Botany: Understanding Plant Biology

Botany: Understanding Plant Biology

Diago Bell

Callisto Reference

www.callistoreference.com

Callisto Reference,
118-35 Queens Blvd., Suite 400,
Forest Hills, NY 11375, USA

Visit us on the World Wide Web at:
www.callistoreference.com

ISBN: 978-1-64116-241-8 (Hardback)

Printed in China.

Cataloging-in-Publication Data

Botany : understanding plant biology / Diago Bell.
p. cm.
Includes bibliographical references and index.
ISBN 978-1-64116-241-8
1. Botany. 2. Plants. 3. Biology. I. Bell, Diago.
QK45.2 .B68 2019
580--dc23

Table of Contents

Preface

Botany is a field of biology that is concerned with the study of plant life. This field has its roots in herbalism, which later evolved with the development and applications of techniques such as optical and electron microscopy, live cell imaging, analysis of chromosome number and plant chemistry, among others. The techniques of molecular genetic analysis such as proteomics and genomics have also significantly advanced the frontiers of this science. The study of plant growth, structure and differentiation, primary metabolism, reproduction, diseases, taxonomy, genetics and epigenetics, etc. are significant areas of study in this field. It has applications in agriculture, horticulture, forestry and breeding, as well as in the provision of materials like rubber, drugs, timber, fiber, etc. This book provides comprehensive insights into the field of plant biology. Most of the topics introduced herein cover the principles and applications of botany. Those with an interest in this field would find this textbook helpful.

Given below is the chapter wise description of the book:

Chapter 1- Botany is a branch of biology that is concerned with the study of plants. This is an introductory chapter which will introduce briefly all the significant aspects of botany and the characteristics of plants including their classification.

Chapter 2- Plant ecology is a branch of ecology that is concerned with the study of plants species, their abundance and distribution, and their interaction with other plants and animals. This chapter has been carefully written to provide an easy understanding of the varied facets of plant ecology, such as ecophysiology, hydrophyte, mesophyte, plant perception, hardiness, plant stress, etc.

Chapter 3- Plant biochemistry is the science concerned with the study of chemical processes in and related to plants. The aim of this chapter is to explore the focus areas of plant biochemistry, such as phytochemistry, photosynthesis, plant nutrition, plant lipids, carbon assimilation and catabolism in plants, among others.

Chapter 4- Genes, variation and heredity in plants are studied under a branch of botany called plant genetics. A significant area of plant genetics is plant breeding, which involves modifying the traits of plants to obtain desired yields. This chapter has been written to provide an understanding of plant genetics and breeding through the elucidation of topics related to plant DNA, RNA, doubled haploidy and transformation vector.

Chapter 5- The study of plant evolution is concerned with the investigation of the processes that resulted in the current diversity of plant life on Earth. The diverse topics elaborated in this chapter include the evolution of plants, their corresponding change in seed size and the process of polyploidy, which are vital to the understanding of plant evolution.

Chapter 6- Plant physiology is a branch of botany that is concerned with the study of the physiology and functioning of plants. Some of the important parts of a plant are leaf, root, stem, flower, etc. This chapter discusses in detail about such important parts of the plant for an in-depth understanding of plant physiology, anatomy and morphology.

Chapter 7- Plant reproduction refers to the production of offspring in plants, achieved either through sexual or asexual reproduction. Sexual reproduction involves the fusion of gametes and results in new progeny that are genetically different from the parent or parents. Asexual reproduction does not require a fusion of gametes and the offspring is genetically identical to the parent. This chapter closely examines the key concepts of sexual and asexual reproduction to provide an extensive understanding of plant reproduction.

At the end, I would like to thank all those who dedicated their time and efforts for the successful completion of this book. I also wish to convey my gratitude towards my friends and family who supported me at every step.

Diago Bell

Chapter 1

Introduction to Botany

Botany is a branch of biology that is concerned with the study of plants. This is an introductory chapter which will introduce briefly all the significant aspects of botany and the characteristics of plants including their classification.

Plant

Plants are multicellular organisms in the kingdom Plantae that use photosynthesis to make their own food. There are over 300,000 species of plants; common examples of plants include grasses, trees, and shrubs. Plants have an important role in the world's ecosystems. They produce most of the world's oxygen, and are important in the food chain, as many organisms eat plants or eat organisms which eat plants.

Plant Characteristics

Plants are autotrophs; they produce their own food. They do so via photosynthesis, which is the process of making nutrients such as sugars from light energy and carbon dioxide. Photosynthesis occurs in cell organelles called chloroplasts, which contain chlorophyll and carotenoids, molecules that absorb light energy and change it into a usable form. Heterotrophs, on the other hand, are organisms that cannot make their own food and must eat other organisms to survive. Many heterotrophs eat plants. Other heterotrophs eat animals that have eaten plants. Plants are primary producers in many ecosystems, giving them a vital role in the survival of many other organisms. In addition, oxygen is a byproduct of photosynthesis, and many organisms depend on oxygen to survive. We couldn't live without plants.

Plants are multicellular organisms with eukaryotic cells. A eukaryotic cell is a relatively large cell with a true nucleus and other organelles that perform specific functions. Plants, protists, fungi, and animals all have eukaryotic cells. Plant cells are distinguished by their cell walls containing cellulose, chloroplasts that perform photosynthesis, and a large central vacuole that holds water and keeps the plant turgid. Prokaryotic cells, on the other hand, are small with no true nucleus or organelles except ribosomes, which produce proteins. Bacteria and archaea have prokaryotic cells.

Many plants have vascular tissue, such as xylem and phloem, that carries water and nutrients throughout the plant. This is particularly important for plants that grow upwards; water needs to travel from the roots up the stem to the leaves. Vascular tissue is found in more "complex" plants. Plants are believed to have evolved from algae-like ancestors. Today, most modern-day algae are classified as bacteria, not plants. However, green algae, which also have cellulose in their cell walls and have chloroplasts that perform photosynthesis, are sometimes grouped with plants.

Plants reproduce both sexually and asexually and have what is known as alternation of generations. A haploid stage alternates with a diploid stage. Haploid is when cells contain one set of chromosomes, while diploid is when cells contain two sets. (For reference, humans are diploid but their gametes—sperm and eggs—are haploid). In plants, two haploid gametes join to form a diploid zygote. This diploid zygote divides through mitosis to become a multicellular organism. It is called the sporophyte, and at maturity, it asexually produces haploid spores. The haploid spores then germinate into multicellular organisms called gametophytes. Gametophytes produce haploid gametes, which fuse to make a diploid organism, and the alternation between diploid and haploid starts all over again.

This image depicts the alternation of generations.

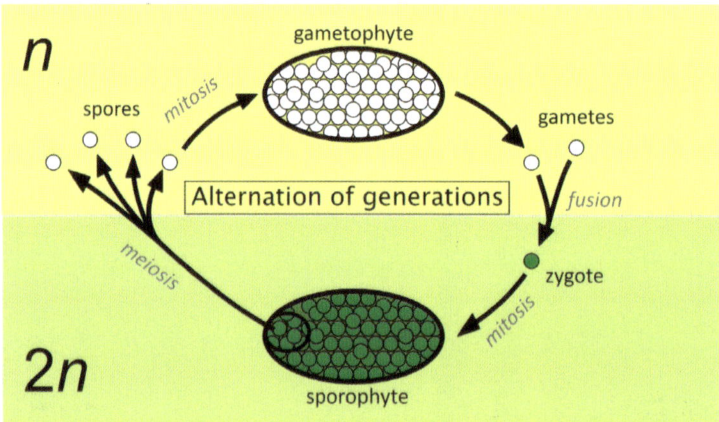

Types of Plants

Charophytes

Charophytes are complex green algae such as stoneworts. They have cells with chloroplasts, cell walls containing cellulose, and store starch, as plants do. They reproduce sexually and some have sperm with flagella (tails that allow them to move), just like some plants do. Some fossil stoneworts are very similar to modern day ones.

Bryophytes

Bryophytes are nonvascular land plants. They do not have vascular tissue, which is tissue that transports water and nutrients. They are found both on land and in water. Common examples of bryophytes are mosses, liverworts, and hornworts. Bryophytes are generally very similar to algae in their lack of a vascular system. They do have parts similar to roots, stems, and leaves, but these are not the true roots, stems, and leaves found in vascular plants. Liverworts were probably the first land plants to evolve. Hornworts have features of both algae and plants, and mosses, the most well-known bryophytes, are the members of this group that are most similar to vascular plants.

Seedless Vascular Plants

Seedless vascular plants produce embryos that are not protected by seeds. Instead, they reproduce via spores. Members of this group include ferns, horsetails, quillworts, clubmosses, and spikemosses. These plants used to be called pteridophytes, but this turned out to be an inaccurate group because ferns and horsetails are more closely related to seed plants than to quillworts, clubmosses, and spikemosses. Seedless vascular plants flourished during the Devonian period and in Carboniferous forests.

Gymnosperms

Gymnosperms include conifers and related plants like ginkgoes and cycads. Gymnosperms have "naked seeds"; their seeds are not contained within an ovary as in flowering plants. Instead, their seeds grow on the surface of leaves, or in the case of conifers, modified structures like cones. The most common example of a gymnosperm is probably the pine tree and its pinecones. Ginkgoes are also well known for being essentially unchanged from ancient ginkgo plants found in fossils from 270 million years ago.

Angiosperms

Angiosperms are flowering plants. They are the most widespread plants today, and over 295,000 different species are known. Their reproductive organs are flowers, which have male parts like stamen and pollen, and female parts like the pistil. When flowers are pollinated, fruits develop containing seeds. Angiosperms have more complex vascular tissue than gymnosperms do.

Plant Classification

All living organisms in this ecosystem are divided into a few major kingdoms. Similarly, plants are also classified into different sub-kingdoms and are mainly based on their certain characteristic features and plant taxonomy.

The classification of plants are mainly based on two features:

- Botanical type.
- Values or geographical or ecological communities.

Importance of Plants Classification

The main purpose of classifying plants is to ensure that the right plants are correctly named grouped and are identified regardless of where it is on earth. Plants are classified into:

1. The evergreens are plants that retain leaves at all times (all year round).
2. Woody plants can also be grouped as deciduous or evergreen.
3. Deciduous plants are seasonal plants which shed its leaves at the end of the growing season, either during the winter season in the temperate climate or during the dry season in the tropical climate.

Apart from the above features, there are some plant species which can be classified based on their life cycle.

Annuals

These are plants that complete their life cycle, seed – plant – seed during a single season. They are normally herbaceous. Few examples like Corn, rice, wheat, and pulses are annuals plant.

Biennials

These are plants that require two years to complete their life cycle. They are normally herbaceous. For example Carrot, cabbage, onions, and beetroot are biennials plant.

Perennials

These are plants that are long-lived, more than two years and are usually woody or herbaceous. Rose, lavender, dianthus, and lilies are few examples of perennials.

Plant Taxonomy

Plant taxonomy is a system for classifying plants based on their genetic and evolutionary relationship. Plant taxonomy is not a fixed science it continues to change as new information becomes available. Plants are classified within a kingdom called as the plant kingdom.

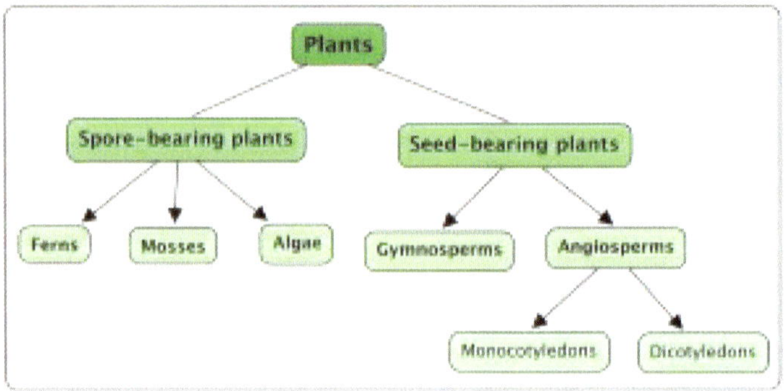

This current system of classification of plants is a natural system which is based on the evolutionary relationship amid other plants.

Coniferophyta

The conifers are remarkably successful as a group, and are among the most common plants in many temperate, alpine, and sub-arctic environment. The conifers generally have strong monopodial vegetative growth. Likely, the monopodial growth habit characteristic of many conifers served them in the evolutionary arms race with huge sauropods like *Brachiosaurus*. The stems and roots have an active vascular cambium, and produce large amounts of wood. This attribute, and their tendency to grow in almost pure stands of uniform age has made them economically important wood producers. The wood of conifers is relatively soft and is the wood of choice for framing homes and the production of plywood. Their leaves vary from needle-like to scale-like to strap-like. And most, but not all, are evergreen.

Their reproductive structures almost always occur in strobili. Some of the earliest conifers like *Cordaites* (Cordaitopsida) had both compound staminate and ovulate strobili; however, *Lebachia* (Voltziopsida) had compound ovulate but simple staminate strobili. Like the voltziopsids, most modern conifers have compound ovulate cones and simple staminate cones. Today, taxa in the Taxales have simplified the ovulate cone to single scales.

The life histories of the extant conifers are variations on that of *Pinus*. Early in the spring small simple strobili with abaxialsporangia (microsporangia) on microsporophylls. The microspore mother cells undergo meiosis to form microspores within the microsporangium. A microspore becomes a pollen grain when the microspore divides to form a generative cell and a prothallial cell. The spore wall which is made of two layers, the entine and the exine, separates and the external wall becomes two balloon-like extensions (giving the pollen grain a Mickey Mouse appearance). The prothallial cell divides again to form two cells. This is what the gametophyte thallus has been reduced to in the conifers (and other seed plants). The generative cell divides to form a tube cell and a generative cell (this will divide once more to form the two sperm). The tube cell grows slowly through the megasporangial wall of the seed until 14 months after pollination the pollen tube enters the archegonial chamber.

Months earlier, the ovule begins to develop on the adaxial surface of the megastrobilus cone scale. Before the formation of the megagametophyte, the ovule, which has its micropyle oriented toward the cone axis, exudes a droplet that catches pollen in the surface tension. When the droplet is reabsorbed, the pollen grains caught by it are pulled through the micropyle and into the pollen chamber (this step is called pollination). The megasporangium has a single megaspore mother cell which undergoes meiosis to form a linear tetrad of megaspores, three of which are aborted. The megaspore begins to undergo free-nuclear enlargement after which it becomes cellular. Some of the cells closest to the micropylar end differentiate into eggs (*Pinus* can have up to 11 eggs in a single ovule). Fertilization occurs 14 months later when the pollination tube enters the archegonial chamber and one of the two sperm fuse with the egg and a zygote is formed. The zygote begins to develop into an embryonic plant with multiple cotyledons. The mature seed has a seed coat formed by the integuments (parental 2N), a megasporangium (parental 2N), a megagametophyte (1N), and an embryo (daughter 2N). When released, the seed has an elongation of the integument which forms a propeller allowing for greater range of dispersal.

Like ferns, conifers have been among the dominant plants almost from the time of their origin in the Carboniferous. They increased in importance well into the Mesozoic and were important members of the landscape when flowering plants appeared. The abundant conifers left remains of many fossils of their wood and leaves. In addition, the resin produced by most conifers has fossilized to form amber.

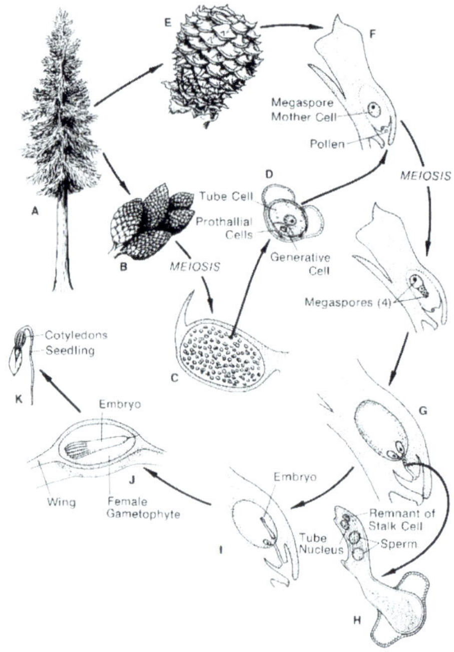

Figure: *Pinus* Life History

A. Mature plant, B. Staminate cones, C. Abaxial microsporangium, D. Mature pollen grain, E. Ovulate cone, F. Pine ovule on the adaxial surface of the ovulate cone scale illustrating pollination, G. Fertilization after the pollen tube enters the archaegonial chamber, H. Detail of the pollen tube with the two sperm, I. Maturing seed with endosperm and embryo, J. Mature seed with developed embryo, K. A germinating seedling.

Systematics of the Coniferophyta

The origin of conifers has been a puzzle. Their similarities to the progymnosperms have led authors like Pearson to place them near the root of the gymnosperm phylogeny. Crane and Tudge show the conifers emerging with the ginkgophytes and cycads as sisters to the gnetophyte-flowering plant clade. Chaw et al. and Bowe present similar analyses based on multiple genes (genomes: nuclear, mitochondrial, and chloroplast) in which *Ginkgo* is a sister to the gnetophyte-conifer clade. Their analyses show the conifers separating into two groups. Conifer group 1 is made of the Pinaceae and the Gnetophytes. Conifer group 2 is all of the other extant conifers. The relative position of the gnetophytes appears to be key not only to understanding the confers, but also the topology of the spermophyte branch of the plant tree. The gne-pine hypothesis suggests that the gnetophytes are sisters to Pinaceae and emerge within the conifers. Other analyses suggest the gne-cup hypothesis in which the gnetophytes are sisters to the Cupressopsida clade of the conifers. Other analyses show that the gnetophytes are sisters to the conifers or sisters to all living seed-bearing plants. Fossil and morphological analyses associate the gnetophytes with the flowering plants.

Doyle, Hilton and Bateman, and Tomescu summarize molecular and morphological cladistic analyses which point the conifers in two different directions. The molecular work embeds the gnetophytes in a conifer clade that includes the Pinales. The morphological analyses based on ovule position, sperm transfer support the molecular trees. However, Tomescu, whose analysis is centered on leaf organization, suggests that the gnetophytes are separate from the conifers.

Figure is an integration of molecular, anatomical, developmental, and fossil work. In our system, the conifers are monophyletic with the Pinaceae sisters to the rest of the conifers (the Cupressopsida). The gnetophytes are sisters to the conifers and the flowering plants are basal to the conifers+gnetophytes+ginkgophytes. Based on this topology, we have organized the living conifers into two classes: Pinopsida (1 family) and Cupressopsida (6 families). This is very different from the classical systematic treatments of the conifers in which the conifers are separated according to the ovulate cones: Coniferopsida (compound ovulate cones) and Taxopsida (simple, reduced isolated ovules on sporophylls).

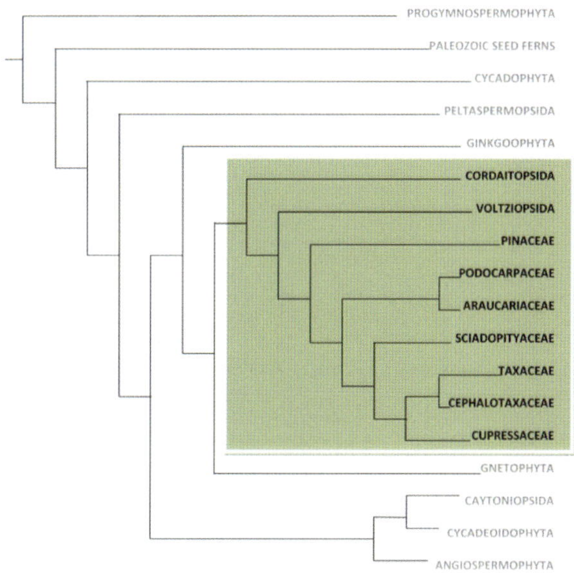

Figure: The relationships between spermophytes (seed plants) is an integration of molecular studies, anatomy and fossil evidence. In this cladogram, the conifers (taxa in shaded box) are monophyletic and include the extinct Cordaitopsida and Voltziopsida. The gnetophytes are sisters to the conifers.

Anthophyta

The Anthophyta is the angiosperms or flowering plants which is the largest and youngest phylum of plants and the one whose members dominate the vegetation of the modern world. The origin of the angiosperms is an enigma, but from all evidence they probably arose sometime during the Late Jurassic or Early Cretaceous, but were first fossilized in the Cretaceous. As they evolved over the ensuing millenia, they developed adaptations that made them ever more successful in the competition for a place in the sun. At the same time the angiosperms were diversifying, the terrestrial animals, too, were diversifying and taking advantage of the new food source. The extraordinary success of the angiosperms is not just that they have flowers—which certainly helps—but because they have, as well, a combination of other structural, developmental, and ecological features not found in their entirety in other groups.

The short list of characteristics of angiosperms includes:

- Flowers, the means of reproduction.
- Ovules surrounded by two integuments.
- Double fertilization, which leads to formation of polyploid endosperm tissue.

- Simple microgametophytes and megagametophytes. The microgametophyte is a three-nucleate structure, the megagametophyte an eight-nucleate one.
- Stamens with two pairs of pollen sacs.
- Sieve tubes and companion cells in the phloem, vessels in the xylem.

Evolution of flowers: With a spotty, incomplete fossil record of the early flowers, much of the understanding of flower evolution is inferred from modern flowers. Taxonomists for a century have defined angiosperm families on floral structure and separated "primitive" from "advanced" features. In this assessment, early, primitive flower characteristics are: an undifferentiated perianth with sepals and petals alike and separate; an indefinite number of parts in each floral whorl; spiral attachment superior ovaries; radial symmetry; and so forth.

Early carpels were leaf-like and seeds were borne on the edges. In advanced flowers, the carpel is folded inward and the seeds are enclosed. Closed carpels have differentiated stigmas, styles, and ovaries. The pollen does not land on the ovules directly.

Pollination: Flowers and their pollinators coevolved; that is, two or more species act as selective forces on one another and each undergoes evolutionary change. Early flowers probably were wind pollinated, but the selective advantages of cross-fertilization by animal pollinators must have been a powerful selective evolutionary force from the very beginning.

Specializations to ensure cross-fertilization and attract pollinators include: colors in wavelengths visible to the pollinators; nectaries placed so that access requires passage across pollen sacs; odors; structural changes such as long corolla tubes and spurs filled with nectar.

Dispersal: Concomitant with the changes to insure fertilization are those that insure dispersal of the products of fertilization, such as the seeds and fruits. Fruits can be dry or fleshy, remain closed or split open at maturity, have hooks or spines that attach to fur or feathers. Seeds can have hard coats, colors, wings, plumes, and all manner of other clever ways to move the new generation away from the old—which is the underlying point of the whole process. Dispersal not only permits colonization of new areas by a species, but also prevents competition for water and minerals between parent and offspring at the home site.

Secondary metabolites (products): Chemical compounds produced by plants are either: 1.) primary products found in all plant cells that are necessary for life, such as amino acids, or 2.) secondary products found in some cells that are important for the survival or propagation of the plants that produce them. When the secondary products were first discovered they were thought to be waste products that plants neither were able to utilize nor get rid of so they were stored out of the way in the vacuoles. With further research it became apparent that the materials were not simply wastes, but had a purpose—to ward off insect attacks, to stop herbivores from eating the plants, or as a response to bacterial and other pathogens.

The toxicity of many of the products is not confined to insect attackers; humans who consume the plants also are affected. Alkaloids produced as secondary metabolites include: cocaine, caffeine, morphine, nicotine, and atropine—a potent pharmacological arsenal. Terpenoids are another class among which are the hydrocarbons, which plants release from their leaves in prodigious amounts and which contribute to air pollution. Terpenoids form the haze that makes the Great

Smoky Mountains "smoky." Rubber is a terpenoid as are taxol and menthol; so are the carotenoids of the plastids and sterols of the cell membranes. Phenolics are imporant secondary metabolites whose plant roles are still being discovered. The evolution of secondary metabolites gave flowering plants a biochemical means to cope with the environment—and added still another improvement over their neighbors.

Systematics

Phylogeny: There are more questions than answers in the phylogeny of the angiosperms. Part of the problem lies in the lack of an adequate fossil record. The first clearly angiosperm fossil is from the Early Cretaceous and is an impression of a fully developed flower. Molecular RNA/DNA sequencing currently is being applied in new phylogenetic (cladistic) analyses to answer the question of angiosperm origins. As yet, no generally accepted answer exists, but several hypothesis are being hotly debated. The molecular data indicate the seed plants most closely related to the angiosperms are the gnetophytes and bennettitaleans, which, incidentally, is the same conclusion reached by some botanists using morphological and anatomical features 50 years ago. Others at the time favored the "seed ferns" as angiosperm ancestors. A second debate revolves around the nature of the first angiosperms. Were they woody or herbaceous? There are no clear answers in that debate either. Cladists in general favor a woody origin, but there are equally vociferous advocates for the herb hypothesis.

Classification: The long-held separation of the angiosperms into two groups on the basis of the number of cotyledons in their seeds—monocots (one) and dicots (two)—is an artificial classification now being abandoned in favor of one based on molecular data, which recognizes evolutionary relationships. The 235,000 species of angio-sperms are separated into three groups:

- Eudicots: 165,000 species; two cotyledons, leaves with net venation, primary vascular bundles in a ring, vascular cambium with secondary growth, pollen with three pores; flower parts primarily in fours or fives or multiples of four or five.
- Monocots: 65,000 species; one cotyledon, leaves with parallel venation, primary bundles scattered, vascular cambium rare, pollen with one pore; flower parts in threes or multiples of three.
- Magnoliids: 5,000 species; primitive characters, pollen with one pore, cells with ether-containing oils; two subgroups: woody magnoliids and paleoherbs; most have fused carpels.

The monocots are a monophyletic group with a common ancestor based on their single cotyledon and a few other features. So, too, are the eudicots with their triaperturate pollen. The magnoliids, however, have no uniting feature and their evolutionary relationships are still being worked out.

Monocotyledon

Monocotyledons commonly referred to as monocots, (Lilianae *sensu* Chase & Reveal) are flowering plants (angiosperms) whose seeds typically contain only one embryonic leaf, or cotyledon. They constitute one of the major groups into which the flowering plants have traditionally been divided, the rest of the flowering plants having two cotyledons and therefore classified as dicotyledons, or dicots. However, molecular phylogenetic research has shown that while the monocots form a monophyletic group or clade (comprising all the descendants of a common ancestor), the dicots do not. Monocots have almost always been recognized as a group, but with various taxonomic ranks and under several different names. The APG III system of 2009 recognises a clade called "monocots" but does not assign it to a taxonomic rank.

Wheat

The monocots include about 60,000 species. The largest family in this group (and in the flowering plants as a whole) by number of species are the orchids (family Orchidaceae), with more than 20,000 species. About half as many species belong to the true grasses (Poaceae), which are economically the most important family of monocots. In agriculture the majority of the biomass produced comes from monocots. These include not only major grains (rice, wheat, maize, etc.), but also forage grasses, sugar cane, and the bamboos. Other economically important monocot crops include various palms (Arecaceae), bananas and plantains (Musaceae), gingers and their relatives, turmeric and cardamom (Zingiberaceae), asparagus (Asparagaceae), pineapple (Bromeliaceae), water chestnut (Cyperaceae), and leeks, onion and garlic (Amaryllidaceae). Many houseplants are monocot epiphytes. Additionally most of the horticultural bulbs, plants cultivated for their blooms, such as lilies, daffodils, irises, amaryllis, cannas, bluebells and tulips, are monocots.

Description

Allium crenulatum (Asparagales), an onion, with typical monocot perianth and parallel leaf venation.

Onion slice: Parallel veins in cross section.

The monocots or monocotyledons have, as the name implies, a single (mono-) cotyledon, or embryonic leaf, in their seeds. Historically, this feature was used to contrast the monocots with the dicotyledons or dicots which typically have two cotyledons; however modern research has shown that the dicots are not a natural group, and the term can only be used to indicate all angiosperms that are not monocots and is used in that respect here. From a diagnostic point of view the number of cotyledons is neither a particularly useful characteristic (as they are only present for a very short period in a plant's life), nor is it completely reliable. The single cotyledon is only one of a number of modifications of the body plan of the ancestral monocotyledons, whose adaptive advantages are poorly understood, but may have been related to adaption to aquatic habitats, prior to radiation to terrestrial habitats. Nevertheless, monocots are sufficiently distinctive that there has rarely been disagreement as to membership of this group, despite considerable diversity in terms of external morphology. However, morphological features that reliably characterise major clades are rare.

Thus monocots are distinguishable from other angiosperms both in terms of their uniformity and diversity. On the one hand the organisation of the shoots, leaf structure and floral configuration are more uniform than in the remaining angiosperms, yet within these constraints a wealth of diversity exists, indicating a high degree of evolutionary success. Monocot diversity includes perennial geophytes such as ornamental flowers including (orchids (Asparagales), tulips and lilies) (Liliales), rosette and succulent epiphytes (Asparagales), mycoheterotrophs (Liliales, Dioscoreales, Pandanales), all in the lilioid monocots, major cereal grains (maize, rice, barley, rye and wheat) in the grass family and forage grasses (Poales) as well as woody tree-like palm trees (Arecales), bamboo, reeds and bromeliads (Poales), bananas and ginger (Zingiberales) in the commelinid monocots, as well as both emergent (Poales, Acorales) and aroids, as well as floating or submerged aquatic plants such as seagrass (Alismatales).

Vegetative

Organisation, Growth and Life Forms

The most important distinction is their growth pattern, lacking a lateral meristem (cambium) that allows for continual growth in diameter with height (secondary growth), and therefore this characteristic is a basic limitation in shoot construction. Although largely herbaceous, some arboraceous monocots reach great height, length and mass. The latter include agaves, palms, pandans, and bamboos. This creates challenges in water transport that monocots deal with in various ways. Some, such as species of *Yucca*, develop anomalous secondary growth, while palm trees utilise an anomalous primary growth form described as establishment growth. The axis undergoes primary thickening, that progresses from internode to internode, resulting in a typical inverted conical shape of the basal primary axis. The limited conductivity also contributes to limited branching of the stems. Despite these limitations a wide variety of adaptive growth forms has resulted from epiphytic orchids (Asparagales) and bromeliads (Poales) to submarine Alismatales (including the reduced Lemnoideae) and mycotrophic Burmanniaceae (Dioscreales) and Triuridaceae (Pandanales). Other forms of adaptation include the climbing vines of Araceae (Alismatales) which use negative phototropism (skototropism) to locate host trees (*i.e.* the darkest area), while some palms such as *Calamus manan* (Arecales) produce the longest shoots in the plant kingdom, up to 185 m long. Other monocots, particularly Poales, have adopted a therophyte life form.

Leaves

The cotyledon, the primordial Angiosperm leaf consists of a proximal leaf base or hypophyll and a distal hyperphyll. In monocots the hypophyll tends to be the dominant part in contrast to other angiosperms. From these, considerable diversity arises. Mature monocot leaves are generally narrow and linear, forming a sheathing around the stem at its base, although there are many exceptions. Leaf venation is of the striate type, mainly arcuate-striate or longitudinally striate (parallel), less often palmate-striate or pinnate-striate with the leaf veins emerging at the leaf base and then running together at the apices. There is usually only one leaf per node because the leaf base encompasses more than half the circumference. The evolution of this monocot characteristic has been attributed to developmental differences in early zonal differentiation rather than meristem activity (leaf base theory).

Roots and Underground Organs

The lack of cambium in the primary root limits its ability to grow sufficiently to maintain the plant. This necessitates early development of roots derived from the shoot (adventitious roots). In addition to roots, monocots develop runners and rhizomes, which are creeping shoots. Runners serve vegetative propagation, have elongated internodes, run on or just below the surface of the soil and in most case bear scale leaves. Rhizomes frequently have an additional storage function and rhizome producing plants are considered geophytes. Other geophytes develop bulbs, a short axial body bearing leaves whose bases store food. Additional outer non-storage leaves may form a protective function. Other storage organs may be tubers or corms, swollen axes. Tubers may form at the end of underground runners and persist. Corms are short lived vertical shoots with terminal inflorescences and shrivel once flowering has occurred. However, intermediate forms may occur such as in *Crocosmia* (Asparagales). Some monocots may also produce shoots that grow directly down into the soil, these are geophilous shoots (Tillich) that help overcome the limited trunk stability of large woody monocots.

Reproductive

Flowers

In nearly all cases the perigone consists of two alternating trimerous whorls of tepals, being homochlamydeous, without differentiation between calyx and corolla. In zoophilous (pollinated by animals) taxa, both whorls are corolline (petal-like). Anthesis (the period of flower opening) is usually fugacious (short lived). Some of the more persistent perigones demonstrate thermonastic opening and closing (responsive to changes in temperature). About two thirds of monocots are zoophilous, predominantly by insects. These plants need to advertise to pollinators and do so by way of phaneranthous (showy) flowers. Such optical signalling is usually a function of the tepal whorls but may also be provided by semaphylls (other structures such as filaments, staminodes or stylodia which have become modified to attract pollinators). However, some monocot plants may have aphananthous (inconspicuous) flowers and still be pollinated by animals. In these the plants rely either on chemical attraction or other structures such as coloured bracts fulfill the role of optical attraction. In some phaneranthous plants such structures may reinforce floral structures. The production of fragrances for olfactory signalling are common in monocots. The perigone also functions as a landing platform for pollinating insects.

Fruit and Seed

The embryo consists of a single cotyledon, usually with two vascular bundles.

Comparison with "Dicots"

Comparison of a monocot (grass: Poales) sprouting (left) with a dicot (right)

Yucca brevifolia (Joshua Tree: Asparagales)

The traditionally listed differences between monocots and "dicots" are as follows. This is a broad sketch only, not invariably applicable, as there are a number of exceptions. The differences indicated are more true for monocots versus eudicots.

Feature	**In Monocots**	**In "Dicots"**
Growth form	Mostly herbaceous, occasionally arboraceous	Herbaceous or arboraceous
Leaves	Leaf shape oblong or linear, often sheathed at base, petiole seldom developed, stipules absent. Major leaf veins usually parallel	Broad, seldom sheathed, petiole common often with stipules. Veins usually reticulate (pinnate or palmate)
Roots	Primary root of short duration, replaced by adventitial roots forming fibrous or fleshy root systems	Develops from the radicle. Primary root often persists forming strong taproot and secondary roots
Plant stem: Vascular bundles	Numerous scattered bundles in ground parenchyma, cambium rarely present, no differentiation between cortical and stelar regions	Ring of primary bundles with cambium, differentiated into cortex and stele (eustelic)
Flowers	Parts in threes (trimerous) or multiples of three (*e.g.* 3, 6 or 9 petals)	Fours (tetramerous) or fives (pentamerous)
Pollen: Number of apertures (furrows or pores)	Monocolpate (single aperture or colpus)	Tricolpate (three)
Embryo: Number of cotyledons (leaves in the seed)	One, endosperm frequently present in seed	Two, endosperm present or absent

Comparison of Monocots and "Dicots"

A number of these differences are not unique to the monocots, and while still useful no one single feature, will infallibly identify a plant as a monocot. For example, trimerous flowers and monosulcate pollen are also found in magnoliids, of which exclusively adventitious roots are found in some of the Piperaceae. Similarly, at least one of these traits, parallel leaf veins, is far from universal among the monocots. Monocots with broad leaves and reticulate leaf veins, typical of dicots, are found in a wide variety of monocot families: for example, *Trillium*, *Smilax* (greenbriar), and *Pogonia* (an orchid), and the Dioscoreales (yams). *Potamogeton* are one of several monocots with tetramerous flowers. Other plants exhibit a mixture of characteristics. Nymphaeaceae (water lilies) have reticulate veins, a single cotyledon, adventitious roots and a monocot like vascular bundle. These examples reflect their shared ancestry. Nevertheless, this list of traits is a generally valid set of contrasts, especially when contrasting monocots with eudicots rather than non-monocot flowering plants in general.

Apomorphies

Monocot apomorphies (characteristics that are derived during radiation rather than inherited from an ancestral form) include herbaceous habit, leaves with parallel venation and sheathed base, embryo with a single cotyledon, atactostele stele, numerous adventitious roots, sympodial growth, and trimerous (3 parts per whorl) flowers that are pentacyclic (5 whorled) with 3 sepals, 3 petals, 2 whorls of 3 stamens each and 3 carpels. In contrast monosculate pollen is considered an ancestral trait, probably plesiomorphic.

Synapomorphies

The distinctive features of the monocots have contributed to the relative taxonomic stability of the group. Douglas E. Soltis and others identify thirteen synapomorphies (shared characteristics that unite monophyletic groups of taxa);

1. Calcium oxalate raphides
2. Absence of vessels in leaves
3. Monocotyledonous anther wall formation
4. Successive microsporogenesis
5. Syncarpous gynoecium
6. Parietal placentation

7. Monocotyledonous seedling
8. Persistent radicle
9. Haustorial cotyledon tip
10. Open cotyledon sheath
11. Steroidal saponins
12. Fly pollination
13. Diffuse vascular bundles and absence of secondary growth

Vascular System

Roystonea regia palm (Arecales) stems showing anomalous secondary growth in monocots, with characteristic fibrous roots.

Monocots have a distinctive arrangement of vascular tissue known as an atactostele in which the vascular tissue is scattered rather than arranged in concentric rings. Collenchyma is absent in monocot stems, roots and leaves. Many monocots are herbaceous and do not have the ability to increase the width of a stem (secondary growth) via the same kind of vascular cambium found in non-monocot woody plants. However, some monocots do have secondary growth, and because it does not arise from a single vascular cambium producing xylem inwards and phloem outwards, it is termed "anomalous secondary growth".Examples of large monocots which either exhibit secondary growth, or can reach large sizes without it, are palms (Arecaceae), screwpines (Pandanaceae), bananas (Musaceae), *Yucca*, *Aloe*, *Dracaena*, and *Cordyline*.

Ecology

Emergence

Some monocots, such as grasses, have hypogeal emergence, where the mesocotyl elongates and pushes the coleoptile (which encloses and protects the shoot tip) toward the soil surface. Since elongation occurs above the cotyledon, it is left in place in the soil where it was planted. Many dicots have epigeal emergence, in which the hypocotyl elongates and becomes arched in the soil. As the hypocotyl continues to elongate, it pulls the cotyledons upward, above the soil surface.

Conservation

The IUCN Red List describes four species as extinct, four as extinct in the wild, 626 as possibly extinct, 423 as critically endangered, 632 endangered, 621 vulnerable, and 269 near threatened of 4,492 whose status is known.

Uses

Monocots are among the most important plants, economically and culturally and account for most of the staple foods of the world, such as cereal grains and starchy root crops, palms, orchids and lilies, building materials, and many medicines. Of the monocots, the grasses are of enormous economic importance as a source of animal and human food, and form the largest component of agricultural species in terms of biomass produced.

Dicotyledon

Dicotyledon, or dicot for short, refers to one of two main groups into which flowering plants (angiosperms) are categorized. Most flowering plants are traditionally divided into two different categories: monocots and dicots. Members of each group tend to share similar features.

Dicots, as their name implies, are named for the number of cotyledons, or embryonic leaves, found in the seed embryo—they have two (di-) cotyledons. Unlike monocots, dicots are not a monophyletic group—meaning that the evolutionary history of dicot plants cannot be traced to a single most common recent ancestor. Instead, a number of lineages diverged earlier than the monocots did.

A word of caution: when classifying flowers into monocots or dicots, remember that there are always exceptions to the rule. Some of the early-diverging dicots seem to have typical monocot characteristics such as scattered vascular bundles, trimerous flowers, and monosulcate pollen grains. Some flowering plants (approximately 2%) don't fit into either category.

Features used to distinguish Monocots from Dicots

Dicots differ from monocots in six distinct structural features. Five of these features are easily observed in the mature angiosperm: the flowers, leaves, roots, stems, and pollen grains. However, the root of these differences stems from the very early embryonic stages of the angiosperm, providing the biggest difference of all between monocots and dicots: the seed.

Flowers

Flowers usually arrange their parts in circles, with the reproductive parts in the middle surrounded by petals and sepals. In dicots, these flower parts are pentamerous. In other words, the flower parts of a dicot are arranged, structured, or numbered in multiples of five, or sometimes four. This is not reliable, however, and is not the easiest characteristic to look for in flowers that have either reduced or numerous parts.

Leaf Venation

Venation refers to the pattern of veins in a leaf blade. These veins are responsible for the transport of water and carbohydrates throughout the plant. In dicots, these veins are arranged in a net-like, or reticulated, pattern. The veins in such leaves appear to look like a finely branched network throughout the leaf blade, with finer veins reticulating between the major veins. However, as with the number of flower parts, leaf venation is also an unreliable characteristic upon which to base your classification assessment. Some flowers may display reticulated venation but are actually monocots such as the aroids and dioscoreales families.

Roots

In plants, the radicle, or the embryonic root, is the first part to emerge from the seed. It shoots down into the ground and begins taking up nutrients and water from the soil. The radicle of a dicot plant develops into the root of the plant. More specifically, the root of a dicot is known as a tap root. Tap root systems have a long and deep primary root, with smaller secondary root growths laterally branching off of the primary root.

Stems

Dicots exhibit secondary growth, which is the ability to increase their diameter via the production of wood and bark. This is the result of two lateral meristems: the cork cambium and the vascular cambium. These lateral meristems continue to produce new cells throughout the life of the woody dicot plant, ultimately increasing the girth of the plant. The rigidity of wood and bark provide mechanical support against gravity and desiccation to dicots, allowing them to grow large, tall, and solid.

In a cross-section of a dicot stem, you will find an epidermis, hypodermis, endodermis, ground tissues, and vascular bundles. Typically, dicot stems have the following characteristics: multicellular epidermal hairs all over the epidermis; chollenchymatous hypodermis; pith; differentiated ground tissues; and a limited number of vascular bundles in a concentric arrangement. The vascular bundles are typically limited to numbers of four or eight, and arranged near the perimeter of the stem in one or two rings.

Pollen Grains

Pollen grains are like the male sex cells of plant; they are the male gametophytes that produce the plant's sperm cells (male gametes). The pollen grains of dicot plants have physical characteristics that distinguish them from the pollen grains of dicot plants. If you look closely, dicot pollen grains are tricolpate, meaning that they have three ridges that go through the outer layer. This structure is derived from the first angiosperms, which had monosulcate pollen grains (having one ridge). The monosulcate form, however, was not retained over the course of divergent evolution.

Embryos

The plant embryo is the part of the seed that contains all of the precursor tissues of the plant and one or more cotyledon. As the name suggests, dicots are characterized by having two (di-) cotyledons in the seed, and two embryonic leaves emerging from the cotyledons.

The seed pods of a dicot are variable in size, shape, texture, and structure. Dicot seed pods can have almost any number of chambers, including zero. More often than not, dicot seed pods contain more seeds than a monocot seed pod.

The cotyledon is the first part of the plant to emerge from the seed, and is the actual basis for distinguishing the two main groups of angiosperms. Cotyledons are important in food absorption and are responsible for absorbing nutrients from the environment until the plant can photosynthesize its own nutrients.

Examples of Dicot

Oak Trees

Although we generally don't think of these trees as flowering plants, they do in fact have insignificant, inconspicuous flowers. These flowers are often overlooked because they are small and yellow-green, often just blending in with their surroundings.

Perhaps a more obvious indication that the oak tree is a dicot is the presence of wood and bark as a result of secondary growth—a characteristic not found in monocots. In fact, all true trees that have wood and bark are dicots, including maple trees, apple trees, and sycamores. Less obviously, their roots do develop from the radicle, which is typical of dicot plants.

Daisies

The daisy is an herbaceous plant without secondary growth. Normally, secondary growth is a telltale sign of a dicot, but the daisy's lack of secondary growth does not mean it is not a dicot; in fact, only about half of dicot species are woody. Instead of using the presence or absence of secondary growth to determine whether the daisy is a dicot, there are other characteristics that you can look for. If you counted all the petals on a daisy, you would find that the daisy's floral parts appear in multiples of four or five, which is characteristic of a dicot. Additionally, the leaves of the daisy have veins that branch in a net-like, or reticulated, pattern—also telling us that the daisy is a dicot.

Roses

Rosa is a genus containing over 100 species of perennial shrubs in the rose family, and roses are, in fact, woody. In the wild, roses usually grow five petals, consistent with the pentamerous pattern seen in all dicots. When cultivated in gardens, however, roses can have double or triple the number of petals while still keeping with the pentamerous pattern. The leaves show a reticulated pattern of venation and the cross-section of a rose stem would reveal concentrically arranged vascular bundles.

Cacti

When we think of flowering plants, cacti may not be the first or even second thing to come to mind; however, cacti do have flowers that bloom under just the right conditions. And when they bloom, their flowers are large, showy, and pentamerous (although it may be difficult to count because in many genera, these flowers have dozens of petals and stamens).

The leaves of cacti are modified to reduce water loss, making it difficult to check for venation pattern. However, the root systems of the cactus plants have not been modified: they have retained the ever-familiar tap root system characteristic of dicot plants.

Legumes

Peas, beans, lentils, and peanuts are all dicots with flowers that grow in clusters. It may be difficult to count the number of floral parts, but they do appear in multiples of four or five. Stems of legumes vary in woodiness and size, but their leaves are all distinctly reticular.

Botany

Pinguicula grandiflora

Botany is the branch of biology dealing with the scientific study of plant life. It is also sometimes referred to as plant science(s) or plant biology. Botany covers a wide range of scientific disciplines that study the structure, growth, reproduction, metabolism, development, diseases, ecology, and evolution of plants.

Historically, botany covered all organisms not considered to be animals, including some "plant-like" organisms, such as algae and fungi, but algae and fungi generally are no longer considered part of the plant kingdom.

Over 400,000 species of plants have been documented on Earth. This multitude of species plays a critical role in the food web, biogeochemical cycles, and maintaining ecological balance. Plant species are generally good examples of complex relations of interdependence—both among various plant species and between plant and animal species. As with animals, plants also rely heavily on sexual reproduction between male and female parts—often, in plants, however, the male and female parts coexist within a single flower.

Dating back to the Roman Empire, botany is one of the oldest disciplines of biology. The study of botany can contribute toward the development of a more ideal human environment through such diverse aspects as helping to feed the expanding human population, discovering new medicines from plants, developing or improving various plant products, and understanding fundamental life processes, ecological relationships, and environmental changes.

The traditional tools of a botanist.

Among the earliest of botanical works, written around 300 B.C.E., are two large treatises by Theophrastus, a philosopher and disciple of Aristotle: *On the History of Plants (Historia Plantarum)* and *On the Causes of Plants*. Together these books constitute the most important contribution to botanical science during antiquity and on into the Middle Ages. As a result, Theophrastus is considered the founder of botany.

The Roman medical writer Dioscorides in the first century C.E., provided important evidence on Greek and Roman knowledge of medicinal plants. He categorized plants based on their medicinal, culinary, or aromatic value.

In 1665, using an early microscope, Robert Hooke discovered cells in cork and a short time later in living plant tissue. The German Leonhart Fuchs, the Swiss Conrad von Gesner, and the British authors Nicholas Culpeper and John Gerard published information on the medicinal uses of plants.

In 1753, Carl Linnaeus published *Species Plantarum,* which included 6,000 plant species. He established the binomial nomenclature, which has been used in the naming of living things ever since.

Sub-disciplines of Botany

Form and Function

Botany can be considered to comprise eight key sub-disciplines each studying a different aspect of plants. These disciplines are genetics, systematics, cytology, anatomy, morphology, physiology, pathology, and ecology.

Genetics is the study of heredity, genes, and gene function. Much modern botany has put to use plant DNA and genomic information to study plants more rigorously than was previously possible. Molecular biology has allowed taxonomists to categorize plant species based on DNA. Plants have been classified into different families and renamed as a result. For this reason, older botanical guides may contain outdated names and classifications. A considerable amount of new knowledge today is being generated from studying model plants like *Arabidopsis thaliana* (mustard weed). Mustard weed was one of the first plants to have its genome sequenced. The sequencing of the rice genome and a large international research community have made rice the de facto cereal/grass/monocot model. Another grass species, *Brachypodium distachyon* is also emerging as an experimental model for understanding the genetic, cellular and molecular biology of temperate grasses. Other commercially important staple foods like wheat, maize, barley, rye, millet, and soybean are also having their genomes sequenced. Some of these are challenging to sequence because they have more than two haploid (n) sets of chromosomes, a condition known as polyploidy, common in the plant kingdom. The "Green Yeast" *Chlamydomonas reinhardtii* (a single-celled, green algae) is another plant model organism the extensive study of which has yielded important insights into cell biology.

Botanical systematics is the study of plant characteristics, especially for the purpose of discerning their evolutionary relationships and establishing different plants' phylogenetic associations. The term "systematics" may or may not overlap with "taxonomy," which concerns itself with scientific classification of species and other taxa. Recent developments are cladistics and molecular systematics.

Cytology is the study of cells, including their function, structure, and life history.

Anatomy is the study of the interior structure of living things.

Morphology is the study of the exterior form of plants, including the placement of stems and leaves on a stem (i.e. alternate or opposite), and also the study of life histories and evolutionary development. Botanical field guides often rely on plant morphology to help biologists identify plant species in the field.

Plant physiology is the study of the function of plants and their cells and tissues. Examples of physiological research includes the study of photosynthetic pathways in different plants and mineral uptake by plants.

Plant pathology is the study of diseases and the structural and functional changes that occur with diseases. This can be important for range of fields, including conservation biology, ecology, agriculture, and horticulture.

Ecology is the study of interactions between organisms and their biotic and abiotic environment as an integrated system.

Other Sub-disciplines

Other sub-disciplines of botany include ethnobotany, the study of how a particular culture, or region has made use of local and indigenous plants, including their use for food, shelter, medicine, clothing, hunting and religion. While ethnobotany has its roots in botany, ethnobotanists also receive training in anthropology, sociology, history, religion, or mythology.

Paleobotany is the study of fossil plants. Palynology, the study of modern and fossil pollen, is also often grouped with paleobotany. Paleobotany and palynology are both instrumental in studying paleoclimatology.

Bryology is the study of mosses, liverworts, and hornworts; phycology is the study of algae; pteridolgy is the study of ferns; and mycology is the study of fungi.

Scope and Importance of Botany

Hibiscus

As with other life forms, plant life can be studied from different perspectives, from the molecular, genetic and biochemical level through organelles, cells, tissues, organs, individuals, plant populations, communities of plants, and entire ecosystems. At each of these levels a botanist might be concerned with the classification (taxonomy), structure (anatomy), or function (physiology) of plant life.

Historically, botany covered all organisms that were not considered to be animals. Some of these "plant-like" organisms include fungi (studied in mycology), bacteria and viruses (studied in microbiology), and algae (studied in phycology). Most algae, fungi, and microbes are no longer considered to be in the plant kingdom.

Plants are a fundamental part of life on earth. They generate the oxygen, food, fibers, fuel, and medicine that allow higher life forms to exist. Plants also absorb carbon dioxide, a significant greenhouse gas, through photosynthesis. A good understanding of plants is crucial to the future of human societies as it allows us to:

- Feed the world
- Understand fundamental life processes
- Utilize medicine and materials
- Understand environmental changes
- Maintain ecological, biodiversity, and ecosystem function

Feed the World

Nearly all the food we eat comes (directly and indirectly) from plants like this American long grain rice.

Virtually *all* of the food we eat comes from plants, either directly from staple foods and other fruit and vegetables, or indirectly through livestock, which rely on plants for fodder. In other words, plants are at the base of nearly all food chains, or what ecologists call the first trophic level. Understanding how plants produce the food we eat is therefore important to be able to *feed the world* and provide food security for future generations, for example through plant breeding.

Not all plants are beneficial to humans, some weeds are a considerable problem in agriculture, and botany provides some of the basic science in order to understand how to minimize their impact. However, other weeds are pioneer plants, which start an abused environment back on the road to rehabilitation, underlining that the term "weed" is a very relative concept, and that broadly defined a weed is simply an undesirable plant that is too successful.

Gregor Mendel laid the foundations of genetics from his studies of plants.

Understand Fundamental Life Processes

Plants are convenient organisms in which fundamental life processes (like cell division and protein synthesis for example) can be studied, without the ethical dilemmas of studying animals or humans. The genetic laws of inheritance were discovered in this way by Gregor Mendel, who was studying the way pea shape is inherited. What Mendel learned from studying plants has had far reaching benefits outside of botany.

Barbara McClintock discovered transposons, or "jumping genes," by studying maize. These transposons, genes that move from one location to the next on a chromosome, are responsible for the mottled look of maize grains. This sort of research has paved the way for the study of other plant genomes and genome evolution.

Other types of physiological research, including the uptake of carbon by plants through photosynthesis and understanding the physiology behind C3 versus C4 photosynthetic plants, are important for understanding the response of plants to climate change and the feedback mechanisms that occur with increased greenhouse gases in the atmosphere.

These are a few examples that demonstrate how botanical research has an ongoing relevance to the understanding of fundamental biological processes.

Utilize Medicine and Materials

Many of our medicine and recreational drugs, like caffeine and nicotine come directly from the plant kingdom. Aspirin, which originally came from the bark of willow trees, is just one example of a plant derivative used in modern medicine. Pharmacognosy is the study of medicinal and toxic plant derivatives. There may be many novel cures for diseases provided by plants that have not yet been discovered. Popular stimulants like coffee, chocolate, tobacco, and tea also come from plants. Most alcoholic beverages come from fermenting plants such as barley malt and grapes.

Plants also provide us with many natural materials, such as cotton, wood, paper, linen, vegetable oils, some types of rope, and rubber. The production of silk would not be possible without the cultivation of the mulberry plant. Sugarcane and other plants have recently been put to use as sources of biofuels, which are important alternatives to fossil fuels. Plants are extremely valuable as recreation for millions of people who enjoy gardening, horticultural, and culinary uses of plants every day.

Understand Environmental Changes

Plants can also help us understand changes in on our environment in many ways:

- Understanding habitat destruction and species extinction is dependent on an accurate and complete catalog of plant systematics and taxonomy.
- Plant responses to ultraviolet radiation can help us monitor problems like ozone depletion.
- Analyzing pollen deposited by plants thousands or millions of years ago can help scientists to reconstruct past climates and predict future ones, an essential part of climate change research.

- Recording and analyzing the timing of plant life cycles are important parts of phenology used in climate-change research.
- Plants can act a bit like the 'miner's canary', an *early warning system,* alerting us to important changes in our environment. For example, lichen, which are sensitive to atmospheric conditions, have been extensively used as pollution indicators.

References

- Plant: biologydictionary.net, Retrieved 30 June 2018
- The-classification-of-plants: byjus.com, Retrieved 14 May 2018
- Coniferophyta, gymnosperms, viridiplantae, archaeplastida-202: comenius.susqu.edu, Retrieved 26 April 2018
- Phylum-anthophyta-the-flowering-plants, plant-biology, seed-plants, biology: cliffsnotes.com, Retrieved 14 March 2018
- Botany: newworldencyclopedia.org, Retrieved 09 July 2018

Chapter 2

Plant Ecology

Plant ecology is a branch of ecology that is concerned with the study of plants species, their abundance and distribution, and their interaction with other plants and animals. This chapter has been carefully written to provide an easy understanding of the varied facets of plant ecology, such as ecophysiology, hydrophyte, mesophyte, plant perception, hardiness, plant stress, etc.

Plant ecology is a sub-discipline of ecology focussed on the distribution and abundance of plants and their interactions with the biotic and abiotic environment.

Plants occur over Earth's surface in well-defined patterns that are closely correlated with both climate and the history of the planet. Forests are the most important of these natural communities from the standpoint of area, carbon content, annual carbon fixation, the cycling of nutrient elements, and influence on energy and water budgets, as well as being the principal reservoir of biodiversity on land. The most extensive forests are the boreal coniferous forests of North America, Scandinavia, northern Europe, and northern Asia. The moist forests of the tropics are the most diverse, often containing as many as 100 species of trees per hectare (1 hectare = 2.47 acres) and occasionally many more.

Autumn at Allegheny National Forest in Pennsylvania.

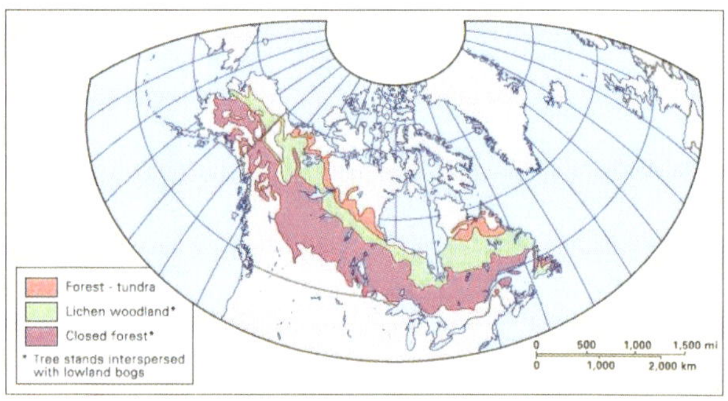

Worldwide distribution of boreal forests.

Forests are commonly distinguished from woodlands and savannas in having a closed canopy of trees that may be 10 metres (about 33 feet) to more than 50 metres (about 165 feet) in height. At their wetter and cooler limits, forests are replaced by tundra, a community of low-growing trees and shrubs with a ground cover of *Sphagnum* and other mosses, dwarfed shrubs, and perennial herbs. The transition from forest to tundra occurs at high latitudes and elevations and may extend over a considerable distance, throughout which trees are scattered, restricted to sheltered areas, and stunted in growth. Such taiga stands are common at the northern limit of trees throughout the Northern Hemisphere. The alpine transition tends to be more abrupt. Both are often determined locally by the distribution of snow.

Vegetation profile of a temperate deciduous forest.

Forests yield at their warmer and drier margins to grassland (called prairie in North America and steppe in Asia). In North America the rich grasslands of the eastern mid-continent, which is affected by moisture moving northward from the Gulf of Mexico, once supported a diversity of herbaceous plants, including grasses that might reach several metres in height, and were known as tall-grass prairie. Westward, where water was less abundant, the grasses diminished in diversity and size to form the short-grass plains. Southward and westward the plains yielded under increasing aridity to desert, as old as any other zone in an evolutionary sense and very diverse in species. Similar patterns of vegetation occur across the broad waist of Asia.

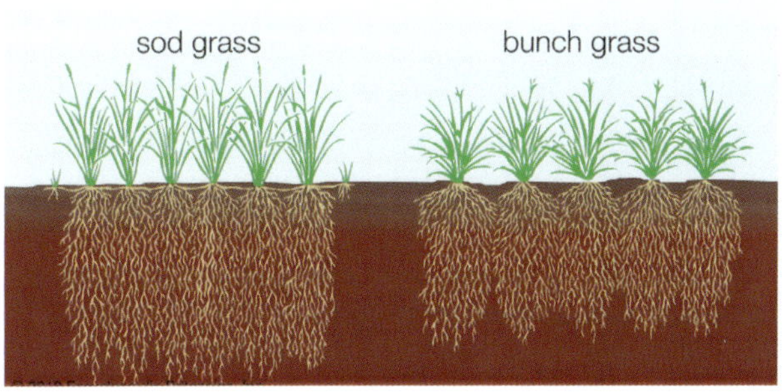

Vegetation profile of a grassland

Edge of prairie and spruce forest

The distinction between grassland and forest is blurred in many places, especially in the semiarid subtropics and in other regions where fires may be common. This transitional zone is savanna, an open grassed woodland. The southeastern coastal plain of North America was originally such a region, a pine savanna that produced plants adapted to, in fact dependent on, burning for survival. The long-leaf pine (*Pinus palustris*), for instance, has a "grass" stage, which lasts for several years of early growth, with the bud protected at the very surface of the ground by a thick tuft of long grasslike leaves that shield it from the heat of a fire. Once sufficient energy has been stored in the taproot and short stem, the tree virtually explodes into growth and passes rapidly through a period of vulnerability to fire. As a tree of 3 metres (10 feet) or more in height, it is safe from all but the hottest of fires. Oaks and other species are replacing pines as savannas disappear from the southeastern United States because of the continuing division of land ownership and protection from fire.

Vegetation profile of a savanna

In contrast, the extensive savannas of eastern Africa have been maintained through climate, fires, and grazing and support an extraordinary diversity of migratory animals that have evolved to use different parts of the ecosystem. The patterns of use are complex and phased in time, and they work to maintain both the diversity and the productivity of the savanna. Those African populations are the last of the large and diverse mammalian and avian populations that once also grazed similar areas of North America, Europe, and Asia.

Acacia tree (*Acacia* species) on a savanna in Zimbabwe

The forests of the moist tropics are among the most complex and fascinating of terrestrial ecosystems. They extend from mountain slopes to river swamps and occurred throughout the tropics wherever precipitation allowed trees to survive. Human activities have virtually eliminated forests from vast areas of the tropics and threaten to destroy the residual forests globally. The largest areas of tropical forest occur now in the Amazon basin of South America and in the Congo basin of west-central Africa. The extensive tropical forests of Southeast Asia have been reduced to fragments of the area they occupied as recently as 1950.

Tropical rainforest in Malaysia.

Vegetation profile of a tropical rainforest.

The rainforests of the Amazon basin have evolved as a part of a river system whose water level fluctuates annually by as much as 15 metres (50 feet) or more along the middle and lower Amazon. There are substantial further differences in the quality of water. The Negro River, for example, drains an area of sands low in nutrient elements, where organic matter has accumulated sufficiently in soils to produce the humic acids that give the river its dark colour and sufficient acidity to affect the plant and animal life it can support. The areas flooded in the annual cycle are forested and are known by the Portuguese word *várzea*. Trees in this zone survive flooding for several weeks annually and provide the basis of a food web that includes fish adapted for grazing on tree fruits and seeds. The grazing fish possess large flat molars adapted for masticating seeds and other coarse organic matter, and they compete for seeds dropped from pods in *várzea* trees on the river.

The distribution of the various zones of vegetation over Earth's surface has changed with climates throughout time. The greatest changes in the recent past have been due to the periods of glacial advance and retreat over the last several hundred thousand years. As recently as 15,000 years ago, glacial ice covered much of eastern North America as far south as the present Hudson estuary and covered all of Scandinavia and northern Europe as well. Sea level was tens to hundreds of feet lower at various times, and extensive coastal areas now flooded were exposed. Such drastic changes in climate have sorted and resorted plant species, allowed the establishment of a diversity of forest types over time, and worked against the establishment or preservation of the mutual dependencies among species so common in the tropics. These higher-latitude forests, while similar in form and genera and sometimes in species around the world, are apparently highly mutable, the antithesis of the stable and self-perpetuating "organism" they were once thought to be. Nonetheless, the concept of the plant community as an organism remains viable and useful in ecology and is the basis of much of the most progressive analysis of how to manage Earth for the successful support of large numbers of people.

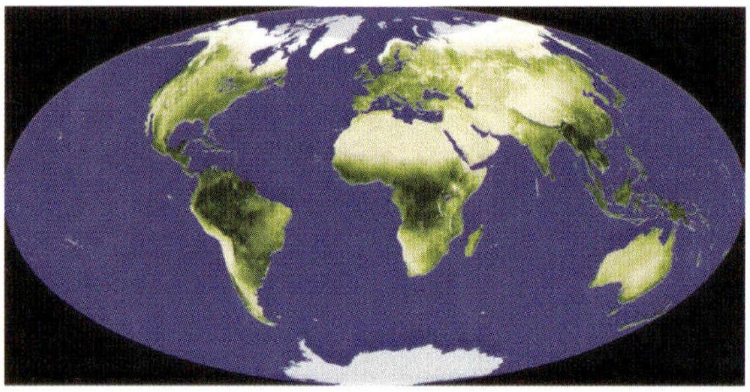

Global vegetation; Normalized difference vegetation index (NDVI)

Scientists map changes in Earth's vegetation over time using a measurement known as the Normalized Difference Vegetation Index (NDVI). NDVI is calculated from satellite data on visible and near-infrared light reflected by vegetation on Earth. By comparing differences between the average NDVI over a given period of time (e.g., a month) and the average NDVI over a span of many years (e.g., two decades), scientists are able to monitor vegetation cover and biomass production and to detect anomalies such as drought.

Succession and Zonation

It is known from studies of plant residues and pollen preserved in the highly acidic sediments of bogs and from observations of contemporary glaciers that the vegetation southward from the glacial front in the Northern Hemisphere was banded in much the same way the vegetation is zoned today: tundra occurred in a zone closest to the ice; coniferous forests occurred in a warmer and drier zone southward; and deciduous forests occurred still farther southward. As the habitat changed—that is, as the glacial ice melted and the glacial front retreated—the vegetation migrated onto the new landscape: first the pioneer species of the tundra—a few hardy low-growing or crustose lichens and mosses of small stature—followed by dwarfed willows and birch and, ultimately, the full panoply of tundra plants. As the climate ameliorated further, the forest followed, always according to pattern, with a few pioneers followed by the full array of species characteristic of the forest.

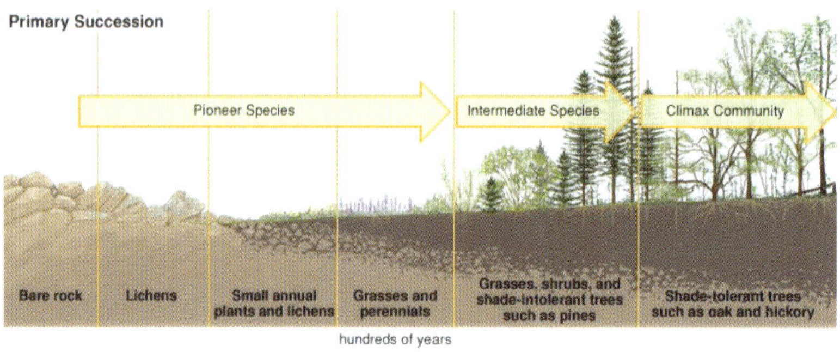

Primary succession

Primary succession begins in barren areas, such as on bare rock exposed by a retreating glacier. The first inhabitants are lichens or plants—those that can survive in such an environment. Over hundreds of years these "pioneer species" convert the rock into soil that can support simple plants such as grasses. These grasses further modify the soil, which is then colonized by other types of plants. Each successive stage modifies the habitat by altering the amount of shade and the composition of the soil. The final stage of succession is a climax community, which is a very stable stage that can endure for hundreds of years.

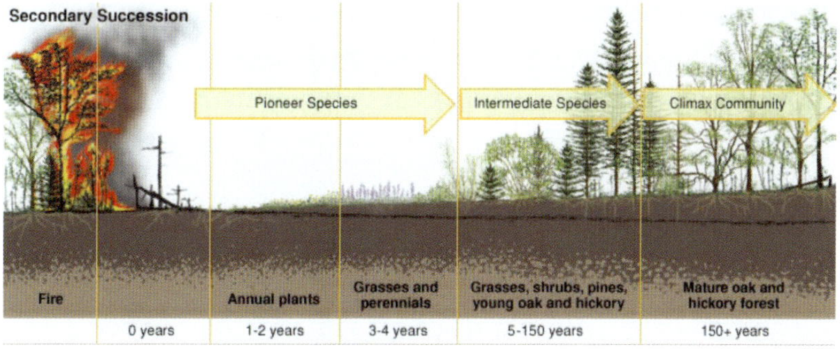

Secondary succession

Secondary succession follows a major disturbance, such as a fire or a flood. The stages of secondary succession are similar to those of primary succession; however, primary succession always begins on a barren surface, whereas secondary succession begins in environments that already possess soil. In addition, through a process called old-field succession, farmland that has been abandoned may undergo secondary succession.

The process of invasion of a new landscape not previously occupied by plants has long been called *primary succession*. In this case the succession was in response to both the availability of a new habitat and a climatic warming that allowed the replacement of tundra by forest and of the coniferous forest ultimately by deciduous forest as the warming continued. Further warming might result in the replacement of deciduous forest by grassland, as has occurred worldwide at the steppe- or prairie-forest border in response to climatic changes over recent centuries. This boundary is strongly influenced by the fires that have swept grasslands throughout time. Disturbances such as clearings for agriculture, wildfires, diseases, and storms severe enough to open gaps in a forest may start secondary successional changes that also reestablish the normal vegetation for that climatic zone.

Throughout much of the tropics, where forests have been destroyed over large areas to make pastures, conditions appropriate for what might be a normal succession within disturbance-caused gaps in an otherwise intact forest do not exist, and the sites are permanently impoverished. The process generates grasslands, shrublands, or sometimes bare earth. More than one-third of the land area of India has been impoverished in this way and is lost to agriculture, forests, or forestry.

Ecosystems and the Biosphere

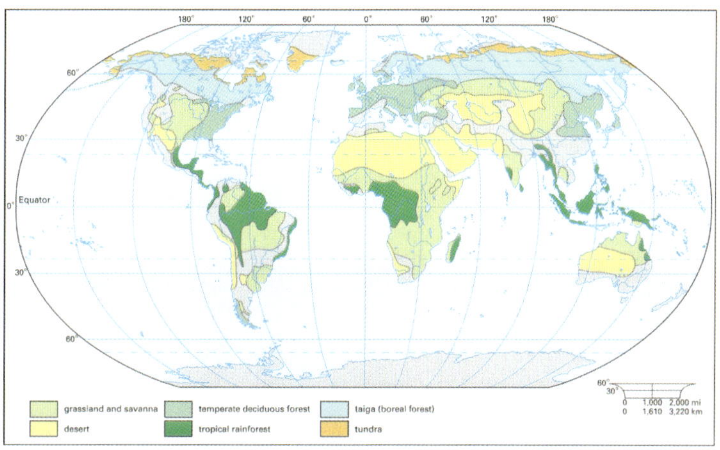

Worldwide distribution of major terrestrial biomes

The thin mantle of the Earth that supports life extends not more than a few feet into the sediments of the abyssal depths of the oceans and to a few tens of thousands of metres into the atmosphere, where pollen grains or other spores may survive. It is this thin layer at Earth's surface that is the biosphere, the place where life occurs. Human activity has changed Earth irreversibly to form the present biosphere, and humans are now a force for further irreversible change.

Dispersal and Colonization

The methods by which plants are distributed over Earth's surface are as diverse and complex as the plants themselves. The most widely occurring plants are the small-bodied rapidly reproducing forms whose spores can be carried by wind and water and by birds and other animals.

Among the seed plants, whose propagules are less mobile, explanations for the current distribution of plants become more complicated and include such profound changes over evolutionary time as the breakup of Pangea some 300 million years ago, the opening of the Atlantic Ocean, and the isolation of North and South America, Australia, and Madagascar from larger continental landmasses. Progressive isolation produced endemism, evolutionary divergence sufficient to generate whole floras peculiar to a particular region, with many species, even genera, not known elsewhere. Volcanic islands are much younger than the continents and support floras derived from chance invaders carried by wind, sea, or animals, including humans. Island floras also come to exhibit endemism. English naturalist Charles Darwin's observations of the fauna and flora of the Galapagos Islands off the western coast of South America led him to recognize the general process of biotic evolution. The islands are diverse in form and climate, isolated in varying degrees from each other and from the continent, and support a highly diversified flora and fauna clearly derived from South America but modified over the time of isolation to contain forms peculiar to the islands.

Diverse mechanisms exist for dispersing spores and seeds at appropriate times and in appropriate habitats to ensure the survival of the new plant. In forests adapted to fire, pinecones may remain closed until a fire occurs. The heat of the fire opens the cones and releases the seeds into a fire-seared habitat, where the seeds germinate, probably stimulated by karrikins, growth regulators in smoke. The seedlings find soil with an abundance of nutrients left by the burning of organic matter and reduced competition from other plants. Seeds of trees frequently have "wings" that permit wider distribution by wind. Other plants, such as thistles and burdock, have awns or other appendages that catch on animals and obtain a wide distribution in areas frequented by them. Still others have elaborate mechanisms for spreading via rhizomes, stems, or other vegetative propagules. Some plants have no significant dispersal mechanisms, and seeds thus end up close to the mother plant.

Especially during the past two centuries, human activities have both deliberately and inadvertently spread the higher plants, and many of their bacterial, fungal, and insect plagues, around Earth's surface. Examples among the higher plants are numerous; while many of the transfers of such plants have been benign and some clearly advantageous, others have been disastrous. Tree species have been moved freely around the world, sometimes with remarkable effectiveness. The Monterey pine (*Pinus radiata*), for instance, is a diminutive and unproductive tree on the coast of California, but it has become a major timber tree as a result of rapid growth in plantations in New Zealand. The rubber tree (*Hevea*) was carried in the 19th century from the moist tropics of Brazil to Java and elsewhere in the South Pacific, where its excellent growth, free of the diseases and competitors that affect it in its native American habitat, nearly destroyed the market for Brazilian rubber. Other inadvertent introductions have been far from benign; some have severely impoverished the landscape. Cheatgrass, or downy brome (*Bromus tectorum*), for example, is an annual grass introduced from Europe and the Transcaspian steppes to the arid intermontane west in North America, probably as a contaminant in fodder in the latter part of the 19th century. It is an aggressive invasive species and spreads rapidly, becoming a continuous ground cover over extensive areas. The plant completes its life cycle early in the season, sets seeds, and stands through the rest of the year as a continuous cover of dry grass not useful for fodder. It does, however, carry fire, which changes and can even destroy the native vegetation.

Plant diseases and pests have been inadvertently spread around the world by human activities. Such diseases continue to exact a high cost in that they progressively impoverish the vegetation. One of the most profound changes in any forested region was produced by the introduction of the chestnut blight (*Cryphonectria parasitica*, formerly *Endothia parasitica*) to North America. The fungus, introduced from Asia, found a home in the Fagaceae of eastern North America but was lethal to the American chestnut (*Castanea dentata*), the dominant species of extensive stands in the southern Appalachians and elsewhere. Once a common and popular shade tree and a source of smooth-grained, easily worked wood and of the abundant mast that was one of the principal foods of indigenous wild turkeys and other animals, the American chestnut never developed resistance to this now widely distributed fungus, which has several hosts in the Fagaceae and which produces wind-borne spores.

The American elm (*Ulmus americana*), a magnificent tree of the moist forests of lowlands throughout eastern North America, has suffered a similar but less-comprehensive loss through the ravages of another exotic fungus, Dutch elm disease (*Ceratocystis ulmi*), which was imported from Europe on infected wood probably early in the 1900s. The fungus is spread by two elm bark beetles. Individual trees sometimes escape the fungus for many years but ultimately succumb before reaching the large sizes that were common throughout its range early in the 20th century.

Human Effects on Plants and Natural Communities

Humans have influenced the structure and development of natural communities for many thousands of years. The major influence has been through fire, which has been used deliberately in herding game, in rejuvenating plants used for fodder, in opening forested plots for agriculture, in maintaining savannas and grasslands, and in keeping forests open for easy passage for hunting. With the rapid expansion of human populations globally in recent decades, however, the changes in vegetation have been profound. Forests have morphed into grasslands and savannas. Succession has replaced pine forests with deciduous forests, as in the southeastern coastal plain and Piedmont regions of North America. In many instances temporary protection from fire has resulted in the accumulation of a large mass of flammable vegetation that, when burning, makes a hot fire, which kills the overstory trees and many species that normally survived the more frequent and less-severe fires of the past. In such circumstances the vegetation may be impoverished, a condition that may prevail indefinitely once established.

Prescribed burn: A fire manager starting a prescribed burn in Yellowstone National Park, in the northwestern United States. Such fires are intentionally set to mimic the natural burn cycles of forests and to prevent the buildup of dead plant matter on the ground, which can fuel dangerous crown fires.

Domestication

The origins of domesticated plants and agriculture are buried in a dim and unrecorded past of 10,000 years and more. Recent experience with Stone Age cultures of the Amazon basin and elsewhere has shown that such cultures have a sophisticated knowledge of plants for such purposes as food, medicine, and the tools and poisons used for hunting and fishing. Amerindian cultures improved a wide range of species by deliberate selection of the more productive and useful forms. Manioc (*Manihot utilissima*) remains a staple of large sections of Latin America, especially Brazil and the Amazon basin. A woody tuberous plant whose origin in the savannas of South America has long been lost, it is propagated vegetatively by planting a piece of tuber or a segment of stem. The tuber is ground to make a flour that must be washed to remove toxic quantities of hydrocyanic acid before being eaten directly or baked into a flat bread called *cazabi* (cassava).

Cassava (*Manihot esculenta*), which is also called manioc, in cultivation in Uganda

Corn, or maize (*Zea mays*), was domesticated about 10,000 years ago either from teosinte (a perennial *Zea* that exists today) or from a lost ancestor that existed in the highlands of what is now central Mexico. Its culture had spread as far north as southern Maine by the time of European settlement of North America. Corn is now the third largest plant-based food source in the world. Other plants domesticated from the region include the common bean, squash, chili, tomato, avocado, papaya, guava, sapodilla, cotton, sisal, and vanilla.

The process of domestication appears to have involved selection and cultivation of the most promising and productive of plants long harvested in the wild. Continued selection brought drastic shifts in the population, shifts in the frequency of genes, and the development of new races totally dependent on humans to maintain them. Races developed in this way have been called cultigens to emphasize their dependence on cultivation. Such is the case with corn: the original stock has probably been lost, although teosinte may offer clues as to the source.

Major crop plants have been domesticated over the last several thousand years from sources identified by the Russian botanist N.I. Vavilov. Common wheat (*Triticum vulgare*) and rye (*Secale cereale*) probably were first domesticated from the grasses of Central Asia. Various millets (*Panicum*) and barley (*Hordeum hexastichum*) originated in the mountainous regions of central and western China, rice (*Oryza sativa*) probably in the Indian region.

Changes in Biosystems: Pollution

There is no plant community anywhere on Earth that has not felt the direct influence of the expansion of the human enterprise. The influence has had the form of direct intrusion through hunting and gathering, cultivation, or extensive harvest of trees or other plants or has had indirect effects through the harvest of fish or other animals or through changes in the chemistry of the environment. The latter changes are now pervasive and range from the introduction of pesticides, such as DDT, and other industrial toxins, such as the PCBs, and acid rain, with substantial modification of the sulfur, nitrogen, and hydrogen ion budgets of ecosystems over large areas. Widespread destruction of forest tree species occurred during the 1980s in eastern North America and throughout Europe because of the combined effects of acid rain and other air pollutants.

Branches from a tree in Germany's Black Forest show needle loss and yellowed boughs caused by acid rain.

Symptoms of incipient damage from air pollution and other chronic disturbances are now common in forests and other types of vegetation around the world, although symptoms may be subtle and difficult to assign as being caused by a particular pollutant. They range from such changes as the elimination of populations of leafy lichens from the bark of trees to the mortality of the trees themselves. Under extreme circumstances, such as in regions around smelters in Sudbury, Ontario, Canada, virtually all higher plants have been eliminated from an area of hundreds of square miles.

One of the most spectacular and informative examples of the patterns of impoverishment of forests exposed to chronic disturbance was induced experimentally in an oak-pine forest at Brookhaven National Laboratory in central Long Island, New York, U.S., using ionizing radiation. A single radiation source was used in the centre of the forest. The exposure, begun in the fall of 1961, was sufficient within months to eliminate all plants from a central area within a few metres of the source and to produce a systematic zonation of the forest at lower exposures. The zones ranged through a lichen and moss community, followed at lower exposures by a continuous ground cover of a single species of sedge (*Carex pensylvanica*). At slightly lower exposures a community of low shrubs occurred, then high shrubs at still lower exposures. An impoverished forest of hardy oaks well within the normal range of variability of the oak-pine forest of the region occurred down the gradient of exposure and beyond that zone; at exposures of one roentgen or less daily, the pines survived and the forest appeared intact and normal. Effects could, however, be detected by careful measurements of growth down to exposures on the order of 0.1 roentgen per day. The experiment offered a specific example of the progressive impoverishment of vegetation where both cause and effect could be measured, something not possible in most instances of pollution. Such experience is useful not only in defining the effects of exposure to ionizing radiation but also in identifying stages in the impoverishment of natural communities as effects accumulate. Similar zonation can be found around the areas contaminated by the 1986 Chernobyl disaster.

Similar changes occur in aquatic systems in response to chronic pollution. The early stages of pollution in bodies of water usually involve enrichment with nutrient elements, especially nitrogen and phosphorus, a process known as eutrophication. This can lead to water blooms (the rapid growth of algae or cyanobacteria), which deplete the water of oxygen and choke aquatic life.

The effects of pollution on land and in water are to favour small-bodied rapidly reproducing organisms that do not depend on complex food webs. The process of simplification and impoverishment is now global, and it affects terrestrial and aquatic communities alike. It is the continuously expanding result of chronic intrusions on natural systems by human influences. The impoverishment threatens all life because it systematically reduces the capacity of Earth to support plants.

Conservation

One of the most serious threats to life is the climate change and the possibility of a rapid warming of Earth's lower atmosphere from the accumulation of heat-trapping gases. The most important of the gases is carbon dioxide, but methane, nitrous oxide, and the chlorofluorocarbon compounds (CFCs) also contribute significantly. The largest sources of carbon dioxide are the combustion of fossil fuels, the destruction of forests, changes in land-use patterns, and, probably, the warming itself. Methane is a product of the use of fossil fuels and of the decay of organic matter in soils and swamps, rice paddies, and bogs. Its rate of evolution is accelerated by warming.

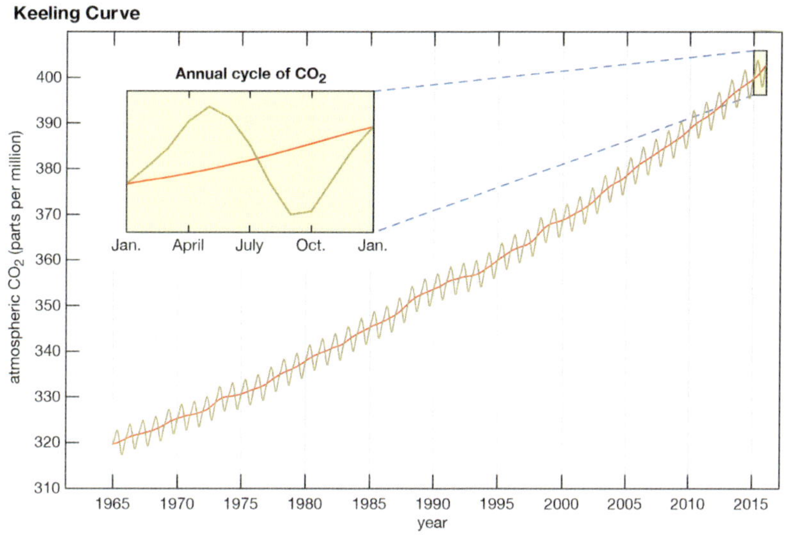

Keeling Curve, named after American climate scientist Charles David Keeling, tracks changes in the concentration of carbon dioxide (CO_2) in Earth's atmosphere at a research station on Mauna Loa in Hawaii. Although these concentrations experience small seasonal fluctuations, the overall trend shows that CO_2 is increasing in the atmosphere.

The effects of the warming are expected to be profound: if global warming has occurred as rapidly as some climatologists have suggested, the mean temperature of Earth's lower atmosphere will increase by 1.8 to 4°C (3.2 to 7.2°F) by 2100 over baseline temperatures established in 2000. A change of 1°C in temperature is equivalent to 96.5–160 km (60–100 miles) of latitudinal change in the middle latitudes. Such high rates of change in climatic zones would not only disrupt agriculture across large areas (e.g., northern India) but also destroy forests and other natural communities at their warmer and drier margins more rapidly than they could be regenerated elsewhere. One of the effects would be to speed the decay of organic matter in soils and accelerate the accumulation of carbon dioxide and methane in the atmosphere, an effect that would further accelerate the warming. Avoiding such change will require abandoning fossil fuels as the major source of energy for industrial societies and replacing these energy sources with renewables, such as solar and wind energy.

Classical approaches to conservation involve specific efforts at preserving species that are endangered. By far the most desirable approach to the preservation of Earth's 5 million–30 million or more species is through management of natural ecosystems. Studies of tropical forests have emphasized that the area required to ensure the habitat of bird populations as well as populations of other animals and plants is very large indeed, on the order of 10,000 hectares (25,000 acres). There is evidence to support earlier experience that the best way to ensure the preservation of the indigenous biota is to maintain a matrix of the indigenous vegetation with more than 50 percent of the land area undisturbed and an additional 25 to 50 percent in successional stages. This type of land-use planning exists in very few places around the world but offers the advantage of a landscape that maintains itself with stable near-maximum flows of energy through living systems and offers, regionally, maximum sustainable support for people at low cost. The difficulty is that the apparently high stocks of land under light use—the forests, the water, and energy in various forms—are an open invitation to immediate short-term exploitation, and their existence may suggest that the landscape can be more intensively occupied.

Ecophysiology

Plant ecophysiology deals with the science of plant interaction with its surrounding environment and its acclimation and adaptation processes. Environmental stress are limiting factors for plant productivity. Plant ecophysiology can be described as an experimental science where the science discusses about the physiological mechanisms with reference to ecological observations. It deals with the control over growth, reproduction, abundance, survival and distribution of plants, these process of plants are affected by interactions of plants with their biotic and abiotic environment. These patterns of ecophysiology help us to understand significance of particular plant traits and their evolutionary heritage.

Plant ecophysiology discusses about two criteria like how plants sense and respond to environmental changes and it responses to variable conditions like gradients of sunlight within tree canopies. As plants are unable to move around like animals, they must survive in these adverse conditions or perish. Plants have a phenotype which can change to their environmental conditions and also they have a vast array of genes that help them to adapt to changing conditions.

Light

Light is an abitoic factor, the intensity of light or its irradiance can be below the optimum or might be excessive. Intensity of light is an important factor to dertermine the temperature of plant organs. The photosynthesis irradiance curve or the light response curve is useful to distinguish plants tolerance to different light intensities. Suboptimal light or shade occurs at the base of the canopy or with the underlying vegetation.

Plants which are tolerant to shade have a wide range of adaptations to help them survive in conditions of shady environment with altered quantity and quality of light. Plants with higher canopies and plants on the open grounds where the the cloud cover is low and the sun's zenith angle is low receive excess of light. This is seen typically in the tropics and high altitude regions. Excess light on the leaves can result in photoinhibition and photodestruction. Plants that are adopted to environments with high incidence of light have a wide range of adaptations to avoid or to scatter the excess light; they also have mechanisms to reduce the amount of injury caused.

Wind

Impact of the winds is caused due to its influence on the canopy which influences the leaves with regulation of moisture, heat and carbon dioxide. In the absence of winds there is a layer of still air present around each leaf which is known as the boundary layer, it insulates the leaf from the environment and provides an atmosphere that is rich in moisture and less susceptible to convective heating or cooling. If there is an increase in wind speed, the leaf environment becomes closely linked to its surrounding environment and it may be difficult for the plant to retain moisture if exposed to dry air. Moderately high winds also allow the plant to cool its leaves more easily when exposed to full sunlight. Plants are entirely passive in their interaction with wind. Plant coat their leaves in fine hair like structures called trichomes to break up the air flow to make themselves less susceptible to changes in wind speed. The dimensions of leaf and canopy are often finely controlled to manipulate the boundary layer depending on the surrounding environmental conditions.

Temperature

Plants produce various proteins in response to extreme temperatures. These proteins protect the leaf from damaging effects of ice formation, reduced rates of enzyme catalysis at lower temperatures, enzyme denaturation and increased photorespiration due to increased temperatures. If temperature decreases below optimum, the plants produces antifreeze proteins and increases the rate of dehydrins. Increase in temperatures results in production of heat shock proteins. Extreme temperatures results in metabolic imbalances which leads to the build-up o reactive oxygen species which are encountered by antioxidant systems. Temperature changes affect cell membranes, causing them to lose its fluid properties which can affect the movement of compounds across the membrane. To prevent this plants can change the composition of their membranes, in lower temperauture conditions, more of unsaturated fatty acids are placed in the membrane and in hot conditions more saturated fatty acids are placed. Plants can also avoid overheating by minimising the sunlight absorbed and by increasing cooling effects of wind and transpiration. Absorption of sulight is reduced by plants by using refletive leaf hairs, scales and waxes. Some species of Macropitilium purpurerum can move their leaves so that they are oriented to avoid the sun. Plants can avoid the impact of low temperatures by altering their microclimate.

Water

Plants can be damaged if there is too much or too little water. Too little water can cause the tissues to dehydrate or even be fatal. If the soil becomes water logged the soil will become anoxic which can kill the roots of the plant.

Plants have the ability to access water depends on the structure of their roots and water potential of the root cells. If the water content of soil is low plants can alter their water potential to maintain a flow of water into the roots and the leaves, this gradient is created by transpiration from the leaves.

In dry soil conditions, plants close their stomata to reduce transpiration and to prevent water loss. The closing of stomata is mediated by chemical signals from the root. If drought conditions continue, the tissues of the plant dehydrate which results in the loss of turgor pressure and leads to wilting. Most plants also respond to drought by altering their water potential and increases root growth. Plants like the xerophytes are adapted to dry environments and have a range of other specialized mechanisms to maintain water and protect tissues from dessication.

CO_2 Concentration

Carbon dioxide play a vital role in plant growth and is important component of photosynthesis. Plants take in carbon dioxide through stomata on their leaves, there is trade-off between CO_2 gain and water loss in plant productivity. The enzyme Rubisco used to capture carbon dioxide is more efficient only when there is high concentration of carbon dioxide in the leaf. Some plants concentrate CO_2 within their leaves using C_4 carbon fixation. Most plant species use C_3 carbon fixation.

Hydrophyte

Hydrophytes are those plants which live in water and adjust with their surroundings. They either remain fully submerged in the water like Hydrilla, Vallisneria, etc. or most of their body parts remain under the water like trapa, lotus, etc. water lilies, sedges, crow foots are other important water plants.

A plant that is adapted to living either in waterlogged soil or partly or wholly submerged in water. Many hydrophytes absorb water and gases over the whole surface and have no stomata like the spiked water milfoil (Myriophyllum spicata), which is completely submerged in water.

The mechanical and vascular tissue of many hydrophytes is reduced and supports them. They often have large intercellular air spaces in their stems, roots, and leaves to overcome the difficulty of obtaining gases from the water. Hydrophytes that are partially submerged have floating leaves with stomata through which gases can be exchanged as in land plants.

However, to prevent the leaves being flooded with water, the petioles may be very long to adjust easily to changes in water level. In the giant water lily (Victoria regia), the enormous leaves have a vertical rim to prevent them from being flooded. Some species like water crowfoot (Ranunculus aquatilis) have both finely divided submerged leaves and floating leaves with stomata.

Aquatic plants are plants that have adapted to living in aquatic environments (saltwater or freshwater). They are also referred to as hydrophytes or macrophytes. These plants require special adaptations for living submerged in water, or at the water's surface – the most common adaptation is aerenchyma, but floating leaves and finely dissected leaves are also common. Aquatic plants can only grow in water or in soil that is permanently saturated with water. They are therefore a common component of wetlands.

Aquatic Plants can usually be Categorized into 4 Main Types:

i. Floating plants

ii. Deep water plants

iii. Marginal plants

iv. Oxygenating plants.

Characteristics of Hydrophytes:

Hydrophytes are aquatic plants that are especially suited for living in aquatic environments. In order to survive, a hydrophyte, also known as an aquatic macrophyte, must either be completely submerged in water, or in some cases be allowed to float on the surface of the water.

(i) Water Retention:

Plants require water to survive, plants usually possess the ability to absorb and retain water to keep the plant alive between watering cycles. Because aquatic plants are literally submerged in water at all times, there is no need for the plant to water retention like non-aquatic plants. As such, there is no need for aquatic plants to expend energy to regulate transpiration, which is the loss of water in the plant due to evaporation.

(ii) Flat Leaves:

Some aquatic plants float on the surface of water; most aquatic plants have flat leaves which act as floation to a portion of the plant. An example of such a plant is the water lily, Nymphaea Attraction, which is a beautiful bright red water lily that can have up to a 12 inch diameter leaf.

(iii) Feathery Roots:

Hydrophytes are supported by water as compared to roots and stem structures; most hydrophytes have roots that are small and feathery. These are designed to take in oxygen from the water, and since the plant is fully submersed in water at all times, there is less need for a long and thick root structure. One plant that has a feathery root is the Salvinia, which is a free-floating aquatic fern.

(iv) Air Sacks:

Many hydrophytes have air sacks (chambers) that help the plant float on the surface of the water. It is important to note that some aquatic plants will float slightly submerged in the water, such as the buttercup. Others, such as water lilies, will float up on top of the surface since their leaves distribute the weight across the surface of the water.

Ecological Adaptations in Hydrophytes:

A. Roots:

1. Due to availability of plenty of water root system is secondary importance and least significant.
2. Roots absent in Wolffia, Ceratophyllum.
3. Poorly developed roots are seen in Hydrilla.
4. Submerged leaves compensate for roots in Salvania.
5. Root caps are absent in Hydrophytes.
6. Amphibious plants growing in mud will have distinct root caps.
7. Root caps are replaced by root pockets in Pistia.
8. If present roots are generally fibrous adventitious reduced in length unbranched or poorly branched.
9. Balancing roots are present in Pistia, Eichhornia.

B. Stems:

1. Stem is long slender and flexible in submerged plants e.g.: Hydrilla, Potamogeton.
2. Stem is slender or thick, short and spongy in free-floating forms e.g.; Eichhornia.
3. Stem is a rhizome in rooted plants with free-floating leaves e.g.: Nymphaea and Nelumbo.

C. Leaves:

1. In submerged forms leaves are thin long and ribbon like (e.g.: Vallisneria) long and linear. (Potamogeton) finely dissected (e.g.; Ceratophyllum).
2. Floating leaves are large and flat with wax coating e.g.: Nymphaea, Nelumbo and Victoria regia.
3. Largest simple leaves in plant kingdom are present in Victoria regia.
4. Petioles are long flexible and covered with mucilage.
5. Petioles are swollen and spongy in ego: Eichhornia, Trapa.
6. Heterophily with submerged, floating and aerial leaves seen in Limnophila, Ranunculus and Sagittaria.

Anatomical Features of Hydrophytes:

1. Cuticle is completely absent in submerged parts of the plants.
2. Cuticle may be present as a thin film on surface of parts exposed to atmosphere.
3. Epidermal cells are with chloroplast useful for absorption and assimilation.
4. Stomata are totally absent in submerged hydrophytes.
5. Exchange of gases takes place through diffusion.
6. Non-functional stomata are seen in Potamogeton.
7. Epistomatous leaves (stomata found only on upper surface) are present in hydrophytes with floating leaves e.g.; Nelumbo.
8. Mechanical tissues like collenchyma and sclerenchyma are more or less absent.
9. Xylem is poorly developed in Hydrophytes as the water absorption takes place all over surface of the plant body.
10. Hydrophytes have aerenchyma in all parts of the plants. Aerenchyma proves provides buoyancy to the hydrophytes.

Nymphaea Leaf c.s.

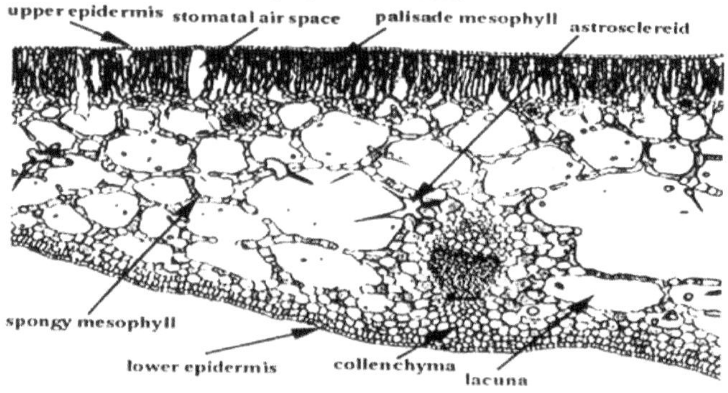

Potamogeton Floating Leaf c.s.

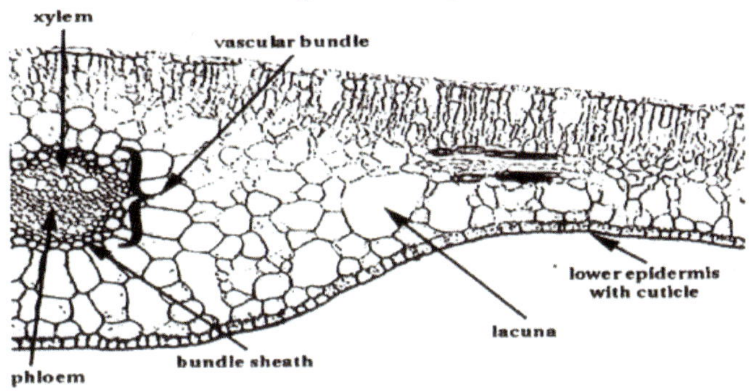

Mesophyte

Mesophytes are the largest group of terrestrial plants that live in a climate of a moderate to hot temperatures and soil that is not dry and is not too wet. Mesophytes are found in many places such as shaded forest areas or sunny meadows or fields. Most plants found in household gardens, parks, streets and general public areas are mesophytes.

Mesophytes can deal with an array of weather conditions, such as humidity, dryness, moisture, and wind. This is because they are adapted to survive throughout all seasons, and have no specific adaptations to deal with extreme cold or heat.

Mesophytes are active during the day, and during the night as well. They are constantly respiring during day and night, but only photosynthesising during the day when light can be turned into solar energy.

Mesophytes are constantly in competition with other plants for water, minerals, and sunlight. Mesophytes would also have to protect themselves from predators and other organisms living in their habitat.

General Adaptations to Maintain Water Balance

- Roots of Mesophytes

The roots of Mesophyte plants are important structures that are adapted to absorb as much water as possible. Root systems can be fibrous, which are thin roots and are only penetrate the soil at a shallow level, meaning that they can absorb water quickly from the surface of the soil. Roots assist in maintaining plant water balance by being able to transport water into the plant.

Roots can also be formed in a taproot system. In a taproot system here is one thick primary root, with several smaller roots branching off the primary root. Taproots penetrate deeply into the soil, meaning that they can absorb moisture from the deep soil. The image below shows the difference between the two roots systems. Fibrous roots have an advantage of absorbing water straight away, but won't be able to reach water in the deep soil, whereas taproots can reach both areas because of their deep roots.

All roots are able to absorb water from soil by osmosis. This is how water enters plants and leaves, through the roots. The water from the high water concentrated soil enters the roots by osmosis, which has a lower concentration of water. Water that has diffused through the roots will then enter the xylem vessels, to be used for other purposes.

The roots supply a large amount of water used for transpiration, and without roots, a large portion of water would not be able to enter the plant and the plant would die due to lack of water.

- Hairs (root and leaf)

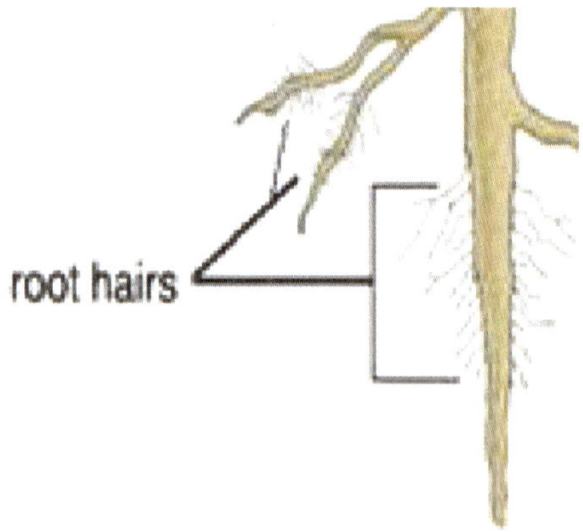

Root hairs

Some roots of mesophytic plants have short thick root hairs present, which helps with obtaining water and water absorption as it increases the surface area for diffusion of the water by osmosis. The water is absorbed by the root hairs, and then the water is then absorbed by the roots to be transported through the plant for further use.

Leaves also sometimes have hairs present on the epidermis'. Hairs on the leaves help create a microclimate. Hairs on leaves trap and condense the water on the surface of the leaf, reducing the concentration gradient between the inside of the leaf and the outside, therefore making it less possible for water to diffuse out of the leaf by osmosis, reducing water loss. An example of leaf hairs on leaves are on mesophytic plant of the Rudbeckia hirta. Mesophytes without root hairs may have a higher rate of transpiration.

- Stomata

Most mesophytic plants only have stomata present on their lower epidermis. Stomata are cells that control the exchange of water and gases from inside and outside of the leaf. Stomata also control transpiration. Water diffusion through stomata is affected by several factors such as, wind, temperature, and light. It is for these reasons that the stomata are present on the lower epidermis (underside) of the leaf. When stomata are underneath the leaf, there is less light exposure to the stomata, a cooler temperature is provided, and the stomata will not be exposed to as high winds as they would be if they were present on the top side of the leaf. Having the stomata on the bottom of the leaf reduces evaporation and transpiration, therefore meaning that water balance is being maintained by preventing water loss.

Stomata control the exit and entrance of water. Without the stomata being able to open, water would not be able to evaporate, causing overhydration, which can result in death of the organism. Stomata can also remain closed during certain periods of weather, especially heat or dryness, as this weather encourages evaporation. The stomata will remain closed until the right amount of water is present in the plant again. This is important for mesophytes, as they live in an environment where many weather conditions are possible, and at times of intense heat, dryness or humidity, the stomata will need to stay closed to prevent water loss through transpiration and to preserve the water it already has. At times of extreme wetness, the stomata will to open to allow water to flow through and exit the plant without causing overhydration. Having the stomata on the bottom of the leaf may limit the gas supply, as leaves may not be in direct contact with a supply of gas, and may find it hard to retrieve these gases into the plant by the stomata on the bottom of the leaf, compared to if they were on the top of the leaf.

- Waxy cuticle

Mesophytic leaves have a waxy cuticle present on both epidermis', but it is thicker on the upper epidermis. The waxy cuticle prevents water loss as it traps water inside the leaf, keeping moisture contained, and it prevents water from easily diffusing through the leaf to the outside. It is thicker on the upper epidermis because the upper side of a leaf is exposed to harsher conditions such as temperature, wind, humidity and light exposure compared to the underside of the leaf. Mesophytes are exposed to all of these conditions, hence why most mesophytic plants possess the waxy cuticle. The waxy cuticle helps to prevent water loss therefore maintaining the right amount of water inside the plant. In the image to the left, of the Ilex or Holly plant, you can see that the waxy cuticle is shiny and water sits on top of the upper epidermis. A waxy cuticle may restrict the rate at which gas is diffused into the plant at, making it harder to obtain essential gases.

- Structure

Mesophyte plants often are supported and rely on turgidity of the cell for structure. The stem cell vacuoles are filled with water, and therefore become turgid, creating a stable and strong stem supporting the plant. Water needs to be supplied to the plant for this reason, without water, the vacuole would not be turgid and would pull away from the cell wall causing plasmolysis, causing the mesophyte plant to wilt and fall over. Mesophyte cells are also lignified, meaning that the presence of lignin in cells supports the structure by strengthening the cells. Mesophyte plants live in areas of high winds sometimes, and often have to stand up for protection, so water must be provided to the plants in order for them to stay standing and be protected from weathering. Plant structure is a reason why plants must obtain a certain amount of water at all times, and contributes to their water balance process. Their strong structure may be harmful during windy periods, as they may snap or break due to their woody structure, harming the organism.

- Vascular System

Most mesophytic plants have Xylem and Phloem vessels. The xylem vessels carry water from the roots into the leaves by transpiration. Xylem vessels are made of dead cells and resemble a long tunnel which is able to transport fluid from the roots to the leaves. These vessels are important as they carry water from the roots, and transports minerals essential for life processes into the plant. Phloem vessels transport water around the whole plant. Phloem vessels allow water to travel up and down the plant, whereas xylem vessels only carry water upwards for transpiration. Without the xylem vessels, the mesophytic plants would have no way of obtaining water from the roots, making the plant water deprived. Without the phloem vessels, water would not be able to reach the whole of the plant, and areas of the plant would become mineral and water deprived.

Xerophytes

A xerophyte (xero meaning dry, phyte meaning plant) is a plant which is able to survive in an environment with little availability of water or moisture. Plants like the cacti and other succulents are typically found in deserts where low rainfall is the normal phenomen, but few xerophytes can also be found in moist habitats such as tropical forests, exploiting niches where water supplies are limited or too intermittent for mesophytic plants.

Plants that live under arctic conditions may also have a need for xerophytic adaptations, as water is unavailable for uptake when the ground is frozen. Their leaves are covered with silvery hairs.

Adaptations of xerophytes include reduced permeability of the epidermal layer, stomata and cuticle to maintain optimal amounts of water in the tissues by reducing transpiration, adaptations of the root system to acquire water from deep underground sources or directly from humid atmospheres and succulence, or storage of water in swollen stems, leaves or root tissues. The typical morphological consequences of these adaptations are collectively called xeromorphisms.

Types of Xerophytic Plants:

These are plants adapted to grow in dry habitats.

They are classified into four categories on the basis of their morphology and life cycle pattern:

a. Ephemeral Annuals:

These plants are also called as drought evaders or drought escapers. They are annuals and complete their life cycle within a very short period. They do not withstand dry seasons but actually avoid them. Argemone mexicana, Solatium xanthocarpum.

b. Succulent:

These plants grow in habitats with less or no water. They store water whenever it is available. They have succulent and fleshy organs like stems, leaves and roots which serve as water storage organs and accumulate large amounts of water during the brief rainy seasons. Euphorbia and Opuntia.

c. Non-Succulent Perennials:

These are drought resistant and called as true xerophytes. They possess a number of morphological, anatomical and physiological characteristics which enable them to withstand critical dry conditions. Calotropis, Acacia, Casuarina and Nerium.

d. Succulent Plants:

Succulent plants typically store water in stems or leaves. They include the Cactaceae families which typically have stems that are round and store a lot of water. Often, as in cacti where the leaves are reduced to spines, their leaves are vestigial, or they do not have leaves.

Characteristics of Xerophytes

(i) Reduction in Air Flow:

Some xerophytes have tiny hairs on their surface to provide a wind break and reduce air flow, thereby reducing the rate of evaporation. When a plant surface is covered with tiny hairs, it is called tomentose. In a still environment, the areas under the leaves/spines where transpiration is taking place form a small localized environment that is more saturated than normal with water vapour.

If this is not blown away by wind, the water vapour potential gradient is reduced and so is transpiration. Thus, in a windier situation, this localization is not held and so the gradient remains high, which aids the loss of water vapour. Spines trap a layer of moisture and also slow air movement over tissues.

(ii) Reflectivity:

The color of a plant, or of the waxes or hairs on its surface, may serve to reflect sunlight and reduce evaporation. An example is the white chalky wax (epicuticular wax) coating of Dudleya brittonii, which has the highest ultraviolet light (UV) reflectivity of any known naturally occurring biological substance.

(iii) Physiological:

Some plants can store water in root structures, trunk structures, stems, and leaves. Water storage in swollen parts of the plant is known as succulence. A swollen trunk or root at the ground level of a plant is called a caudex and plants with swollen bases are called caudiciforms. Tiny pores on the surface of a xerophytic plant called stomata may open only at night, so as to reduce evaporation.

Plants may secrete resins and waxes (epicuticular wax) on their surfaces, which reduce evaporation. Examples are the heavily scented and flammable resins (volatile organic compounds) of some chaparral plants, such as Malosma laurina, or the chalky wax of Dudleya pulverulenta.

Plants may drop their leaves in times of dryness (drought deciduous), or modify the leaves produced so that they are smaller.

During dry times, xerophytic plants may stop growing and go dormant, change the kind of photosynthesis or change the allocation of the products of photosynthesis from growing new leaves to the roots.

Ecological Adaptation in Xerophytes:

1. Plants growing in habitats where water supply is absent or physiologically dry are called Xerophytes.
2. Xerophytes classified based on their (a) Morphology (b) Physiology (c) Life cycle pattern.
3. Plants growing in dry or arid zones are called Ephimerals or Drough evaders or drought escapers. Eg; Tribulus.
4. Ephemerals are Annuals and complete their life cycles in 6-8 weeks.
5. Xerophytic plants absorbing more water during rainy season and storeing them in different body parts are called Succulents or drought avoiding plants.
6. Succulents store water in the form of mucilage.
7. Leaf succulents: Bryophylum, Aloe, Agave.
8. Root succulents: Asparagus.
9. Perennial plants which can withstand prolonged period of drought are called Non-succulents or true xerophytes Eg: Casurina, Nerium, Ziziphus, Calotropis etc.
10. Ecological adaptations of Xerophytes.
11. The all three major groups of xerophytes have some common adaptations to survive in very dry conditions.

a. Roots:

1. Root system is very well developed with extensive branching and often longer than shoot system.
2. Root hairs and root caps are very well developed.

b. Stems:

1. Mostly they are stunted, woody hard and covered with thick bark.
2. In some xerophytes stem becomes underground.
3. In some plants stem becomes fleshy, green, leaf-like phylloclades covered with spines, Eg: Opuntia.
4. Stems are usually covered by hairs and or waxy coatings.

c. Leaves:

1. Leaves are very much reduced small scale like and sometimes modified in to spines to reduce the rate of transpiration. Lamina may be long narrow needle-like or divided in to many leaflets as Eg: Acacia.
2. Foliage leaves become thick fleshy and succulent or tough and leathery in texture. Eg: Aloe.
3. Leaf surfaces are shiny glazed to reflect light and heat. Eg. Calotropis.

Anatomical features of Xerophyte:

1. Epidermis is covered thick cuticle to reduce the rate of transpiration.
2. Epidermal cells may have silica crystals.
3. Epidermis is multilayered Eg: Nerium.
4. Waxy coating is present on leaves and stem Eg: Calotropis.
5. Stomata are generally confined to lower epidermis of leaves called hypostomatous.
6. Stomata are present in pits called sunken stomata. They are lined with hairs Eg: Nerium.
7. Mesophyll is differentiated in to palisade and spongy parenchyma.
8. Mechanical & vascular tissues are well developed.

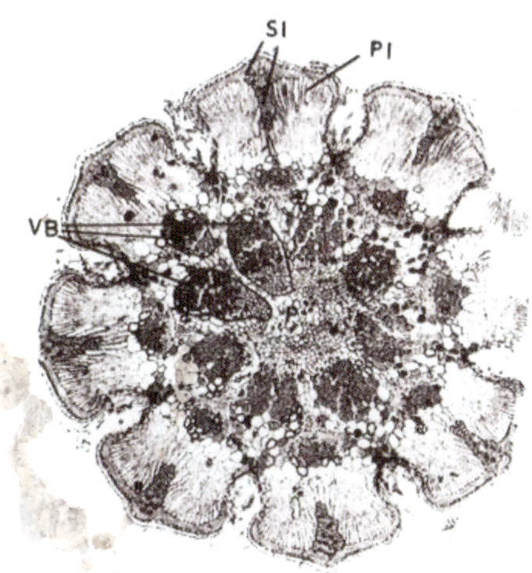

CASUARINA

Plant Perception

In botany, plant perception is the ability of plants to sense the environment and adjust their morphology, physiology and phenotype accordingly.Research draws on the fields of plant physiology, ecology and molecular biology.

Examples of stimuli which plants perceive and can react to include chemicals, gravity, light, moisture, infections, temperature, oxygen and carbon dioxide concentrations, parasite infestation, physical disruption, and touch. Plants have a variety of means to detect such stimuli and a variety of reaction responses or behaviors.

Plant perception occurs on a cellular level and its concomitant reactive behavior is mediated by phytochromes, kinins, hormones, antibiotic or other chemical release, changes of water and chemical transport, and other means. These responses are generally slow, taking at minimum a number of hours to accomplish, and can best be observed with time-lapse cinematography, but rapid movements can occur as well.

Plants have many strategies to fight off pests. For example, they can produce different toxins (phytoalexins) against invaders or they can induce rapid cell death in invading cells to hinder the pests from spreading out. These strategies depend on quick and reliable recognition-systems.

Aspects of Perception

Light

Many plant-organs contain photo-sensitive compounds (phototropins, cryptochromes and phytochromes) each reacting very specifically to certain wavelengths of light. These light-sensors tell the plant if it's day or night, how long the day is,how much light is available and from where the light comes. Shoots grow towards light and roots usually grow away from light. These responses are called phototropism and skototropism respectively. They are brought about by light sensitive pigments like phototropins and phytochromes and the plant hormone auxin. Many plants exhibit certain phenomena at specific times of the day, for example certain flowers open only in the mornings. Plants keep track of the time of the day with a molecular clock. This internal clock is set to the solar clock every day using sunlight. The internal clock coupled with the ability to perceive light also allows plants to measure the time of the day and so find the season of the year. This is how many plants know when to flower. The seeds of many plants sprout only after they are exposed to light. This response is carried out by phytochrome signalling. Plants are also able to sense the quality of light and respond appropriately, for example in low light conditions plants produce more photosynthetic pigments whereas when the light is very bright and/or if the levels of harmful UV increase, plants produce more of their protective pigments that act as sunscreens.

Gravity

To orient themselves correctly, plants must have an adequate sense of the direction of gravity's unidirectional pull. The subsequent response plant movement is known as gravitropism. Typically, in the root this works as gravity is sensed and translated in the root tip, and subsequently roots grow towards gravity via elongation of the cells. In the shoot, similar effects are happening, but gravity is perceived and then growth occurs in the opposite direction, as the above ground part of the plant experiences negative gravitropism.

At the root tip, there are amyloplasts containing starch granules that fall in the direction of gravity. This weight activates secondary receptors, which signal to the plant the direction of gravitational pull. After this occurs, auxin is redistributed through polar auxin transport and differential growth towards gravity begins. In the shoots, auxin redistribution occurs in a way to produce differential growth away from gravity.

For perception to happen, the plant must sense, perceive and translate the direction of gravity. Without gravity, proper orientation will not occur and the plant will not effectively grow. The root will not be able to uptake nutrients or water, and the shoot will not grow towards the sky to maximize photosynthesis.

Processes

Detection

Displacement can be detected by plants. Poplar stems can detect reorientation and inclination (equilibrioception).

Pathway Signals

Wounded tomatoes are known to produce the volatile odour methyl jasmonate as an alarm signal. Neighbouring plants can then detect the chemical and prepare for the attack by producing chemicals that defend against insects or attract predators.

Plants systematically use hormonal signalling pathways to coordinate their own development and morphology.

Neurochemicals:

Plants produce several proteins found in the animal neuron systems such as acetylcholine esterase, glutamate receptors, GABA receptors, and endocannabinoid signaling components. They also use ATP, NO, and ROS like animals for signaling.

Electrophysiology:

Although plant cells are not neurons, they can be electrically excitable and can display rapid electrical responses (action potentials) to environmental stimuli. These action potentials can influence processes such as actin-based cytoplasmic streaming, plant organ movements, wound responses, respiration, photosynthesis, and flowering. These electrical responses can cause the synthesis of numerous organic molecules, including ones that act as neuroactive substances in other organisms. Thus, plants accomplish behavioural responses in environmental, communicative, and ecological contexts.

Signal Response

A plant's concomitant reactive behavior is mediated by phytochromes, kinins, hormones, antibiotic or other chemical release, changes of water and chemical transport, and other means. These responses are generally slow, taking at minimum a number of hours to accomplish, and can best be observed with time-lapse cinematography, but rapid movements can occur as well. Plants respond to volatile signals produced by other plants. Jasmonate levels also increase rapidly in response to mechanical perturbations such as tendril coiling.

Plants have many strategies to fight off pests. For example, they can produce different toxins (phytoalexins) against invaders or they can induce rapid cell death in invading cells to hinder the pests from spreading out.

Some plants are capable of rapid movement: the mimosa plant (*Mimosa pudica*) makes its thin leaves point down at the slightest touch and carnivorous plants such as the Venus flytrap snap shut by the touch of insects. In the Venus flytrap, touch is detected by cilia that generate an action potential that stimulates motor cells, by which movement occurs.

In plants, the mechanism responsible for adaptation is signal transduction.

Adaptive responses include:

- Active foraging for light and nutrients. They do this by changing their architecture, physiology and phenotype.
- Leaves and branches are positioned and oriented in response to light source.
- Ability to detect soil volume and adapt growth accordingly independently of nutrient availability.
- Adaptively defend against herbivores.

Plant Intelligence

Plants do not have a brain or neuronal network, but reactions within signalling pathways may provide a biochemical basis for learning and memory in addition to computation and problem solving. Controversially, the brain is used as a metaphor in plant intelligence to provide an integrated view of signalling.

Plants respond to environmental stimuli by movement and changes in morphology. They communicate while actively competing for resources. In addition, plants accurately compute their circumstances, use sophisticated cost–benefit analysis and take tightly controlled actions to mitigate and control diverse environmental stressors. Plants are also capable of discriminating positive and negative experiences and of learning by registering memories from their past experiences. Plants use this information to update their behaviour in order to survive present and future challenges of their environment.

Plant physiology studies the role of signalling to integrate data obtained at the genetic, biochemical, cellular and physiological levels to understand plant development and behaviour. The neurobiological view sees plants as information-processing organisms with rather complex processes of communication occurring throughout the individual plant organism. It studies how environmental information is gathered, processed, integrated and shared (sensory plant biology) to enable these adaptive and coordinated responses (plant behaviour); and how sensory perceptions and behavioural events are 'remembered' in order to allow predictions of future activities upon the basis of past experiences. Plants, it is claimed by some plant physiologists, are as sophisticated in behaviour as animals but this sophistication has been masked by the time scales of plants' response to stimuli, many orders of magnitude slower than animals.

It has been argued that although plants are capable of adaptation, it should not be called intelligence, as plant neurobiologists are relying primarily on metaphors and analogies to argue that complex responses in plants can only be produced by intelligence. A bacterium can monitor its environment and instigate developmental processes appropriate to the prevailing circumstances, but is that intelligence? Such simple adaptation behaviour might be bacterial intelligence but is clearly not animal intelligence. However, plant intelligence fits a definition of intelligence proposed by David Stenhouse in a book about evolution and animal intelligence where he described it as adaptively variable behaviour during the lifetime of the individual. Critics of the concept have also argued that a plant cannot have goals once it is past the development stage of seedling because, as a modular organism, each module seeks its own survival goals and the resulting organism level behavior is not centrally controlled. This view, however, necessarily accommodates the possibility that a tree is a collection of individually intelligent modules cooperating, competing and influencing each other to determine behavior in a bottom-up fashion. The development into a larger organism whose modules must deal with different environmental conditions and challenges is not universal across plant species however, as smaller organisms might be subject to the same conditions across their bodies, at least, when the below and above ground parts are considered separately. Moreover, the claim that central control of development is completely absent from plants is readily falsified by apical dominance.

The Italian botanist Federico Delpino wrote on the idea of plant intelligence in 1867. Charles Darwin studied the movement of plants and in 1880 published a book. In the book he concludes:

It is hardly an exaggeration to say that the tip of the radicle thus endowed acts like the brain of one of the lower animals; the brain being situated within the anterior end of the body, receiving impressions from the sense-organs, and directing the several movements.

In philosophy, there are few studies of the implications of plant perception. Michael Marder put forth a phenomenology of plant life based on the physiology of plant perception. Paco Calvo Garzon offers a philosophical take on plant perception based on the cognitive sciences and the computational modeling of consciousness.

Comparison to Neurobiology

A plant's sensory and response system has been compared to the neurobiological processes of animals. Plant neurobiology, an unfamiliar misnomer, concerns mostly the sensory adaptive behaviour of plants and plant electrophysiology. Indian scientist J. C. Bose is credited as the first person to research and talk about neurobiology of plants. Many plant scientists and neuroscientists, however, view this as inaccurate, because plants do not have neurons.

The ideas behind plant neurobiology were criticised in a 2007 article by Amedeo Alpi and 35 other scientists, including such eminent plant biologists as Gerd Jürgens, Ben Scheres, and Chris Sommerville. The breadth of fields of plant science represented by these researchers reflects the fact that the vast majority of the plant science research community reject plant neurobiology. Their main arguments are that:

- Plant neurobiology does not add to our understanding of plant physiology, plant cell biology or signaling.
- There is no evidence for structures such as neurons, synapses or a brain in plants.
- The common occurrence of plasmodesmata in plants which poses a problem for signaling from an electrophysiological point of view since extensive electrical coupling would preclude the need for any cell-to-cell transport of a 'neurotransmitter-like' compounds.

The authors call for an end to superficial analogies and questionable extrapolations if the concept of plant neurobiology is to benefit the research community.

There were several responses to the criticism clarifying that the term plant neurobiology is a metaphor and metaphors have proved useful on several previous occasions. Plant ecophysiology describes this phenomenon.

Parallels in other Taxa

As described above in the case of a plant, similar mechanisms exist in a bacterial cell, a choanoflagellate, a fungal hypha, or a sponge, among the many other examples. All of these individual organisms of the respective taxa, despite being devoid of a brain or nervous system, are capable of sensing their immediate and momentary environment and responding accordingly. In the case of single-celled life, the sensory pathways are even more primitive in the sense that they take place on the surface of a single cell, as opposed to a network of many cells.

Hardiness

Hardiness is genetic. That is why some plants are hardier than others, even why some cultivars are hardier than others of the same plant. They may have been bred or selected as being hardier clones. To confuse this further, plants adapt or change genetically over time to their climate. This is why a species growing in a southern location may not be as hardy as the same species growing in a northern location. This may be a consideration when buying plants, or ordering them through the mail.

Roots, stems, leaf buds, and flower buds usually all are hardy to different temperatures. This is why many perennials die to the ground in winter, only to have their roots survive and produce new shoots the following spring. It is also why forsythia often have leaves but no flowers. The flower buds, being less hardy than the leaf buds, are killed by cold. If flowers appear to a particular height, but not above, this represents the depth of snow cover that protected the flower buds during the killing cold.

Hardiness is a function of location in a different sense as well. A particular, plant such as impatiens, may be perennial in a southern or warmer climate, only to be annual if moved to a colder climate.

Mechanisms

Woody plants survive freezing temperatures by suppressing the formation of ice in living cells or by allowing water to freeze in plant parts that are not affected by ice formation. The common mechanism for woody plants to survive up to −40°C (−40°F) is supercooling. Woody plants that survive lower temperatures are dehydrating their cells, allowing water to freeze between cell walls and the cells to survive. Plants normally considered hardy may not survive freezing if they are not readily acclimated, which renders them unable to use these mechanisms.

Hardiness Ratings

Heather, *Calluna vulgaris*, one of Britain's hardiest plants (North York Moors)

Various hardiness ratings are published. In the USA, the most widely used is the USDA system of hardiness zones based on average minimum yearly temperatures. This system was developed specifically for the extremely diverse range of conditions in the United States (US), from baking desert to frozen tundra. Another commonly used system is the Sunset Climate Zone system. This system is much more specific to climates (i.e. precipitation, temperature, and humidity based) and less dependent on the yearly minimum.

In contrast the United Kingdom (UK) and Western Europe have an oceanic climate, and experience a narrower range of temperatures that is tempered by the presence of the Gulf Stream. This results in areas like western Scotland experiencing relatively mild winter conditions that enable the growing of subtropical plants, despite being well to the north of subtropical climate areas. The Royal Horticultural Society has published a set of hardiness ratings applicable to the UK. The ratings range from H1a to H7. H1a, higher than 15°C (59°F), applies to tropical plants permanently under glass in heat; while H7, below −20°C (−4°F), applies to very cold-tolerant plants such as heathers. Most outdoor plants in the UK fall within the range H4, −10 to −5°C (14 to 23°F) (hardy in the average winter) to H5, −15 to −10°C (5 to 14°F) (hardy in a cold winter). Also, the average minimum temperature in the UK is much warmer than the average minimums in most of the US. This means that the coldest areas in the UK would be considered USDA Zone 7, plants considered 'Fully Hardy' in the UK may not be hardy below Zone 7 in the US.

In addition to cold tolerance, plant hardiness has been observed to be linked to how much stress specific plants are undergoing into the winter, or even how fast the onset of cold weather is in a specific year. This means that often stressed plants will exhibit less cold tolerance than plants that have been well maintained. Plants may also die if the winter changes from balmy to exceptionally cold in a short period of time.

Plant Stress

Any change in the surrounding environment may disrupt homeostasis. Environmental modulation of homeostasis may be defined as biological stress. Thus, it follows that plant stress implies some adverse effect on the physiology of a plant induced upon a sudden transition from some optimal environmental condition where homeostasis is maintained to some suboptimal condition which disrupts this initial homeostatic state. Thus, plant stress is a relative term since the experimental design to assess the impact of a stress always involves the measurement of a physiological phenomenon in a plant species under a suboptimal, stress condition compared to the measurement of the same physiological phenomenon in the same plant species under optimal conditions.

Plants Respond to Stress in Several Different Ways

Plant stress can be divided into two primary categories. Abiotic stress is a physical (e.g., light, temperature) or chemical insult that the environment may impose on a plant. Biotic stress is a biological insult, (e.g., insects, disease) to which a plant may be exposed during its lifetime. Some plants may be injured by a stress, which means that they exhibit one or more metabolic dysfunctions. If the stress is moderate and short term, the injury may be temporary and the plant may recover when the stress is removed. If the stress is severe enough, it may prevent flowering, seed formation, and induce senescence that leads to plant death. Such plants are considered to be susceptible. Some plants escape the stress altogether, such as ephemeral, or short-lived, desert plants.

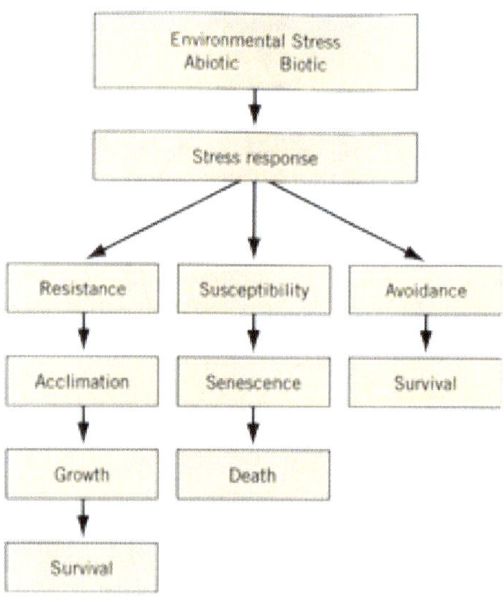

Figure: Effect of environmental stress on plant survival

Ephemeral plants germinate, grow, and flower very quickly following seasonal rains. They thus complete their life cycle during a period of adequate moisture and form dormant seeds before the onset of the dry season. In a similar manner, many arctic annuals rapidly complete their life cycle during the short arctic summer and survive over winter in the form of seeds. Because ephemeral plants never really experience the stress of drought or low temperature, these plants survive the environmental stress by stress avoidance. Avoidance mechanisms reduce the impact of a stress, even though the stress is present in the environment. Many plants have the capacity to tolerate a particular stress and hence are considered to be stress resistant. Stress resistance requires that the organism exhibit the capacity to adjust or to acclimate to the stress.

Stress Resistance requires that the Organism Exhibit the Capacity to Adjust or to Acclimate to the Stress

A plant stress usually reflects some sudden change in environmental condition. However, in stress-tolerant plant species, exposure to a particular stress leads to acclimation to that specific stress in a time-dependent manner. Thus, plant stress and plant acclimation are intimately linked with each other. The stress-induced modulation of homeostasis can be considered as the signal for the plant to initiate processes required for the establishment of a new homeostasis associated with the acclimated state. Plants exhibit stress resistance or stress tolerance because of their genetic capacity to adjust or to acclimate to the stress and establish a new homeostatic state over time. Furthermore, the acclimation process in stress-resistant species is usually reversible upon removal of the external stress.

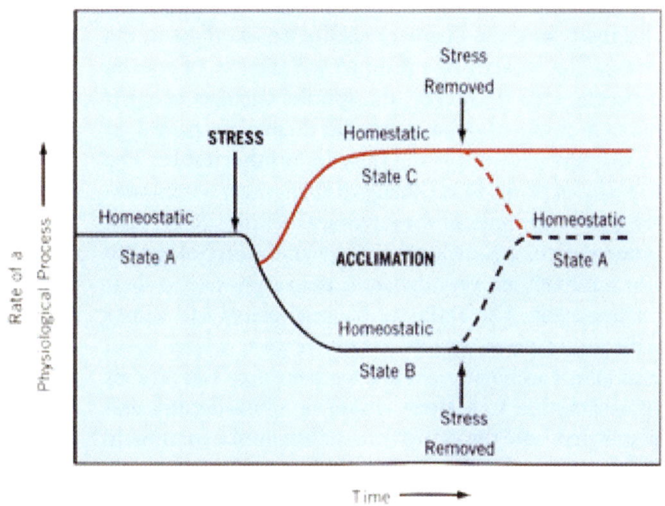

Figure. A schematic relationship between stress and acclimation

The establishment of homeostasis associated with the new acclimated state is not the result of a single physiological process but rather the result of many physiological processes that the plant integrates over time, that is, integrates over the acclimation period. Plants usually integrate these physiological processes over a short-term as well as a long-term basis. The *short-term processes* involved in acclimation can be initiated within seconds or minutes upon exposure to a stress but may be transient in nature. That means that although these processes can be detected very soon after the onset of a stress, their activities also disappear rather rapidly. As a consequence, the life-time of these processes is rather short. In contrast, *long-term processes* are less transient and thus usually exhibit a longer lifetime. However, the lifetimes of these processes overlap in time such that the short-term processes usually constitute the initial responses to a stress while the long-term processes are usually detected later in the acclimation process. Such a hierarchy of short- and long-term responses indicates that the attainment of the acclimated state can be considered a complex, time-nested response to a stress. Acclimation usually involves the differential expression of specific sets of genes associated with exposure to a particular stress. The remarkable capacity to *regulate gene expression* in response to environmental change in a time-nested manner is the basis of plant plasticity.

Adaptation and Phenotypic Plasticity

Plants have various mechanisms that allow them to survive and often prosper in the complex environments in which they live. Adaptation to the environment is characterized by genetic changes in the entire population that have been fixed by natural selection over many generations. In contrast, individual plants can also respond to changes in the environment, by directly altering their physiology or morphology to allow them to better survive the new environment. These responses require no new genetic modifications, and if the response of an individual improves with repeated exposure to the new environmental condition then the response is one of acclimation. Such responses are often referred to as phenotypic plasticity, and represent nonpermanent changes in the physiology or morphology of the individual that can be reversed if the prevailing environmental conditions change.

Individual Plants may also show Phenotypic Plasticity that allows them to Respond to Environmental Fluctuations

In addition to genetic changes in entire populations, individual plants may also show phenotypic plasticity; they may respond to fluctuations in the environment by directly altering their morphology and physiology. The changes associated with phenotypic plasticity require no new genetic modifications, and many are reversible. Both genetic adaptation and phenotypic plasticity can contribute to the plant's overall tolerance of extremes in their abiotic environment. As a consequence, a plant's physiology and morphology are not static but are very dynamic and responsive to their environment. The ability of biennial plants and winter cultivars of cereal grains to survive over winter is an example of acclimation to low temperature. The process of acclimation to a stress is known as hardening and plants that have the capacity to acclimate are commonly referred to as hardy species. In contrast, those plants that exhibit a minimal capacity to acclimate to a specific stress are referred to as nonhardy species.

Imbalances of Abiotic Factors

Plants may experience physiological stress when an abiotic factor is deficient or in excess (referred to as an imbalance). The deficiency or excess may be chronic or intermittent. Abiotic conditions to which native plants are adapted may cause physiological stress to non-native plants. Most agricultural crops, for example, are cultivated in regions to which they are not highly adapted. Field crops are estimated to produce only 22% of their genetic potential for yield because of suboptimal climatic and soil conditions.

Imbalances of abiotic factors in the environment cause *primary and secondary effects* in plants.

Primary effects such as reduced water potential and cellular dehydration directly alter the physical and biochemical properties of cells, which then lead to secondary effects. These secondary effects, such as reduced metabolic activity, ion cytotoxicity, and the production of reactive oxygen species, initiate and accelerate the disruption of cellular integrity, and may lead ultimately to cell death. Different abiotic factors may cause similar primary physiological effects because they affect the same cellular processes. This is the case for water deficit, salinity, and freezing, all of which cause reduction in hydrostatic pressure (turgor pressure, Ψp) and cellular dehydration. Secondary physiological effects caused by different abiotic imbalances may overlap substantially. It is evident that imbalances in many abiotic factors reduce cell proliferation, photosynthesis, membrane integrity, and protein stability, and induce production of *reactive oxygen species (ROS)*, oxidative damage, and cell death.

Light-dependent Inhibition of Photosynthesis

As photoautotrophs, plants are dependent upon – and exquisitely adapted to – visible light for the maintenance of a positive carbon balance through photosynthesis. Higher energy wavelengths of electromagnetic radiation, especially in the ultraviolet range, can inhibit cellular processes by damaging membranes, proteins, and nucleic acids. However, even in the visible range, irradiances far above the light saturation point of photosynthesis cause high light stress, which can disrupt chloroplast structure and reduce photosynthetic rates, a process known as photoinhibition.

Photoinhibition by high light leads to the production of destructive forms of oxygen:

Excess light excitation arriving at the PSII reaction center can lead to its inactivation by the direct damage of the D1 protein. Excess absorption of light energy by photosynthetic pigments also produces excess electrons outpacing the availability of $NADP^+$ to act as an electron sink at PSI. The excess electrons produced by PSI lead to the production of reactive oxygen species (ROS), notably superoxide (O_2). Superoxide and other ROS are low-molecular-weight molecules that function in signaling and, in excess, cause oxidative damage to proteins, lipids, RNA, and DNA. The oxidative stress generated by excessive ROS destroys cellular and metabolic functions and leads to cell death.

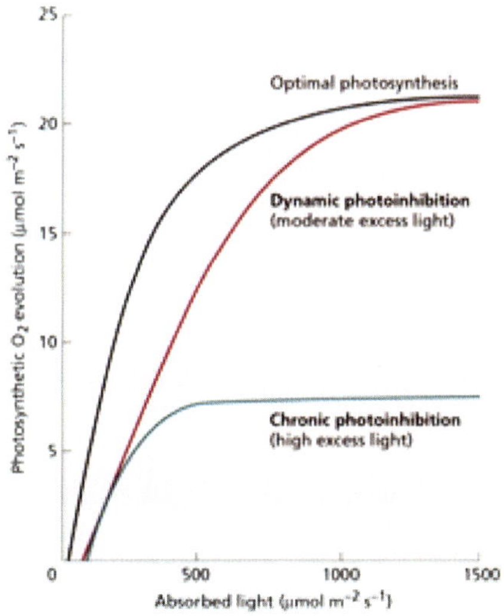

Figure: Changes in the light-response curves of photosynthesis caused by photoinhibition

Temperature Stress

Mesophytic plants (terrestrial plants adapted to temperate environments that are neither excessively wet nor dry) have a relatively narrow temperature range of about 10°C for optimal growth and development. Outside of this range, varying amounts of damage occur, depending on the magnitude and duration of the temperature fluctuation. In this topic we will discuss three types of temperature stress: high temperatures, low temperatures above freezing, and temperatures below freezing. Most actively growing tissues of higher plants are tillable to survive extended exposure to temperatures above 45°C or even short exposure to temperatures of 55°C or above. However, nongrowing cells or dehydrated tissues (e.g., seeds and pollen) remain viable at much higher temperatures. Pollen grains of some species can survive 70°C and some dry seeds can tolerate temperatures as high as 120°C.

Most plants with access to abundant water are able to maintain leaf temperatures below 45°C by evaporative cooling, even at elevated ambient temperatures. However, high leaf temperatures combined with minimal evaporative cooling causes heat stress. Leaf temperatures can rise to 4 to 5°C above ambient air temperature in bright sunlight near midday, when soil water deficit causes partial stomatal closure or when high relative humidity reduces the gradient driving evaporative cooling. Increases in leaf temperature during the day can be more pronounced in plants experiencing drought and high irradiance from direct sunlight.

Temperature Stress can Result in Damaged Membranes and Enzymes

Plant membranes consist of a lipid bilayer interspersed with proteins and sterols, and any abiotic factor that alters membrane properties can disrupt cellular processes. The physical properties of the lipids greatly influence the activities of the integral membrane proteins, including H+-pumping ATPases, carriers, and channel-forming proteins that regulate the transport of ions and other solutes. High temperatures cause an increase in the fluidity of membrane lipids and a decrease in the strength of hydrogen bonds and electrostatic interactions between polar groups of proteins within the aqueous phase of the membrane. High temperatures thus modify membrane composition and structure, and can cause leakage of ions. High tempeatures can also lead to a loss of the three-dimensional structure required for correct function of enzymes or structural cellular components, thereby leading to loss of proper enzyme structure and activity. Misfolded proteins often aggregate and precipitate, creating serious problems within the cell.

Temperature Stress can Inhibit Photosynthesis

Photosynthesis and respiration are both inhibited by temperature stress. Typically, photosynthetic rates are inhibited by high temperatures to a greater extent than respiratory rates. Although chloroplast enzymes such as rubisco, rubisco activase, NADP-G3P dehydrogenase, and PEP carboxylase become unstable at high temperatures, the temperatures at which these enzymes began to denature and lose activity are distinctly higher than the temperatures at which photosynthetic rates begin to decline. This would indicate that the early stages of heat injury to photosynthesis are more directly related to changes in membrane properties and to uncoupling of the energy transfer mechanisms in chloroplasts.

This imbalance between photosynthesis and respiration is one of the main reasons for the deleterious effects of high temperatures. On an individual plant, leaves growing in the shade have a lower temperature compensation point than leaves that are exposed to the sun (and heat). Reduced photosynthate production may also result from stress-induced stomatal closure, reduction in leaf canopy area, and regulation of assimilate partitioning.

Freezing Temperatures cause Ice Crystal Formation and Dehydration

Freezing temperatures result in intra- and extracellular ice crystal formation. Intracellular ice formation physically shears membranes and organelles. Extracellular ice crystals, which usually form before the cell contents freeze, may not cause immediate physical damage to cells, but they do cause cellular dehydration. This is because ice formation substantially lowers the water potential (Ψw) in the apoplast, resulting in a gradient from high Ψw in the symplast to low Ψw in the apoplast. Consequently, water moves from the symplast to the apoplast, resulting in cellular dehydration. Cells that are already dehydrated, such as those in seeds and pollen, are relatively less affected by ice crystal formation. Ice usually forms first within the intercellular spaces and in the xylem vessels, along which the ice can quickly propagate. This ice formation is not lethal to hardy plants, and the tissue recovers fully if warmed. However, when plants are exposed to freezing temperatures for an extended period, the growth of extracellular ice crystals leads to physical destruction of membranes and excessive dehydration.

Imbalances in Soil Minerals

Imbalances in the mineral content of soils can affect plant fitness either indirectly, by affecting plant nutritional status or water uptake, or directly, through toxic effects on plant cells.

Soil mineral content can result in plant stress in various ways:

Several anomalies associated with the elemental composition of soils can result in plant stress, including high concentrations of salts (e.g., Na^+ and Cl^-) and toxic ions (e.g., As and Cd), and low concentrations of essential mineral nutrients, such as Ca^{2+}, Mg^{2+}, N, and P. The term salinity is used to describe excessive accumulation of salt in the soil solution. Salinity stress has two components: nonspecific osmotic stress that causes water deficits, and specific ion effects resulting from the accumulation of toxic ions, which disturb nutrient acquisition and result in cytotoxicity. Salt-tolerant plants genetically adapted to salinity are termed *halophytes*, while less salt-tolerant plants that are not adapted to salinity are termed *glycophytes*.

Soil Salinity Occurs Naturally and as the Result of Improper Water Management Practices

In natural environments, there are many causes of salinity. Terrestrial plants encounter high salinity close to the seashore and in estuaries where seawater and freshwater mix or replace each other with the tides. The movement of seawater upstream into rivers can be substantial, depending on the strength of the tidal surge. Far inland, natural seepage from geologic marine deposits can wash salt into adjoining areas. Evaporation and transpiration remove pure water (as vapor) from the soil, concentrating the salts in the soil solution. Soil salinity is also increased when water droplets from the ocean disperse over land and evaporate.

Human activities also contribute to soil salinization. Improper water management practices associated with intensive agriculture can cause substantial salinization of croplands. In many areas of the world, salinity threatens the production of staple foods. Irrigation water in semiarid and arid regions is often saline. Only halophytes, the most salt-tolerant plants, can tolerate high levels of salts. Glycophytic crops cannot be grown with saline irrigation water.

Saline soils are often associated with high concentrations of NaCl, but in some areas Ca^{2+}, Mg^{2+}, and $SO4^{-}$ are also present in high concentrations in saline soils. High Na^{+} concentrations that occur in sodic soils (soils in which Na^{+} occupies $\geqslant$10% of the cation exchange capacity) not only injure plants but also degrade the soil structure, decreasing porosity and water permeability. Salt incursion into the soil solution causes water deficits in leaves and inhibits plant growth and metabolism.

High Cytosolic Na+ and Cl- denature Proteins and Destabilize Membranes

The most widespread example of a specific ion effect is the cytotoxic accumulation of Na^{+} and Cl- ions under saline conditions. Under non-saline conditions, the cytosol of higher plant cells contains about 100 mM K^{+} and less than 10 mM Na^{+}, an ionic environment in which enzymes are optimally functional. In saline environments, cytosolic Na^{+} and Cl- increase to more than 100 mM, and these ions become cytotoxic. High concentrations of salt cause protein denaturation and membrane destabilization by reducing the hydration of these macromolecules. However, Na^{+} is a more potent denaturant than K^{+}.

At high concentrations, apoplastic Na^{+} also competes for sites on transport proteins that are necessary for high-affinity uptake of K^{+}, an essential macronutrient. Further, Na^{+} displaces Ca^{2+} from sites on the cell wall, reducing Ca^{2+} activity in the apoplast and resulting in greater Na^{+} influx, presumably through nonselective cation channels. Reduced apoplastic Ca^{2+} concentrations caused by excess Na^{+} may also restrict the availability of Ca^{2+} in the cytosol. Since cytosolic Ca^{2+} is necessary to activate Na^{+} detoxification via efflux across the plasma membrane, elevated external Na^{+} has the ability to block its own detoxification.

Developmental and Physiological Mechanisms against Environmental Stress

Plants can Modify their Life Cycles to Avoid Abiotic Stress

One way plants can adapt to extreme environmental conditions is through modification of their life cycles. For example, annual desert plants have short life cycles: they complete them during the periods when water is available, and are dormant (as seeds) during dry periods. Deciduous trees of the temperate zone shed their leaves before the winter so that sensitive leaf tissue is not damaged by cold temperatures. During less predictable stressful events (e.g., a summer of significant but erratic rainfall) the growth habits of some species may confer a degree of tolerance to these conditions. For example, plants that can grow and flower over an extended period (*indeterminate growth*) are often more tolerant to erratic environmental extremes than plants that develop preset numbers of leaves and flower over only very short periods (*determinate growth*).

Phenotypic Changes in Leaf Structure and Behavior

Because of their roles in photosynthesis, leaves (or their equivalent) are crucial to the survival of a plant. To function, leaves must be exposed to sunlight and air, but this also makes them particularly vulnerable to environmental extremes. Plants have thus evolved various mechanisms that enable them to avoid or mitigate the effects of abiotic extremes to leaves. Such mechanisms include changes in leaf area, leaf orientation, trichomes, and the cuticle.

Turgor reduction is the earliest significant biophysical effect of water deficit. As a result, turgor-dependent processes such as *leaf expansion* and root elongation are the most sensitive to water deficits. When water deficit develops slowly enough to allow changes in developmental processes, it has several effects on growth, one of which is a limitation of leaf expansion. Because leaf expansion depends mostly on cell expansion, the principles that underlie the two processes are similar. Inhibition of cell expansion results in a slowing of leaf expansion early in the development of water deficits. The resulting smaller leaf area transpires less water, effectively conserving a limited water supply in the soil over a longer period. Altering *leaf shape* is another way that plants can reduce leaf area. Under conditions of water, heat, or salinity extremes, leaves may be narrower or may develop deeper lobes during development. The result is a reduced leaf surface area and therefore, reduced water loss and heat load (defined as amount of heat loss [cooling] required to maintain a leaf temperature close to air temperature). For protection against overheating during water deficit, the leaves of some plants may orient themselves away from the sun. *Leaf orientation* may also change in response to low oxygen availability.

Figure. Altered leaf shape can occur in response to environmental changes: leaf from outside (left) and inside (right) of a tree canopy

Plants can Regulate Stomatal Aperture in Response to Dehydration Stress

The ability to control stomatal aperture allows plants to respond quickly to a changing environment, for example to avoid excessive water loss or limit uptake of liquid or gaseous pollutants through stomata. Stomatal opening and closing is modulated by uptake and loss of water in guard cells, which changes their turgor pressure. Although guard cells can lose turgor as a result of a direct loss of water by evaporation to the atmosphere, stomatal closure in response to dehydration is almost always an active, energy-dependent process rather than a passive one. Abscisic acid (ABA) mediates the solute loss from guard cells that is triggered by a decrease in the water content of the leaf. Plants constantly modulate the concentration and cellular localization of ABA, and this allows them to respond quickly to environmental changes, such as fluctuations in water availability.

Plants Adjust Osmotically to Drying Soil by Accumulating Solutes

Osmotic adjustment is the capacity of plant cells to accumulate solutes and use them to lower Ψw during periods of osmotic stress. The adjustment involves a net increase in solute content per cell that is independent of the volume changes that result from loss of water. The decrease in ΨS (= osmotic potential) is typically limited to about 0.2 to 0.8 MPa, except in plants adapted to extremely dry conditions.

There are two main ways by which osmotic adjustment can take place. A plant may *take up ions* from the soil, or *transport ions* from other plant organs to the root, so that the solute concentration of the root cells increases. For example, increased uptake and accumulation of $K+$ will lead to decreases in ΨS due to the effect of the potassium ions on the osmotic pressure within the cell. This is a common event in saline areas, where ions such as potassium and calcium are readily available to the plant. The accumulation of ions during osmotic adjustment is predominantly restricted to the vacuoles, where the ions are kept out of contact with cytosolic enzymes or organelles.

When ions are compartmentalized in the vacuole, other solutes must accumulate in the cytoplasm to maintain water potential equilibrium within the cell. These solutes are called *compatible solutes* (or *compatible osmolytes*). Compatible solutes are organic compounds that are osmotically active in the cell, but do not destabilize the membrane or interfere with enzyme function, as high concentrations of ions can. Plant cells can hold large concentrations of these compounds without detrimental effects on metabolism. Common compatible solutes include amino acids such as proline, sugar alcohols such as mannitol, and quaternary ammonium compounds such as glycine betaine.

Phytochelatins Chelate Certain Ions, Reducing their Reactivity and Toxicity

Chelation is the binding of an ion with at least two ligating atoms within a chelating molecule. Chelating molecules can have different atoms available for ligation, such as sulfur (S), nitrogen (N), or oxygen (O), and these different atoms have different affinities for the ions they chelate. By wrapping itself around the ion it binds to form a complex, the chelating molecule renders the ion less chemically active, thereby reducing its potential toxicity. The complex is then usually translocated to other parts of the plant, or stored away from the cytoplasm (typically in the vacuole). Phytochelatins are low-molecular-weight thiols consisting of the amino acids glutamate, cysteine, and glycine, with the general form of $(\gamma\text{-Glu-Cys})n\text{Gly}$. The thiol groups act as ligands for ions of trace elements such as Cd and As. Once formed, the phytochelatin-metal complex is transported into the vacuole for storage.

Many Plants have the Capacity to Acclimate to Cold Temperature

The ability to tolerate freezing temperatures under natural conditions varies greatly among tissues. Seeds and other partially dehydrated tissues, as well as fungal spores, can be kept indefinitely at temperatures near absolute zero (0 K, or -273°C), indicating that these very low temperatures are not intrinsically harmful. Hydrated, vegetative cells can also retain viability at freezing temperatures, provided that ice crystal formation can be restricted to the intercellular spaces and cellular dehydration is not too extreme.

Temperate plants have the capacity for *cold acclimation* – a process whereby exposure to low but nonlethal temperatures (typically above freezing) increases the capacity for low temperature survival. Cold acclimation in nature is induced in the early autumn by exposure to short days and nonfreezing, chilling temperatures, which combine to stop growth. A diffusible factor that promotes acclimation, most likely ABA, moves from leaves via the phloem to overwintering stems. ABA accumulates during cold acclimation and is necessary for this process.

Plants Survive Freezing Temperatures by Limiting Ice Formation

During rapid freezing, the protoplast, including the vacuole, may supercool; that is, the cellular water remains liquid because of its solute content, even at temperatures several degrees below its theoretical freezing point. Supercooling is common to many species of the hardwood forests. Cells can supercool to only about -40°C, the temperature at which ice forms spontaneously. Spontaneous ice formation sets the low-temperature limit at which many alpine and subarctic species that undergo deep supercooling can survive. It may also explain why the altitude of the timberline in mountain ranges is at or near the -40°C minimum isotherm. Several specialized plant proteins, termed antifreeze proteins, limit the growth of ice crystals through a mechanism independent of lowering of the freezing point of water. Synthesis of these antifreeze proteins is induced by cold temperatures. The proteins bind to the surfaces of ice crystals to prevent or slow further crystal growth.

Cold-resistant Plants Tend to have Membranes with more Unsaturated Fatty Acids

As temperatures drop, membranes may go through a phase transition from a flexible liquid-crystalline structure to a solid gel structure. The phase transition temperature varies with species (tropical species: 10-12°C; apples: 3-10°C) and the actual lipid composition of the membranes. Chilling-resistant plants tend to have membranes with more unsaturated fatty acids. Chilling-sensitive plants, on the other hand, have a high percentage of saturated fatty acid chains, and membranes with this composition tend to solidify into a semicrystalline state at a temperature well above 0°C. Prolonged exposure to extreme temperatures may result in an altered composition of membrane lipids, a form of acclimation. Certain transmembrane enzymes can alter lipid saturation, by introducing one or more double bonds into fatty acids. This modification lowers the temperature at which the membrane lipids begin a gradual phase change from fluid to semicrystalline form and allows membranes to remain fluid at lower temperatures, thus protecting the plant against damage from chilling.

A Large variety of Heat Shock Proteins can be Induced by different Environmental Conditions

Under environmental extremes, protein structure is sensitive to disruption. Plants have several mechanisms to limit or avoid such problems, including osmotic adjustment for maintenance of hydration and chaperone proteins that physically interact with other proteins to facilitate protein folding, reduce misfolding and aggregation, and stabilize protein tertiary structure. In response to sudden 5 to 10°C increases in temperature, plants produce a unique set of chaperone proteins referred to as heat shock proteins (HSPs). Cells that have been induced to synthesize HSPs show improved thermal tolerance and can tolerate subsequent exposure to temperatures that otherwise would be lethal. Heat shock proteins are also induced by widely different environmental conditions, including water deficit, ABA treatment, wounding, low temperature, and salinity. Thus, cells that have previously experienced one condition may gain cross-protection against another.

During Mild or short-term Water Shortage, Photosynthesis is Strongly Inhibited, but Phloem Translocation is Unaffected Until the Shortage becomes Severe

Changes in the environment may stimulate shifts in metabolic pathways. When the supply of O2 is insufficient for aerobic respiration, roots first begin to ferment pyruvate to lactate through the action of lactate dehydrogenase; this recycles NADH to NAD^+, allowing the maintenance of ATP production through glycolysis. Production of lactate (lactic acid) lowers the intracellular pH, inhibiting lactate dehydrogenase and activating pyruvate decarboxylase. These changes in enzyme activity quickly lead to a switch from lactate to ethanol production. The net yield of ATP in fermentation is only 2 moles of ATP per mole of hexose sugar catabolized (compared with 36 moles of ATP per mole of hexose respired in aerobic respiration). Thus, injury to root metabolism by O2 deficiency originates in part from a lack of ATP to drive essential metabolic processes such as root absorption of essential nutrients.

Water shortage decreases both photosynthesis and the consumption of assimilates in the expanding leaves. As a consequence, water shortage indirectly decreases the amount of photosynthate exported from leaves. Because phloem transport depends on pressure gradients, decreased water potential in the phloem during water deficit may inhibit the movement of assimilates. The ability to continue translocating assimilates is a key factor in almost all aspects of plant resistance to drought.

Plant stress measurement is the quantification of environmental effects on plant health. When plants are subjected to less than ideal growing conditions, they are considered to be under stress. Stress factors can affect growth, survival and crop yields. Plant stress research looks at the response of plants to limitations and excesses of the main abiotic factors (light, temperature, water and nutrients), and of other stress factors that are important in particular situations (e.g. pests, pathogens, or pollutants). Plant stress measurement usually focuses on taking measurements from living plants. It can involve visual assessments of plant vitality, however, more recently the focus has moved to the use of instruments and protocols that reveal the response of particular processes within the plant (especially, photosynthesis, plant cell signalling and plant secondary metabolism)

- Determining the optimal conditions for plant growth, e.g. optimising water use in an agricultural system.
- Determining the climatic range of different species or subspecies.
- Determining which species or subspecies are resistant to a particular stress factor.

Instruments used to measure Plant Stress

Measurements can be made from living plants using specialised equipment. Among the most commonly used instruments are those that measure parameters related to photosynthesis (chlorophyll content, chlorophyll fluorescence, gas exchange) or water use (porometer, pressure bomb). In addition to these general purpose instruments, researchers often design or adapt other instruments tailored to the specific stress response they are studying.

Photosynthesis Systems

Photosynthesis systems use infrared gas analyzers (IRGAS) for measuring photosynthesis. CO_2 concentration changes in leaf chambers are measured to provide carbon assimilation values for leaves or whole plants. Research has shown that the rate of photosynthesis is directly related to the amount of carbon assimilated by the plant. Measuring CO_2 in the air, before it enters the leaf chamber, and comparing it to air measured for CO_2 after it leaves the leaf chamber, provides this value using proven equations. These systems also use IRGAs, or solid state humidity sensors, for measuring H_2O changes in leaf chambers. This is done to measure leaf transpiration, and to correct CO_2 measurements. The light absorption spectrum for CO_2 and H_2O overlap somewhat, therefore, a correction is necessary for reliable CO_2 measuring results. The critical measurement for most plant stress measurements is designated by "A" or carbon assimilation rate. When a plant is under stress, less carbon is assimilated. CO_2 IRGAs are capable of measuring to approximately +/- 1 μmol or 1ppm of CO_2.

Because these systems are effective in measuring carbon assimilation and transpiration at low rates, as found in stressed plants, they are often used as the standard to compare to other types of instruments. Photosynthesis instruments come in field portable and laboratory versions. They are also designed to measure ambient environmental conditions, and some systems offer variable microclimate control of the measuring chamber. Microclimate control systems allow adjustment of the measuring chamber temperature, CO_2 level, light level, and humidity level for more detailed investigation.

The combination of these systems with fluorometers, can be especially effective for some types of stress, and can be diagnostic, e.g. in the study of cold stress and drought stress.

Chlorophyll Fluorometers

Chlorophyll fluorescence emitted from plant leaves, gives an insight into the health of the photosynthetic systems within the leaf. Chlorophyll fluorometers are designed to measure variable fluorescence of photosystem II. This variable fluorescence can be used to measure the level of plant stress. The most commonly used protocols include those aimed at measuring the photosynthetic efficiency of photosystem II, both in the light (ΔF/Fm') and in a dark-adapted state (Fv/Fm). Chlorophyll fluorometers are, for the most part, less expensive tools than photosynthesis systems, they also have a faster measurement time and tend to be more portable. For these reasons they have become one of the most important tools for field measurements of plant stress.

Fv/Fm

Fv/Fm tests whether or not plant stress affects photosystem II in a dark adapted state. Fv/Fm is the most used chlorophyll fluorescence measuring parameter in the world. The majority of fluorescence measurements are now made using modulated fluorometers with the leaf poised in a known state.

Light that is absorbed by a leaf follows three competitive pathways. It may be used in photochemistry to produce ATP and NADPH used in photosynthesis, it can be re-emitted as fluorescence, or dissipated as heat. The Fv/Fm test is designed to allow the maximum amount of the light energy to take the fluorescence pathway. It compares the dark-adapted leaf pre-photosynthetic fluorescent state, called minimum fluorescence, or Fo, to maximum fluorescence called Fm. In maximum fluorescence, the maximum number of reaction centers have been reduced or closed by a saturating light source. In general, the greater the plant stress, the fewer open reaction centers available, and the Fv/Fm ratio is lowered. Fv/Fm is a measuring protocol that works for many types of plant stress.

In Fv/Fm measurements, after dark adaption, minimum fluorescence is measured, using a modulated light source. This is a measurement of antennae fluorescence using a modulated light intensity that is too low to drive photosynthesis. Next, an intense light flash, or saturation pulse, of a limited duration, is used, to expose the sample, and close all available reaction centers. With all available reaction centers closed, or chemically reduced, maximum fluorescence is measured. The difference between maximum fluorescence and minimum fluorescence is Fv, or variable fluorescence. Fv/Fm is a normalize ratio created by dividing variable fluorescence by maximum fluorescence. It is a measurement ratio that represents the maximum potential quantum efficiency of Photosystem II if all capable reaction centers were open. An Fv/Fm value in the range of 0.79 to 0.84 is the approximate optimal value for many plant species, with lowered values indicating plant stress (Maxwell K., Johnson G. N. 2000), (Kitajima and Butler, 1975). Fv/Fm is a fast test that usually takes a few seconds. It was developed in and around 1975 by Kitajima and Butler. Dark adaptation times vary from about fifteen minutes to overnight. Some researchers will only use pre-dawn values.

Y(II) or ΔF/Fm' and ETR

Y(II) is a measuring protocol that was developed by Bernard Genty with the first publications in 1989 and 1990. It is a light adapted test that allows one to measure plant stress while the plant is undergoing the photosynthetic process at steady-state photosynthesis lighting conditions. Like FvFm, Y(II) represents a measurement ratio of plant efficiency, but in this case, it is an indication of the amount of energy used in photochemistry by photosystem II under steady-state photosynthetic lighting conditions. For most types of plant stress, Y(II) correlates to plant carbon assimilation in a linear fashion in C_4 plants. In C_3 plants, most types of plant stress correlate to carbon assimilation in a curve-linear fashion. According to Maxwell and Johnson, it takes between fifteen and twenty minutes for a plant to reach steady-state photosynthesis at a specific light level. In the field, plants in full sunlight, and not under canopy, or partly cloudy conditions, are considered to be at steady state. In this test, light irradiation levels and leaf temperature must be controlled or measured, because while the Y(II) parameter levels vary with most types of plant stress, it also varies with light level and temperature. Y(II) values will be higher at lower light levels than at higher light levels. Y(II) has the advantage that it is more sensitive to a larger number of plant stress types than Fv/Fm.

ETR, or electron transport rate, is also a light-adapted parameter that is directly related to Y(II) by the equation, $ETR = Y(II) \times PAR \times 0.84 \times 0.5$. By multiplying Y(II) by the irradiation light level in the PAR range (400 nm to 700 nm) in μmols, multiplied by the average ratio of light absorbed by the leaf 0.84, and multiplied by the average ratio of PSII reaction centers to PSI reaction centers, 0.50, relative ETR measurement is achieved.

Relative ETR values are valuable for stress measurements when comparing one plant to another, as long as the plants to be compared have similar light absorption characteristics. Leaf absorption characteristics can vary by water content, age, and other factors. If absorption differences are a concern, absorption can be measured with the use of an integrating sphere. For more accurate ETR values, the leaf absorption value and the ratio of PSII reaction centers to PSI reaction centers can be included in the equation. If different leaf absorption ratios are an issue, or they are an unwanted variable, then using Y(II) instead of ETR, may be the best choice. Four electrons must be transported for every CO_2 molecule assimilated, or O_2 molecule evolved, differences from gas exchange measurements, especially in C_3 plants, can occur under conditions that promote photorespiration, cyclic electron transport, and nitrate reduction.

Quenching Measurements

Quenching measurements have been traditionally used for light stress, and heat stress measurements. In addition, they have been used to study plant photoprotective mechanisms, state transitions, plant photoinhibition, and the distribution of light energy in plants. While they can be used for many types of plant stress measurement, the time required, and the additional expense required for this capability, limit their use. These tests commonly require overnight dark adaptation, and fifteen to twenty minutes in lighted conditions to reach steady state photosynthesis before measurement.

Puddle Model and Lake Model Quenching Parameters

Understanding of the organization of plant antennae, or plant light collection structures, and reaction centers, where the photosynthetic light reaction actually takes place, has changed over the years. It is now understood that a single antennae does not link only to a single reaction center, as was previously described in the puddle model. Current evidence indicates that reaction centers are connected with shared antennae in terrestrial plants. As a result, the parameters used to provide reliable measurements have changed to represent the newer understanding of this relationship. The model that represents the newer understanding of the antennae - reaction center relationship is called the lake model.

Lake model parameters were provided by Dave Kramer in 2004. Since then, Luke Hendrickson has provided simplified lake model parameters that allow the resurrection of the parameter NPQ, from the puddle model, back into the lake model. This is valuable because there have been so many scientific papers that have used NPQ for plant stress measurement, as compared to papers that have used lake model parameters.

OJIP or OJIDP

OJIP or OJIDP is a dark adapted chlorophyll fluorescence technique that is used for plant stress measurement. It has been found that by using a high time resolution scale, the rise to maximum fluorescence from minimum fluorescence has intermediate peaks and dips, designated by the OJID and P nomenclature. Over the years, there have been multiple theories of what the rise, time scale, peaks and dips mean. In addition, there is more than one school as to how this information should be used for plant stress testing. Like Fv/Fm, and the other protocols, the research shows that OJIP works better for some types of plant stress than it does for others.

Choosing the best Chlorophyll Fluorescence Protocol and Parameter

When choosing the correct protocol, and measuring parameter, for a specific type of plant stress, it is important to understand the limitations of the instrument, and the protocol used. For example, it was found that when measuring Oak leaves, a photosynthesis system could detect heat stress at 30 °C and above, Y(II) could detect heat stress at 35 °C and above, NPQ could detect heat stress at 35 °C and above, and Fv/Fm could only detect heat stress at 45 °C and above. (Haldiman P, & Feller U. 2004) OJIP was found to detect heat stress at 44 °C and above on samples tested (Strasser 2004).

The relationship between carbon assimilation measurements made by photosynthesis systems of the dark Calvin cycle, and measurements of variable fluorescence of photosystem II (PSII), made by chlorophyll fluorometers of the light reaction, are not always straightforward. For this reason, choosing the correct chlorophyll fluorescence protocol can also be different for C_3 and C_4 plants. It has been found, for example, that Y(II) and ETR are good tests for drought stress in C_4 plants, but a special assay is required to measure drought stress in most C_3 plants at usable levels. In C_3 plants, photorespiration, and the Mehler reaction, are thought to be a principal cause.

Chlorophyll Content Meters

These are instruments that use light transmission through a leaf, at two wavelengths, to determine the greenness and thickness of leaves. Transmission in the infrared range provides a measurement related to leaf thickness, and a wavelength in the red light range is used to determine greenness. The ratio of the transmission of the two wavelengths provides a chlorophyll content index that is referred to as CCI or alternatively as a SPAD index. CCI is a linear scale, and SPAD is a logarithmic scale. These instruments and scales have been shown to correlate to chlorophyll chemical tests for chlorophyll content except at very high levels.

Chlorophyll content meters are commonly used for nutrient plant stress measurement, that includes nitrogen stress, and sulfur stress. Because research has shown, that if used correctly, chlorophyll content meters are reliable for nitrogen management work, these meters are often the instruments of choice for crop fertilizer management because they are relatively inexpensive. Research has demonstrated that by comparing well fertilized plants to test plants, the ratio of the chlorophyll content index of test plants, divided by the chlorophyll content index of well fertilized plants, will provide a ratio that is an indication of when fertilization should occur, and how much should be used. It is common to use a well fertilized stand of crops in a specific field and sometimes in different areas of the same field, as the fertilization reference, due to differences from field to field and within a field. The research done to date uses either ten and thirty measurements on test and well fertilized crops, to provide average values. Studies have been done for corn and wheat. One paper suggests that when the ratio drops below 95%, it is time to fertigate. The amounts of fertilizer are also recommended.

Crop consultants also use these tools for fertilizer recommendations. However, because strict scientific protocols are more time consuming and more expensive, consultants sometimes use well-fertilized plants located in low-lying areas as the standard well-fertilized plants. They typically also use fewer measurements. The evidence for this approach involves anecdotal discussions with crop consultants. Chlorophyll content meters are sensitive to both nitrogen and sulfur stress at usable levels. Chlorophyll fluorometers require a special assay, involving a high actinic light level in combination with nitrogen stress, to measure nitrogen stress at usable levels. In addition, chlorophyll fluorometers will only detect sulfur stress at starvation levels. For best results, chlorophyll content measurements should be made when water deficits are not present. Photosynthesis systems will detect both nitrogen and sulfur stress.

Vegetation

The term vegetation is used in ecology to describe the overall characteristics of plant cover in an area by:

- Referring to dominant plant growth forms or structural characteristics, e.g., forest vegetation, grassland vegetation;
- Using either colloquial or technical descriptors, e.g., desert vegetation, cliff vegetation, arctic vegetation; or,
- Referring to specific plant communities, e.g., peat bog or trembling aspen vegetation types.

Vegetation is a broader term than flora, which refers specifically to the plant diversity of an area.

Spatially, vegetation can be thought of as the mosaic of plant communities across the landscape.

The diversity of vegetation for an area can be specified using any of the three descriptive categories noted above, e.g.,

- Alberta has forest, shrub, and grassland vegetation;
- Canada has temperate, boreal and arctic vegetation; or,
- The river valley has spruce and cottonwood riparian vegetation communities.

Vegetation integrates the combined influence of a variety of environmental factors. The main factors determining vegetation conditions in a specific area are:

- Climate: macroclimate has a primary influence on the overall flora – the plants that make up the vegetation.
- Organisms: although plants are the key organisms, animals, fungi, and microorganisms influence the resulting vegetation development through various interspecies and habitat relationships.
- Topography: determines the movement of surface and soil water, and therefore the moisture and some nutrient availability; also influences microclimate, e.g., aspect, cold air drainage.
- Soil parent material: is a primary influence on moisture and nutrient conditions in the substrates within which plants grow.
- Time: the length of development time, since either vegetation establishment or significant disturbance of existing vegetation, affects the characteristics of current vegetation.

Vegetation can range from natural to cultural.

- Natural vegetation, the focus is that which develops 'naturally' with little to no influence or modification by humans, except, in some cases, by First Nations peoples (e.g., regular burning to increase certain habitats or food products). Native species characterize natural vegetation.
- Semi-natural vegetation is generally included in the 'natural' landscape, as humans have influenced the natural vegetation in many areas and certain values and services are provided by semi-natural and natural vegetation. Some introduced, i.e., non-native, species are usually a component of semi-natural vegetation.
- Cultural vegetation is planted and/or maintained by humans, often with considerable input of energy. The species involved are often not native to the area.

Relationship between Vegetation and Agriculture

Farmers are responsible for maintaining a majority of the cropland vegetation in arable regions of the world. Vegetation (or the lack of it) is a key factor affecting weather patterns and micro-climates.

Figure. Bradford pear orchard.

The following factors are affected by vegetation:

(1) Plant covers and crops are inherently responsible for a vast majority of the water storage and usage. When irrigation is taken out of the equation, rain, snow, fog, dew, frost and soil water are all used and captured by vegetation. Some of this water is intercepted by the leaves themselves, and some of the water that infiltrates the soil is absorbed by plant roots. When water is available in excess, plants can transpire large volumes of water as a by-product of photosynthesis. This in turn causes evaporative cooling in the atmosphere directly surrounding the crop.

(2) Vegetation contributes to the level of carbon dioxide that is present in the atmosphere. Seasonal variations have shown that higher levels of CO_2 are measured in the early-mid part of the growing season. There are also daily variations in CO_2, with the highest levels measured during the day.

(3) Radiation measurements within a crop or vegetated area reveal a complex relationship between absorption, reflection, transmission and emission. The albedo of a crop is lower than the sum of all of the leaves in the canopy due to differences in crop architecture and angle of solar radiation. Most individual leaves have an albedo of about 0.25, but the albedo for a crop is less and depends upon crop height. For a crop of 1m in height, the surface albedo is between 0.18 and 0.25. However, this value is dependent upon the vegetation type. If leaves are wilted, or if the soil is exposed beneath the canopy this value can be different. Also, this albedo measurement is the one taken at mid-day. Early morning and late day measurements of albedo are much higher and the values follow the position of the sun in the sky. Also, tropical measurements of albedo are usually less that for a similar vegetated surface found at higher latitudes.

References

- Baluška F, Volkmann D, Mancuso S (2006). Communication in Plants: Neuronal Aspects of Plant Life. Springer Verlag. ISBN 978-3-540-28475-8
- Ecology, plant: britannica.com, Retrieved 31 June 2018
- Zimmermann MR, Maischak H, Mithöfer A, Boland W, Felle HH (March 2009). "System potentials, a novel electrical long-distance apoplastic signal in plants, induced by wounding". Plant Physiology. 149 (3): 1593–600. doi:10.1104/pp.108.133884. PMC 2649404. PMID 19129416
- Ecophysiology, plant-kingdom: biology.tutorvista.com, Retrieved 19 July 2018
- Karban R, Baxter KJ (2001). "Induced Resistance in Wild Tobacco with Clipped Sagebrush Neighbors: The Role of Herbivore Behavior". Journal of Insect Behavior. 14 (2): 147–156. doi:10.1023/A:1007893626166
- Hydrophytes-meaning-and-characteristics-plants-botany-75467: biologydiscussion.com, Retrieved 10 June 2018
- Scheel D, Wasternack C (2002). Plant signal transduction. Oxford: Oxford University Press. ISBN 0-19-963879-9
- Mesophytes: mesohydroxero.weebly.com, Retrieved 21 April 2018
- Pickard BG (1973). "Action Potentials in Higher Plants". Botanical Review. 39 (2): 172–201. doi:10.1007/BF02859299. JSTOR 4353850
- Xerophyte-meaning-and-characteristics-plants-botany-75464: biologydiscussion.com, Retrieved 28 May 2018
- Long S.P., and Bernacchi C.J. (2003) Gas exchange measurements, what can they tell us about the underlying limitations to photosynthesis? Procedures and sources of error Journal of Experimental Botany, Page 1 of 9
- Vegetation: climate.ncsu.edu, Retrieved 14 2018

Chapter 3

Plant Biochemistry

Plant biochemistry is the science concerned with the study of chemical processes in and related to plants. The aim of this chapter is to explore the focus areas of plant biochemistry, such as phytochemistry, photosynthesis, plant nutrition, plant lipids, carbon assimilation and catabolism in plants, among others.

Plant biochemistry is the study of chemical processes within and relating to living organisms. By controlling information flow through biochemical signaling and the flow of chemical energy through metabolism, biochemical processes give rise to the complexity of life. Over the last 40 years, biochemistry has become so successful at explaining living processes that now almost all areas of the life sciences from botany to medicine are engaged in biochemical research.

Phytochemistry

Phytochemistry is the chemistry (chemical analysis) of plant products. The chemicals present in leaf powder are analysed chemically by qualitatively and quantitively. Presence or absence of a chemical will gives the criteria to evaluates the drug or to standardize the drug. Estimation of particular chemical or element will also considered as criteria. The quantitive analysis of elements / chemicals like nitrogen, crude proteins, crude fats, crude fibers, carbohydrates etc. may fluctuate with the age of the shrubs plants, season of collection, hence these values are not considered as criteria. But their use in combination roughly gives the idea. Their values with little variation should be accepted. The quantitive analysis of a single chemical or element should not be considered as strict criteria for standardization or evaluation.

Phytochemistry (plant chemistry) also plays an important role in systematic botany. This is based on the assumption that related plants have a similar chemistry.

For example, in lichens chemical methods are used for identification of genera and species. But there is a controversy between Taxonomists as some morphologically similar lichens are chemically different; these two strains are then known as "chemical strains" and not as different species. In recent past at least 80 different lichen substances are identified based on the chemistry of lichens.

Chemical substances found in plants are Betalins, Anthocyanins, Glucosinolates, fatty acids and oils, flavonoides etc. There are three major techniques to study the phytochemical substances.

These are as follows:

(A) Serology,

(B) Electrophoresis, and

(C) Amino acid sequencing.

Serology

Serology is the branch of biology related with the nature of antigenic materials and antibodies. Serological data help in detecting and comparing non- morphological characteristics in taxonomic studies.

Serology is one of the techniques which is used in chemosystematic studies. It relies on the immunological reactions shown by the mammals when they are invaded by foreign proteins or antigens. The phenomenon was first used in 1897. Carl Mez and his associates at Korigsberg in Germany used serological bioassay methods to study angiosperm family relationship.

The fundamental principle of serology is that when plant extract containing antigen (protein) is injected into a mammal, a proteinaceous antibody will be formed in the animal. This antibody is specific to the antigens (E) and capable of coagulating it.

These antibodies are extracted from the blood in the form of antiscrum. Antiserum coagulates further supply of antigen and thus can be used as a standard test against plant extracts. The amount of coagulation is used as a measure of similarity to plant extract. Hence, the similarity of the species corresponding to the plant extracts (E1, E2...) will be compared.

The individual antigen antibody reaction, forming the total precipitation can be recorded separately in two ways:

(a) By following the antigenic material and anti serum to diffuse towards one another in a gel (different proteins travel at different rates so different reaction will take place at different sites in the gel) this method is known as double diffusion serology.

(b) By first separating the antigen unidimensionally in a gel by electrophoresis and then allowing them to diffuse towards a trough of antiserum. This method is called Immuno electrophoresis.

Presently in the double diffusion serology method the antiserum is placed in a small circular well, surrounded by a ring of similar wells containing the samples of antigens. Radio Immuno assay (RIA) and Enzyme Linked Immunosorbent Assay (ELISA) are also used where either the antibodies or the antigens are labelled with radioactive molecules or enzymes which can be detected in even very small quantities.

Most of the serological work is carried out on organs where protein is found as a food reserve, e.g., seeds, stems, tubers etc.

Serology is a useful taxonomic tool at all levels from above the family to below the species. Hawkes worked on the relationships of the tuber bearing species of Solanum to elucidate the ancestors and close relatives of the cultivated variety of potatoes. Jensen worked at family level on Ranunculaceae and Berberdaceae.

The classification of Ranunculaceae at the tribal and generic levels indicated by serological relationship strongly supported that based on classical evidence particularly from the chromosome. Fairbrother worked on Cornaceae and Magnoliaceae.

Smith has the relationship of annual brome grass (Bromus), relating immuno electrophoretic data to these from morphology, cytology, and cytogenetics. He recognised a variety of Bromus secalinus as a new species. B. pseudosecalinus on the basis of his studies.

Serology is not useful in identification of particular proteins. It cannot interpret the significance of the results.

Electrophoresis

Electrophoretic separation of protein is dependent on the amphoteric properties of amino acids. Whereby they are charged negatively or positively according to the pH of the medium and will travel through a gel at various speeds in a voltage gradient. The principle is basically applied and carried out in a polyacrylamide gel. Protein separation depends upon the sieving effect of the gel pores.

A crude separation is brought about in the larger porous zone while a complete separation into discrete bands in the smaller pored zone. After suitable separation time, the current is switched off and the separated polypeptide on the gel are located by stains. General protein stains may be used or specific stain is used for a particular enzyme.

SDS-PAGE System

Polyacrylamide gel electrophoresis when carried out in the presence of SDS – (Sodium Dodecyl Sulphate) the whole system is called as SDS-PAGE system, i.e., sodium dodecyl sulphate polyacrylamide gel electrophoresis system.

SDS is considered as a strong detergent which can denature proteins. Under appropriate conditions all reduced polypeptides bind the same amount of SDS on a weight basis, i.e., 1.4 gm SDS/gm of polypeptide. The separation is dependent on the size of the SDS—polypeptide complexes.

SDS -PAGE may be run on either a disc or a slab gel system. Most widely used is slab gel system because:

(a) Versatility of manipulation as drying, photography etc.

(b) Uniformity with regard to polymerisation, shrinking, swelling during staining and destaining.

(c) In case where molecular weight of the separate polypeptides are to be determined, both standard and unknown portions can be run on the same gel and the accuracy of results is much higher than disc-gel system.

Electrophoresis Apparatus:

The dimension of slab gels are 16 x 14 cm with 1.5 mm thickness containing 10 to 15 slots (25 x 8 x 1.5 mm) Glass plates are assembled and the separating gel recipes are poured between them, and left to polymerize for about 30-40 minutes.

Stack gel solution is then transferred to gel mould with a pasteur pipette. Combs are inserted and the stack gel is left to polymerize for about 20-30 minutes.

Sample Preparation:

Equal volume of 2x sample buffer is added to a protein sample (50 ml). 3 ml of 0.2% bromophenol blue dye is also added to it. The sample is heated in boiling water bath for 2 minutes and the protein denatures completely. This sample is then loaded to gel slot with micropipette. The electrophoresis run is done bromophenol blue dye reaches the bottom of the gel.

Staining is done by using dye as coomasic blue containing coomasic brilliant blue R 250 dissolved in 95% methanol and glacial acetic acid or silver staining containing 0.36% sodium hydroxide, ammonium hydroxide and 19.4% silver nitrate, citric acid, formaldehyde and 95% ethanol.

Data Recording:

After electrophoresis the glass plates are separated and the gel is kept on a clean glass plate. The length of the separation gel and the distance travelled by the bromophenol blue dye are measured. Staining and destaining is done. After destaining the gel is kept on top of a light box and the length of the separation gel and the distance travelled by each polypeptide are measured from top of separation gel.

The Rf of each polypeptide is calculated by:

$$Rf = \frac{\text{Distance of protein migration}}{\text{Gel length after destaining}} \times \frac{\text{Gel length before destaining}}{\text{Distance of dye migration}}$$

Data Analysis:

After calculation of Rf values of each of the protein bands, a data matrix is usually prepared with two rows corresponding to the characters. The score could be '1' for the presence and 'O' for absence of a band or polypeptide.

The data matrix thus prepared is ready for further analysis, essentially through numerical means to compute different similarity or dissimilarity indices and finally to construct a scheme of classification based on such protein characters.

Taxonomic Significance:

The taxonomic significance of the results of electrophoresis are useful mainly at and below generic levels. The most valuable results was obtained in cereals related to wheat in relation to the genomic constitution and ancestors of the tetraploids and Hexaploids.

Johnson worked on storage proteins and concluded that the hexaploid wheat (Triticum aestivum) contains a sum of protein possessed by the diploid species which are ancestors of it.

It had been studied in wild variety of Cajanus cajan while studying total seed storage protein in them SDS-PAGE method. Its protein profile of wild species were found to be very specific and distinct, pointing out towards the polyphyletic origin of the cultigen.

Amino Acid Sequencing

It is a modern way of identifying pure protein down to atomic level. First step is to break the total polypeptide chain into smaller peptide fragments and then they are sequenced separately.

Amino acid sequencing investigates the variation in the precise sequence of amino acids in a single homologous protein throughout a range of organisms. The process depends upon the fact that proteins do not have a single invariable structure, but a proportion of it may without altering its essential function.

Cytochrome 'C' is the molecule that was first used in plants for sequencing purpose. 79 out of 113 amino acids vary from species to species but alternation of even one of the other 34 destroys the functioning of the molecule. Stace determined amino acid.

Sequence of cytochrome in 25 species of vascular plants. Sequence of Plstocyanin, Ferredoxin, Ribulose-I, 5-bis phosphate Carboxylase, etc. is also carried out in various plants. Plastocyanin sequence of over 70 species of angiosperms have been determined so far.

Amino acids sequences of protein are very useful taxonomically at the level of families and order as they are precise, relatively stable and independent of the traditional characters. The construction of a phylogenetic tree based on amino acid sequence and its using the ancestral sequences method permits the reconstruction of a precise, quantitative and objective topology of relationship.

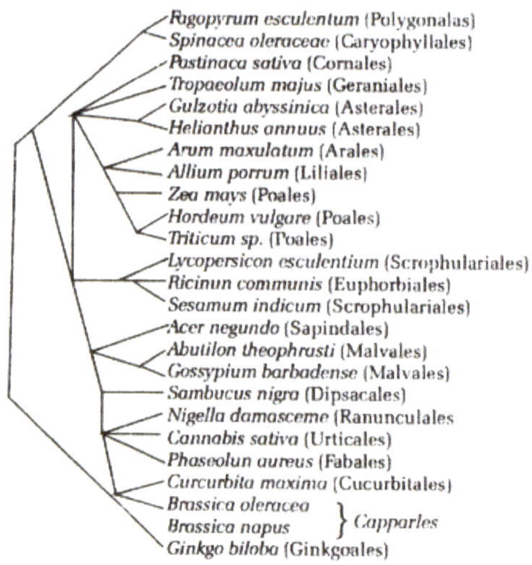

Figure: Cladogram relating the cytochrome amino acid sequences of 25 species of spermatophytes. Ginkgo biloba shows isolated position

Photosynthesis

Photosynthesis is the process by which green plants and certain other organisms transform light energy into chemical energy.

During photosynthesis in green plants, light energy is captured and used to convert water, carbon dioxide, and minerals into oxygen and energy-rich organic compounds.

It would be impossible to overestimate the importance of photosynthesis in the maintenance of life on Earth. If photosynthesis ceased, there would soon be little food or other organic matter on Earth. Most organisms would disappear, and in time Earth's atmosphere would become nearly devoid of gaseous oxygen. The only organisms able to exist under such conditions would be the chemosynthetic bacteria, which can utilize the chemical energy of certain inorganic compounds and thus are not dependent on the conversion of light energy.

Energy produced by photosynthesis carried out by plants millions of years ago is responsible for the fossil fuels (i.e., coal, oil, and gas) that power industrial society. In past ages, green plants and small organisms that fed on plants increased faster than they were consumed, and their remains were deposited in Earth's crust by sedimentation and other geological processes. There, protected from oxidation, these organic remains were slowly converted to fossil fuels. These fuels not only provide much of the energy used in factories, homes, and transportation but also serve as the raw material for plastics and other synthetic products. Unfortunately, modern civilization is using up in a few centuries the excess of photosynthetic production accumulated over millions of years. Consequently, the carbon dioxide that has been removed from the air to make carbohydrates in photosynthesis over millions of years is being returned at an incredibly rapid rate. The carbon dioxide concentration in Earth's atmosphere is rising the fastest it ever has in Earth's history, and this phenomenon is expected to have major implications on Earth's climate.

Requirements for food, materials, and energy in a world where human population is rapidly growing have created a need to increase both the amount of photosynthesis and the efficiency of converting photosynthetic output into products useful to people. One response to those needs—the so-called Green Revolution, begun in the mid-20th century—achieved enormous improvements in agricultural yield through the use of chemical fertilizers, pest and plant-disease control, plant breeding, and mechanized tilling, harvesting, and crop processing. This effort limited severe famines to a few areas of the world despite rapid population growth, but it did not eliminate widespread malnutrition. Moreover, beginning in the early 1990s, the rate at which yields of major crops increased began to decline. This was especially true for rice in Asia. Rising costs associated with sustaining high rates of agricultural production, which required ever-increasing inputs of fertilizers and pesticides and constant development of new plant varieties, also became problematic for farmers in many countries.

A second agricultural revolution, based on plant genetic engineering, was forecast to lead to increases in plant productivity and thereby partially alleviate malnutrition. Since the 1970s, molecular biologists have possessed the means to alter a plant's genetic material (deoxyribonucleic acid, or DNA) with the aim of achieving improvements in disease and drought resistance, product yield and quality, frost hardiness, and other desirable properties. However, such traits are inherently complex, and the process of making changes to crop plants through genetic engineering has turned out to be more complicated than anticipated. In the future such genetic engineering may result in improvements in the process of photosynthesis, but by the first decades of the 21st century, it had yet to demonstrate that it could dramatically increase crop yields.

Another intriguing area in the study of photosynthesis has been the discovery that certain animals are able to convert light energy into chemical energy. The emerald green sea slug (*Elysia chlorotica*), for example, acquires genes and chloroplasts from *Vaucheria litorea*, an alga it consumes, giving it a limited ability to produce chlorophyll. When enough chloroplasts are assimilated, the slug may forgo the ingestion of food. The pea aphid (*Acyrthosiphon pisum*) can harness light to manufacture the energy-rich compound adenosine triphosphate (ATP); this ability has been linked to the aphid's manufacture of carotenoid pigments.

General Characteristics

Development of the Idea:

The study of photosynthesis began in 1771 with observations made by the English clergyman and scientist Joseph Priestley. Priestley had burned a candle in a closed container until the air within the container could no longer support combustion. He then placed a sprig of mint plant in the container and discovered that after several days the mint had produced some substance (later recognized as oxygen) that enabled the confined air to again support combustion. In 1779 the Dutch physician Jan Ingenhousz expanded upon Priestley's work, showing that the plant had to be exposed to light if the combustible substance (i.e., oxygen) was to be restored. He also demonstrated that this process required the presence of the green tissues of the plant.

In 1782 it was demonstrated that the combustion-supporting gas (oxygen) was formed at the expense of another gas, or "fixed air," which had been identified the year before as carbon dioxide. Gas-exchange experiments in 1804 showed that the gain in weight of a plant grown in a carefully weighed pot resulted from the uptake of carbon, which came entirely from absorbed carbon dioxide, and water taken up by plant roots; the balance is oxygen, released back to the atmosphere. Almost half a century passed before the concept of chemical energy had developed sufficiently to permit the discovery (in 1845) that light energy from the sun is stored as chemical energy in products formed during photosynthesis.

Overall Reaction of Photosynthesis

In chemical terms, photosynthesis is a light-energized oxidation–reduction process. (Oxidation refers to the removal of electrons from a molecule; reduction refers to the gain of electrons by a molecule.) In plant photosynthesis, the energy of light is used to drive the oxidation of water (H_2O), producing oxygen gas (O_2), hydrogen ions (H^+), and electrons. Most of the removed electrons and hydrogen ions ultimately are transferred to carbon dioxide (CO_2), which is reduced to organic products. Other electrons and hydrogen ions are used to reduce nitrate and sulfate to amino and sulfhydryl groups in amino acids, which are the building blocks of proteins. In most green cells, carbohydrates—especially starch and the sugar sucrose—are the major direct organic products of photosynthesis. The overall reaction in which carbohydrates—represented by the general formula (CH_2O)—are formed during plant photosynthesis can be indicated by the following equation:

$$CO_2 + 2H_2O \xrightarrow[\text{green plants}]{\text{light}} (CH_2O) + O_2 + H_2O.$$

This equation is merely a summary statement, for the process of photosynthesis actually involves numerous reactions catalyzed by enzymes (organic catalysts). These reactions occur in two stages: the "light" stage, consisting of photochemical (i.e., light-capturing) reactions; and the "dark" stage, comprising chemical reactions controlled by enzymes. During the first stage, the energy of light is absorbed and used to drive a series of electron transfers, resulting in the synthesis of ATP and the electron-donor-reduced nicotine adenine dinucleotide phosphate (NADPH). During the dark stage, the ATP and NADPH formed in the light-capturing reactions are used to reduce carbon dioxide to organic carbon compounds. This assimilation of inorganic carbon into organic compounds is called carbon fixation.

During the 20th century, comparisons between photosynthetic processes in green plants and in certain photosynthetic sulfur bacteria provided important information about the photosynthetic mechanism. Sulfur bacteria use hydrogen sulfide (H_2S) as a source of hydrogen atoms and produce sulfur instead of oxygen during photosynthesis. The overall reaction is:

$$CO_2 + 2H_2S \xrightarrow[\substack{\text{sulfur}\\\text{bacteria}}]{\text{light}} (CH_2O) + S_2 + H_2O.$$

In the 1930s Dutch biologist Cornelis van Niel recognized that the utilization of carbon dioxide to form organic compounds was similar in the two types of photosynthetic organisms. Suggesting that differences existed in the light-dependent stage and in the nature of the compounds used as a source of hydrogen atoms, he proposed that hydrogen was transferred from hydrogen sulfide (in bacteria) or water (in green plants) to an unknown acceptor (called A), which was reduced to H_2A. During the dark reactions, which are similar in both bacteria and green plants, the reduced acceptor (H_2A) reacted with carbon dioxide (CO_2) to form carbohydrate (CH_2O) and to oxidize the unknown acceptor to A. This putative reaction can be represented as:

$$CO_2 + 2H_2A \xrightarrow{\text{light}} (CH_2O) + 2A + H_2O.$$

Van Niel's proposal was important because the popular (but incorrect) theory had been that oxygen was removed from carbon dioxide (rather than hydrogen from water, releasing oxygen) and that carbon then combined with water to form carbohydrate (rather than the hydrogen from water combining with CO_2 to form CH_2O).

By 1940 chemists were using heavy isotopes to follow the reactions of photosynthesis. Water marked with an isotope of oxygen (^{18}O) was used in early experiments. Plants that photosynthesized in the presence of water containing $H_2{}^{18}O$ produced oxygen gas containing ^{18}O; those that photosynthesized in the presence of normal water produced normal oxygen gas. These results provided definitive support for van Niel's theory that the oxygen gas produced during photosynthesis is derived from water.

Basic Products of Photosynthesis

Carbohydrates are the most important direct organic product of photosynthesis in the majority of green plants. The formation of a simple carbohydrate, glucose, is indicated by a chemical equation.

$$\underset{\text{carbon dioxide}}{6CO_2} + \underset{\text{water}}{12H_2A} \xrightarrow[\text{green plants}]{\text{light}} \underset{\text{glucose}}{C_6H_{12}O_6} + \underset{\text{oxygen}}{6O_2} + \underset{\text{water}}{6H_2O}.$$

Little free glucose is produced in plants; instead, glucose units are linked to form starch or are joined with fructose, another sugar, to form sucrose.

Not only carbohydrates, as was once thought, but also amino acids, proteins, lipids (or fats), pigments, and other organic components of green tissues are synthesized during photosynthesis. Minerals supply the elements (e.g., nitrogen, N; phosphorus, P; sulfur, S) required to form these compounds. Chemical bonds are broken between oxygen (O) and carbon (C), hydrogen (H), nitrogen, and sulfur, and new bonds are formed in products that include gaseous oxygen (O_2) and organic compounds. More energy is required to break the bonds between oxygen and other elements (e.g., in water, nitrate, and sulfate) than is released when new bonds form in the products. This difference in bond energy accounts for a large part of the light energy stored as chemical energy in the organic products formed during photosynthesis. Additional energy is stored in making complex molecules from simple ones.

Evolution of the Process

Although life and the quality of the atmosphere today depend on photosynthesis, it is likely that green plants evolved long after the first living cells. When Earth was young, electrical storms and solar radiation probably provided the energy for the synthesis of complex molecules from abundant simpler ones, such as water, ammonia, and methane. The first living cells probably evolved from these complex molecules. For example, the accidental joining (condensation) of the amino acid glycine and the fatty acid acetate may have formed complex organic molecules known as porphyrins. These molecules, in turn, may have evolved further into coloured molecules called pigments—e.g., chlorophylls of green plants, bacteriochlorophyll of photosynthetic bacteria, hemin (the red pigment of blood), and cytochromes, a group of pigment molecules essential in both photosynthesis and cellular respiration.

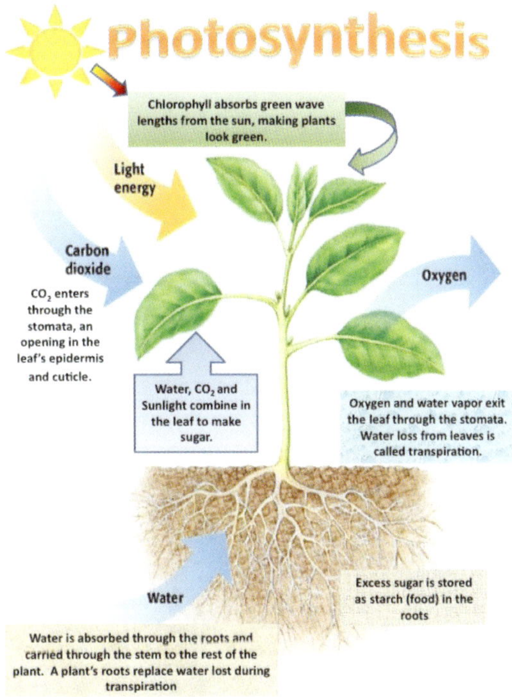

Primitive coloured cells then had to evolve mechanisms for using the light energy absorbed by their pigments. At first, the energy may have been used immediately to initiate reactions useful to the cell. As the process for utilization of light energy continued to evolve, however, a larger part of the absorbed light energy probably was stored as chemical energy, to be used to maintain life. Green plants, with their ability to use light energy to convert carbon dioxide and water to carbohydrates and oxygen, are the culmination of this evolutionary process.

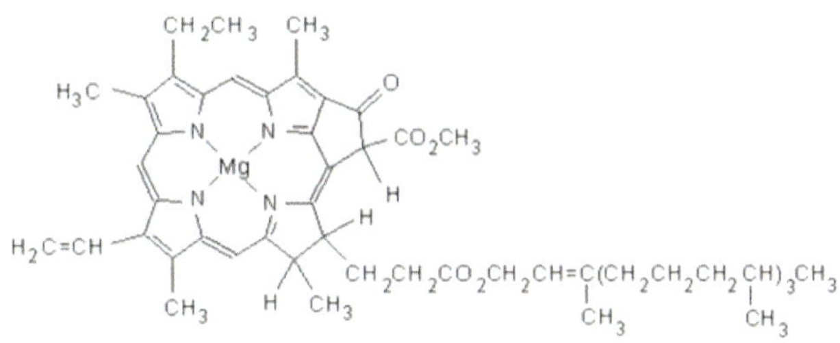

Chlorophyll structure

The first oxygenic (oxygen-producing) cells probably were the blue-green algae (cyanobacteria), which appeared about two billion to three billion years ago. These microscopic organisms are believed to have greatly increased the oxygen content of the atmosphere, making possible the development of aerobic (oxygen-using) organisms. Cyanophytes are prokaryotic cells; that is, they contain no distinct membrane-enclosed subcellular particles (organelles), such as nuclei and chloroplasts. Green plants, by contrast, are composed of eukaryotic cells, in which the photosynthetic apparatus is contained within membrane-bound chloroplasts. The complete genome sequences of cyanobacteria and higher plants provide evidence that the first photosynthetic eukaryotes were likely the red algae that developed when nonphotosynthetic eukaryotic cells engulfed cyanobacteria. Within the host cells, these cyanobacteria evolved into chloroplasts.

There are a number of photosynthetic bacteria that are not oxygenic (e.g., the sulfur bacteria). The evolutionary pathway that led to these bacteria diverged from the one that resulted in oxygenic organisms. In addition to the absence of oxygen production, nonoxygenic photosynthesis differs from oxygenic photosynthesis in two other ways: light of longer wavelengths is absorbed and used by pigments called bacteriochlorophylls, and reduced compounds other than water (such as hydrogen sulfide or organic molecules) provide the electrons needed for the reduction of carbon dioxide.

Factors that Influence the Rate of Photosynthesis

The rate of photosynthesis is defined in terms of the rate of oxygen production either per unit mass (or area) of green plant tissues or per unit weight of total chlorophyll. The amount of light, the carbon dioxide supply, temperature, water supply, and the availability of minerals are the most important environmental factors that affect the rate of photosynthesis in land plants. The rate of photosynthesis is also determined by the plant species and its physiological state—e.g., its health, its maturity, and whether it is in flower.

Light Intensity and Temperature

The complex mechanism of photosynthesis includes a photochemical, or light-harvesting, stage and an enzymatic, or carbon-assimilating, stage that involves chemical reactions. These stages can be distinguished by studying the rates of photosynthesis at various degrees of light saturation (i.e., intensity) and at different temperatures. Over a range of moderate temperatures and at low to medium light intensities (relative to the normal range of the plant species), the rate of photosynthesis increases as the intensity increases and is relatively independent of temperature. As the light intensity increases to higher levels, however, the rate becomes saturated; light "saturation" is achieved at a specific light intensity, dependent on species and growing conditions. In the light-dependent range before saturation, therefore, the rate of photosynthesis is determined by the rates of photochemical steps. At high light intensities, some of the chemical reactions of the dark stage become rate-limiting. In many land plants, a process called photorespiration occurs, and its influence upon photosynthesis increases with rising temperatures. More specifically, photorespiration competes with photosynthesis and limits further increases in the rate of photosynthesis, especially if the supply of water is limited.

Carbon Dioxide

Included among the rate-limiting steps of the dark stage of photosynthesis are the chemical reactions by which organic compounds are formed by using carbon dioxide as a carbon source. The rates of these reactions can be increased somewhat by increasing the carbon dioxide concentration. Since the middle of the 19th century, the level of carbon dioxide in the atmosphere has been rising because of the extensive combustion of fossil fuels, cement production, and land-use changes associated with deforestation. The atmospheric level of carbon dioxide climbed from about 0.028 percent in 1860 to 0.032 percent by 1958 (when improved measurements began) and to 0.040 percent by 2016. This increase in carbon dioxide directly increases plant photosynthesis, but the size of the increase depends on the species and physiological condition of the plant. Furthermore, most scientists maintain that increasing levels of atmospheric carbon dioxide affect climate, increasing global temperatures and changing rainfall patterns. Such changes will also affect photosynthesis rates.

Water

For land plants, water availability can function as a limiting factor in photosynthesis and plant growth. Besides the requirement for a small amount of water in the photosynthetic reaction itself, large amounts of water are transpired from the leaves; that is, water evaporates from the leaves to the atmosphere via the stomata. Stomata are small openings through the leaf epidermis, or outer skin; they permit the entry of carbon dioxide but inevitably also allow the exit of water vapour. The stomata open and close according to the physiological needs of the leaf. In hot and arid climates the stomata may close to conserve water, but this closure limits the entry of carbon dioxide and hence the rate of photosynthesis. The decreased transpiration means there is less cooling of the leaves and hence leaf temperatures rise. The decreased carbon dioxide concentration inside the leaves and the increased leaf temperatures favour the wasteful process of photorespiration. If the level of carbon dioxide in the atmosphere increases, more carbon dioxide could enter through a smaller opening of the stomata, so more photosynthesis could occur with a given supply of water.

Minerals

Several minerals are required for healthy plant growth and for maximum rates of photosynthesis. Nitrogen, sulfate, phosphate, iron, magnesium, calcium, and potassium are required in substantial amounts for the synthesis of amino acids, proteins, coenzymes, deoxyribonucleic acid (DNA) and ribonucleic acid (RNA), chlorophyll and other pigments, and other essential plant constituents. Smaller amounts of such elements as manganese, copper, and chloride are required in photosynthesis. Some other trace elements are needed for various nonphotosynthetic functions in plants.

Internal Factors

Each plant species is adapted to a range of environmental factors. Within this normal range of conditions, complex regulatory mechanisms in the plant's cells adjust the activities of enzymes (i.e., organic catalysts). These adjustments maintain a balance in the overall photosynthetic process and control it in accordance with the needs of the whole plant. With a given plant species, for example, doubling the carbon dioxide level might cause a temporary increase of nearly twofold in the rate of photosynthesis; a few hours or days later, however, the rate might fall to the original level because photosynthesis produced more sucrose than the rest of the plant could use. By contrast, another plant species provided with such carbon dioxide enrichment might be able to use more sucrose, because it had more carbon-demanding organs, and would continue to photosynthesize and to grow faster throughout most of its life cycle.

Photosynthetic Efficiency

The photosynthetic efficiency is the fraction of light energy converted into chemical energy during photosynthesis in plants and algae. Photosynthesis can be described by the simplified chemical reaction:

$$6H_2O + 6CO_2 + \text{energy} \rightarrow C_6H_{12}O_6 + 6O_2$$

where $C_6H_{12}O_6$ is glucose (which is subsequently transformed into other sugars, cellulose, lignin, and so forth). The value of the photosynthetic efficiency is dependent on how light energy is defined – it depends on whether we count only the light that is absorbed, and on what kind of light is used. It takes eight (or perhaps 10 or more) photons to utilize one molecule of CO_2. The Gibbs free energy for converting a mole of CO_2 to glucose is 114 kcal, whereas eight moles of photons of wavelength 600 nm contains 381 kcal, giving a nominal efficiency of 30%. However, photosynthesis can occur with light up to wavelength 720 nm so long as there is also light at wavelengths below 680 nm to keep Photosystem II operating. Using longer wavelengths means less light energy is needed for the same number of photons and therefore for the same amount of photosynthesis. For actual sunlight, where only 45% of the light is in the photosynthetically active wavelength range, the theoretical maximum efficiency of solar energy conversion is approximately 11%. In actuality, however, plants do not absorb all incoming sunlight (due to reflection, respiration requirements of photosynthesis and the need for optimal solar radiation levels) and do not convert all harvested energy into biomass, which results in an overall photosynthetic efficiency of 3 to 6% of total solar radiation. If photosynthesis is inefficient, excess light energy must be dissipated to avoid damaging the photosynthetic apparatus. Energy can be dissipated as heat (non-photochemical quenching), or emitted as chlorophyll fluorescence.

Typical Efficiencies

Plants

Quoted values sunlight-to-biomass efficiency

Plant	Efficiency
Plants, typical	0.1% 0.2–2%
Typical crop plants	1–2%

The following is a breakdown of the energetics of the photosynthesis process from *Photosynthesis* by Hall and Rao:

Starting with the solar spectrum falling on a leaf,

47% lost due to photons outside the 400–700 nm active range (chlorophyll utilizes photons between 400 and 700 nm, extracting the energy of one 700 nm photon from each one).

30% of the in-band photons are lost due to incomplete absorption or photons hitting components other than chloroplasts.

24% of the absorbed photon energy is lost due to degrading short wavelength photons to the 700 nm energy level.

68% of the utilized energy is lost in conversion into d-glucose.

35–45% of the glucose is consumed by the leaf in the processes of dark and photo respiration.

Stated another way:

100% sunlight → non-bioavailable photons waste is 47%, leaving.

53% (in the 400–700 nm range) → 30% of photons are lost due to incomplete absorption, leaving.

37% (absorbed photon energy) → 24% is lost due to wavelength-mismatch degradation to 700 nm energy, leaving.

28.2% (sunlight energy collected by chlorophyll) → 32% efficient conversion of ATP and NADPH to d-glucose, leaving.

9% (collected as sugar) → 35–40% of sugar is recycled/consumed by the leaf in dark and photo-respiration, leaving.

5.4% net leaf efficiency.

Many plants lose much of the remaining energy on growing roots. Most crop plants store ~0.25% to 0.5% of the sunlight in the product (corn kernels, potato starch, etc.).

Photosynthesis increases linearly with light intensity at low intensity, but at higher intensity this is no longer the case. Above about 10,000 lux or ~100 watts/square meter the rate no longer increases. Thus, most plants can only utilize ~10% of full mid-day sunlight intensity. This dramatically reduces average achieved photosynthetic efficiency in fields compared to peak laboratory results. However, real plants (as opposed to laboratory test samples) have lots of redundant, randomly oriented leaves. This helps to keep the average illumination of each leaf well below the mid-day peak enabling the plant to achieve a result closer to the expected laboratory test results using limited illumination.

Only if the light intensity is above a plant specific value, called the compensation point the plant assimilates more carbon and releases more oxygen by photosynthesis than it consumes by cellular respiration for its own current energy demand.

Photosynthesis measurement systems are not designed to directly measure the amount of light absorbed by the leaf. Nevertheless, the light response curves that the class produces do allow comparisons in photosynthetic efficiency between plants.

Algae and other Monocellular Organisms

From a 2010 study by the University of Maryland, photosynthesizing Cyanobacteria have been shown to be a significant species in the global carbon cycle, accounting for 20–30% of Earth's photosynthetic productivity and convert solar energy into biomass-stored chemical energy at the rate of ~450 TW.

Worldwide Figures

According to the cyanobacteria study above, this means the total photosynthetic productivity of earth is between ~1500–2250 TW, or $4.73–7.10 \times 10^{22}$ J (47,300–71,000 exajoules) per year. Using this source's figure of 178,000 TW of solar energy hitting the Earth's surface, the total photosynthetic efficiency of the planet is 0.84% to 1.26%.

Efficiencies of Various Biofuel Crops

Popular choices for plant biofuels include: oil palm, soybean, castor oil, sunflower oil, safflower oil, corn ethanol, and sugar cane ethanol.

An analysis of a proposed Hawaiian oil palm plantation claimed to yield 600 gallons of biodiesel per acre per year. That comes to 2835 watts per acre or 0.7 W/m^2. Typical insolation in Hawaii is around 5.5 $kWh/(m^2day)$ or 230 W/m^2. For this particular oil palm plantation, if it delivered the claimed 600 gallons of biodiesel per acre per year, would be converting 0.3% of the incident solar energy to chemical fuel. Total photosynthetic efficiency would include more than just the biodiesel oil, so this 0.3% number is something of a lower bound.

Contrast this with a typical photovoltaic installation, which would produce an average of roughly 22 W/m^2 (roughly 10% of the average insolation), throughout the year. Furthermore, the photovoltaic panels would produce electricity, which is a high-quality form of energy, whereas converting the biodiesel into mechanical energy entails the loss of a large portion of the energy. On the other hand, a liquid fuel is much more convenient for a vehicle than electricity, which has to be stored in heavy, expensive batteries.

Most crop plants store ~0.25% to 0.5% of the sunlight in the product (corn kernels, potato starch, etc.), sugar cane is exceptional in several ways to yield peak storage efficiencies of ~8%.

Ethanol fuel in Brazil has a calculation that results in: "Per hectare per year, the biomass produced corresponds to 0.27 TJ. This is equivalent to 0.86 W/m^2. Assuming an average insolation of 225 W/m^2, the photosynthetic efficiency of sugar cane is 0.38%." Sucrose accounts for little more than 30% of the chemical energy stored in the mature plant; 35% is in the leaves and stem tips, which are left in the fields during harvest, and 35% are in the fibrous material (bagasse) left over from pressing.

C3 vs. C4 and CAM Plants

C3 plants use the Calvin cycle to fix carbon. C4 plants use a modified Calvin cycle in which they separate Ribulose-1,5-bisphosphate carboxylase oxygenase (RuBisCO) from atmospheric oxygen, fixing carbon in their mesophyll cells and using oxaloacetate and malate to ferry the fixed carbon to RuBisCO and the rest of the Calvin cycle enzymes isolated in the bundle-sheath cells. The intermediate compounds both contain four carbon atoms, which gives C4. In Crassulacean acid metabolism (CAM), time isolates functioning RuBisCo (and the other Calvin cycle enzymes) from high oxygen concentrations produced by photosynthesis, in that O_2 is evolved during the day, and allowed to dissipate then, while at night atmospheric CO_2 is taken up and stored as malic or other acids. During the day, CAM plants close stomata and use stored acids as carbon sources for sugar, etc. production.

The C3 pathway requires 18 ATP for the synthesis of one molecule of glucose while the C4 pathway requires 20 ATP. Despite this reduced ATP efficiency, C4 is an evolutionary advancement, adapted to areas of high levels of light, where the reduced ATP efficiency is more than offset by the use of increased light. The ability to thrive despite restricted water availability maximizes the ability to use available light. The simpler C3 cycle which operates in most plants is adapted to wetter darker environments, such as many northern latitudes. Corn, sugar cane, and sorghum are C4 plants. These plants are economically important in part because of their relatively high photosynthetic efficiencies compared to many other crops. Pineapple is a CAM plant.

Chloroplast: Photosynthetic Units of Green Plants

One of the main functions of the chloroplast is its role in photosynthesis, the process by which light is transformed into chemical energy, to subsequently produce food in the form of sugars. Water (H_2O) and carbon dioxide (CO_2) are used in photosynthesis, and sugar and oxygen (O_2) is made, using light energy. Photosynthesis is divided into two stages—the light reactions, where water is split to produce oxygen, and the dark reactions, or Calvin cycle, which builds sugar molecules from carbon dioxide. The two phases are linked by the energy carriers adenosine triphosphate (ATP) and nicotinamide adenine dinucleotide phosphate ($NADP^+$).

Light Reactions

The light reactions of photosynthesis take place across the thylakoid membranes.

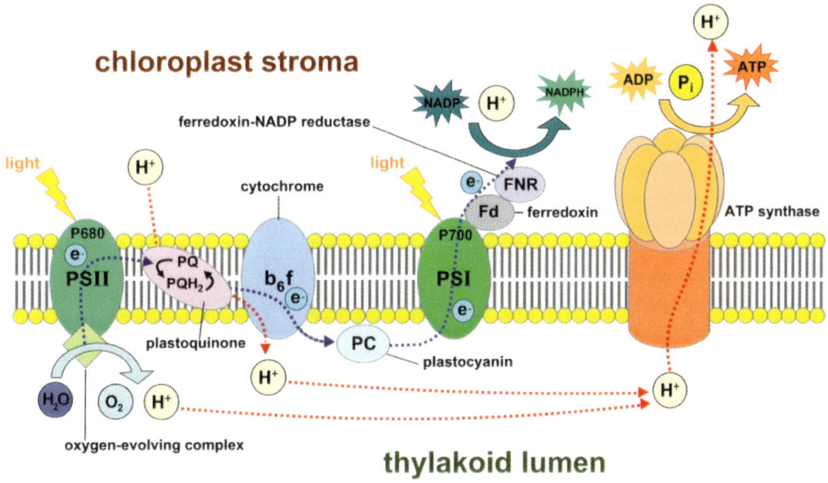

The light reactions take place on the thylakoid membranes. They take light energy and store it in NADPH, a form of $NADP^+$, and ATP to fuel the dark reactions.

Energy Carriers

ATP is the phosphorylated version of adenosine diphosphate (ADP), which stores energy in a cell and powers most cellular activities. ATP is the energized form, while ADP is the (partially) depleted form. $NADP^+$ is an electron carrier which ferries high energy electrons. In the light reactions, it gets reduced, meaning it picks up electrons, becoming NADPH.

Photophosphorylation

Like mitochondria, chloroplasts use the potential energy stored in an H^+, or hydrogen ion gradient to generate ATP energy. The two photosystems capture light energy to energize electrons taken from water, and release them down an electron transport chain. The molecules between the photosystems harness the electrons' energy to pump hydrogen ions into the thylakoid space, creating a concentration gradient, with more hydrogen ions (up to a thousand times as many) inside the thylakoid system than in the stroma. The hydrogen ions in the thylakoid space then diffuse back down their concentration gradient, flowing back out into the stroma through ATP synthase. ATP synthase uses the energy from the flowing hydrogen ions to phosphorylate adenosine diphosphate into adenosine triphosphate, or ATP. Because chloroplast ATP synthase projects out into the stroma, the ATP is synthesized there, in position to be used in the dark reactions.

NADP+ Reduction

Electrons are often removed from the electron transport chains to charge $NADP^+$ with electrons, reducing it to NADPH. Like ATP synthase, ferredoxin-$NADP^+$ reductase, the enzyme that reduces $NADP^+$, releases the NADPH it makes into the stroma, right where it is needed for the dark reactions.

Because $NADP^+$ reduction removes electrons from the electron transport chains, they must be replaced—the job of photosystem II, which splits water molecules (H_2O) to obtain the electrons from its hydrogen atoms.

Cyclic Photophosphorylation

While photosystem II photolyzes water to obtain and energize new electrons, photosystem I simply reenergizes depleted electrons at the end of an electron transport chain. Normally, the reenergized electrons are taken by $NADP^+$, though sometimes they can flow back down more H^+-pumping electron transport chains to transport more hydrogen ions into the thylakoid space to generate more ATP. This is termed cyclic photophosphorylation because the electrons are recycled. Cyclic photophosphorylation is common in C_4 plants, which need more ATP than NADPH.

Dark Reactions

The Calvin cycle incorporates carbon dioxide into sugar molecules.

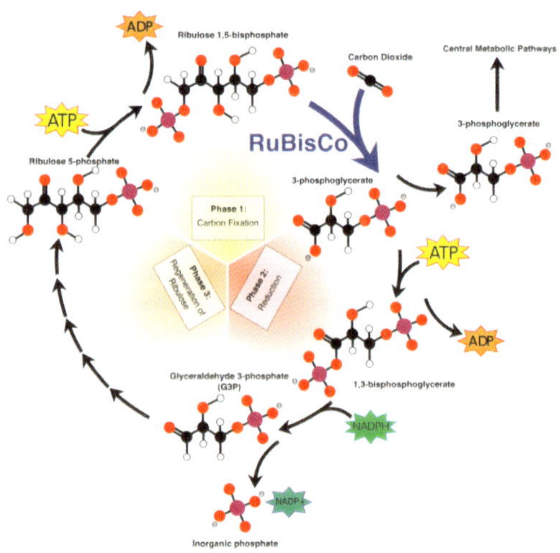

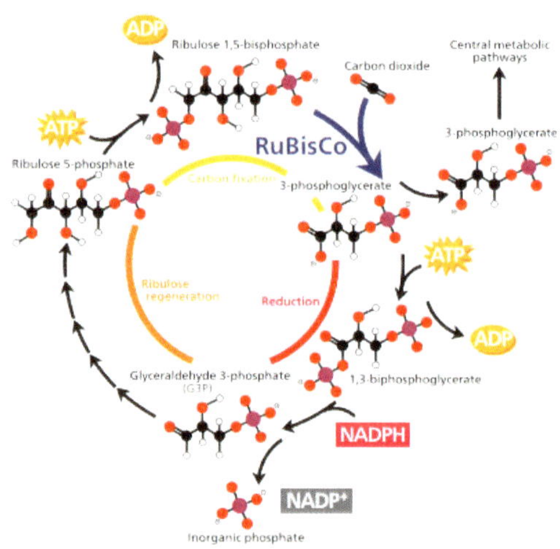

The Calvin cycle, also known as the dark reactions, is a series of biochemical reactions that fixes CO_2 into G3P sugar molecules and uses the energy and electrons from the ATP and NADPH made in the light reactions. The Calvin cycle takes place in the stroma of the chloroplast.

While named *"the dark reactions"*, in most plants, they take place in the light, since the dark reactions are dependent on the products of the light reactions.

Carbon Fixation and G3P Synthesis

The Calvin cycle starts by using the enzyme RuBisCO to fix CO_2 into five-carbon Ribulose bisphosphate (RuBP) molecules. The result is unstable six-carbon molecules that immediately break down into three-carbon molecules called 3-phosphoglyceric acid, or 3-PGA. The ATP and NADPH made in the light reactions is used to convert the 3-PGA into glyceraldehyde-3-phosphate, or G3P sugar molecules. Most of the G3P molecules are recycled back into RuBP using energy from more ATP, but one out of every six produced leaves the cycle—the end product of the dark reactions.

Sugars and Starches

Glyceraldehyde-3-phosphate can double up to form larger sugar molecules like glucose and fructose. These molecules are processed, and from them, the still larger sucrose, a disaccharide commonly known as table sugar, is made, though this process takes place outside of the chloroplast, in the cytoplasm.

Alternatively, glucose monomers in the chloroplast can be linked together to make starch, which accumulates into the starch grains found in the chloroplast. Under conditions such as high atmospheric CO_2 concentrations, these starch grains may grow very large, distorting the grana and thylakoids. The starch granules displace the thylakoids, but leave them intact. Waterlogged roots can also cause starch buildup in the chloroplasts, possibly due to less sucrose being exported out of the chloroplast (or more accurately, the plant cell). This depletes a plant's free phosphate supply, which indirectly stimulates chloroplast starch synthesis. While linked to low photosynthesis rates, the starch grains themselves may not necessarily interfere significantly with the efficiency of photosynthesis, and might simply be a side effect of another photosynthesis-depressing factor.

Photorespiration

Photorespiration can occur when the oxygen concentration is too high. RuBisCO cannot distinguish between oxygen and carbon dioxide very well, so it can accidentally add O_2 instead of CO_2 to RuBP. This process reduces the efficiency of photosynthesis—it consumes ATP and oxygen, releases CO_2, and produces no sugar. It can waste up to half the carbon fixed by the Calvin cycle. Several mechanisms have evolved in different lineages that raise the carbon dioxide concentration relative to oxygen within the chloroplast, increasing the efficiency of photosynthesis. These mechanisms are called carbon dioxide concentrating mechanisms, or CCMs. These include Crassulacean acid metabolism, C_4 carbon fixation, and pyrenoids. Chloroplasts in C_4 plants are notable as they exhibit a distinct chloroplast dimorphism.

Process of Photosynthesis: the Light Reactions

Light Absorption and Energy Transfer

The light energy absorbed by a chlorophyll molecule excites some electrons within the structure of the molecule to higher energy levels, or excited states. Light of shorter wavelength (such as blue) has more energy than light of longer wavelength (such as red), so absorption of blue light creates an excited state of higher energy. A molecule raised to this higher energy state quickly gives up the "extra" energy as heat and falls to its lowest excited state. This lowest excited state is similar to that of a molecule that has just absorbed the longest wavelength light capable of exciting it. In the case of chlorophyll *a*, this lowest excited state corresponds to that of a molecule that has absorbed red light of about 680 nm.

The return of a chlorophyll *a* molecule from its lowest excited state to its original low-energy state (ground state) requires the release of the extra energy of the excited state. This can occur in one of several ways. In photosynthesis, most of this energy is conserved as chemical energy by the transfer of an electron from a special chlorophyll *a* molecule (P_{680} or P_{700}) to an electron acceptor. When this electron transfer is blocked by inhibitors, such as the herbicide dichlorophenylmethylurea (DCMU), or by low temperature, the energy can be released as red light. Such reemission of light is called fluorescence. The examination of fluorescence from photosynthetic material in which electron transfer has been blocked has proved to be a valuable tool for scientists studying the light reactions.

The Pathway of Electrons

The general features of a widely accepted mechanism for photoelectron transfer, in which two light reactions (light reaction I and light reaction II) occur during the transfer of electrons from water to carbon dioxide, were proposed by Robert Hill and Fay Bendall in 1960. This mechanism is based on the relative potential (in volts) of various cofactors of the electron-transfer chain to be oxidized or reduced. Molecules that in their oxidized form have the strongest affinity for electrons (i.e., are strong oxidizing agents) have a low relative potential. In contrast, molecules that in their oxidized form are difficult to reduce have a high relative potential once they have accepted electrons. The molecules with a low relative potential are considered to be strong oxidizing agents, and those with a high relative potential are considered to be strong reducing agents.

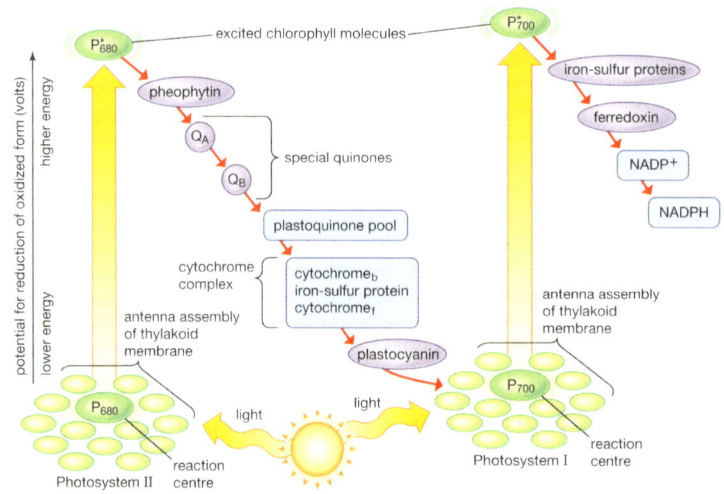

Flow of electrons during the light reaction stage of photosynthesisArrows pointing upward represent light reactions that increase the chemical potential; arrows slanting downward represent flow of electrons via carriers in the membrane.

In diagrams that describe the light reaction stage of photosynthesis, the actual photochemical steps are typically represented by two vertical arrows. These arrows signify that the special pigments P_{680} and P_{700} receive light energy from the light-harvesting chlorophyll-protein molecules and are raised in energy from their ground state to excited states. In their excited state, these pigments are extremely strong reducing agents that quickly transfer electrons to the first acceptor. These first acceptors also are strong reducing agents and rapidly pass electrons to more stable carriers. In light reaction II, the first acceptor may be pheophytin, which is a molecule similar to chlorophyll that also has a strong reducing potential and quickly transfers electrons to the next acceptor. Special quinones are next in the series. These molecules are similar to plastoquinone; they receive electrons from pheophytin and pass them to the intermediate electron carriers, which include the plastoquinone pool and the cytochromes *b* and *f* associated in a complex with an iron-sulfur protein.

In light reaction I, electrons are passed on to iron-sulfur proteins in the lamellar membrane, after which the electrons flow to ferredoxin, a small water-soluble iron-sulfur protein. When $NADP^+$ and a suitable enzyme are present, two ferredoxin molecules, carrying one electron each, transfer two electrons to $NADP^+$, which picks up a proton (i.e., a hydrogen ion) and becomes NADPH.

Each time a P_{680} or P_{700} molecule gives up an electron, it returns to its ground (unexcited) state, but with a positive charge due to the loss of the electron. These positively charged ions are extremely strong oxidizing agents that remove an electron from a suitable donor. The P_{680}^+ of light reaction II is capable of taking electrons from water in the presence of appropriate catalysts. There is good evidence that two or more manganese atoms complexed with protein are involved in this catalysis, taking four electrons from two water molecules (with release of four hydrogen ions). The manganese-protein complex gives up these electrons one at a time via an unidentified carrier to P_{680}^+, reducing it to P_{680}. When manganese is selectively removed by chemical treatment, the thylakoids lose the capacity to oxidize water, but all other parts of the electron pathway remain intact.

In light reaction I, P_{700}^+ recovers electrons from plastocyanin, which in turn receives them from intermediate carriers, including the plastoquinone pool and cytochrome b and cytochrome f molecules. The pool of intermediate carriers may receive electrons from water via light reaction II and the quinones. Transfer of electrons from water to ferredoxin via the two light reactions and intermediate carriers is called noncyclic electron flow. Alternatively, electrons may be transferred only by light reaction I, in which case they are recycled from ferredoxin back to the intermediate carriers. This process is called cyclic electron flow.

Evidence of Two Light Reactions

Many lines of evidence support the concept of electron flow via two light reactions. An early study by American biochemist Robert Emerson employed the algae *Chlorella*, which was illuminated with red light alone, with blue light alone, and with red and blue light at the same time. Oxygen evolution was measured in each case. It was substantial with blue light alone but not with red light alone. With both red and blue light together, the amount of oxygen evolved far exceeded the sum of that seen with blue and red light alone. These experimental data pointed to the existence of two types of light reactions that, when operating in tandem, would yield the highest rate of oxygen evolution. It is now known that light reaction I can use light of a slightly longer wavelength than red (λ = 680 nm), while light reaction II requires light with a wavelength of 680 nm or shorter.

Since those early studies, the two light reactions have been separated in many ways, including separation of the membrane particles in which each reaction occurs. As discussed previously, lamellae can be disrupted mechanically into fragments that absorb light energy and break the bonds of water molecules (i.e., oxidize water) to produce oxygen, hydrogen ions, and electrons. These electrons can be transferred to ferredoxin, the final electron acceptor of the light stage. No transfer of electrons from water to ferredoxin occurs if the herbicide DCMU is present. The subsequent addition of certain reduced dyes (i.e., electron donors) restores the light reduction of $NADP^+$ but without oxygen production, suggesting that light reaction I but not light reaction II is functioning. It is now known that DCMU blocks the transfer of electrons between the first quinone and the plastoquinone pool in light reaction II.

When treated with certain detergents, lamellae can be broken down into smaller particles capable of carrying out single light reactions. One type of particle can absorb light energy, oxidize water, and produce oxygen (light reaction II), but a special dye molecule must be supplied to accept the electrons. In the presence of electron donors, such as a reduced dye, a second type of lamellar particle can absorb light and transfer electrons from the electron donor to ferredoxin (light reaction I).

Photosystems I and II

The structural and photochemical properties of the minimum particles capable of performing light reactions I and II have received much study. Treatment of lamellar fragments with neutral detergents releases these particles, designated photosystem I and photosystem II, respectively. Subsequent harsher treatment (with charged detergents) and separation of the individual polypeptides with electrophoretic techniques have helped identify the components of the photosystems. Each photosystem consists of a light-harvesting complex and a core complex. Each core complex contains a reaction centre with the pigment (either P_{700} or P_{680}) that can be photochemically oxidized, together with electron acceptors and electron donors. In addition, the core complex has some 40 to 60 chlorophyll molecules bound to proteins. In addition to the light absorbed by the chlorophyll molecules in the core complex, the reaction centres receive a major part of their excitation from the pigments of the light-harvesting complex.

Quantum Requirements

The quantum requirements of the individual light reactions of photosynthesis are defined as the number of light photons absorbed for the transfer of one electron. The quantum requirement for each light reaction has been found to be approximately one photon. The total number of quanta required, therefore, to transfer the four electrons that result in the formation of one molecule of oxygen via the two light reactions should be four times two, or eight. It appears, however, that additional light is absorbed and used to form ATP by a cyclic photophosphorylation pathway. (The cyclic photophosphorylation pathway is an ATP-forming process in which the excited electron returns to the reaction centre). The actual quantum requirement, therefore, probably is 9 to 10.

Process of Photosynthesis: Conversion of Light Energy to ATP

The electron transfers of the light reactions provide the energy for the synthesis of two compounds vital to the dark reactions: NADPH and ATP. The previous section explained how noncyclic electron flow results in the reduction of $NADP^+$ to NADPH. In this part, the synthesis of the energy-rich compound ATP is described.

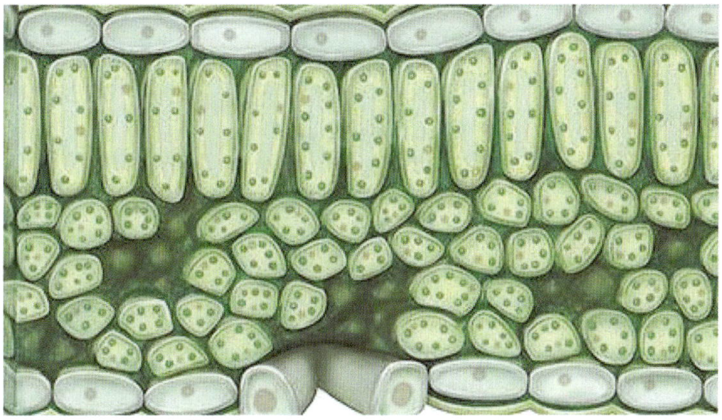

chloroplast; photosynthesis Chloroplasts play a key role in the process of photosynthesis.

ATP is formed by the addition of a phosphate group to a molecule of adenosine diphosphate (ADP)—or to state it in chemical terms, by the phosphorylation of ADP. This reaction requires a substantial input of energy, much of which is captured in the bond that links the added phosphate group to ADP. Because light energy powers this reaction in the chloroplasts, the production of ATP during photosynthesis is referred to as photophosphorylation, as opposed to oxidative phosphorylation in the electron-transport chain in the mitochondrion.

Unlike the production of NADPH, the photophosphorylation of ADP occurs in conjunction with both cyclic and noncyclic electron flow. In fact, researchers speculate that the sole purpose of cyclic electron flow may be for photophosphorylation, since this process involves no net transfer of electrons to reducing agents. The relative amounts of cyclic and noncyclic flow may be adjusted in accordance with changing physiological needs for ATP and reduced ferredoxin and NADPH in chloroplasts. In contrast to electron transfer in light reactions I and II, which can occur in membrane fragments, intact thylakoids are required for efficient photophosphorylation. This requirement stems from the special nature of the mechanism linking photophosphorylation to electron flow in the lamellae.

The theory relating the formation of ATP to electron flow in the membranes of both chloroplasts and mitochondria (the organelles responsible for ATP formation during cellular respiration) was first proposed by English biochemist Peter Dennis Mitchell, who received the 1978 Nobel Prize for Chemistry. This chemiosmotic theory has been somewhat modified to fit later experimental facts. The general features are now widely accepted. A central feature is the formation of a hydrogen ion (proton) concentration gradient and an electrical charge across intact lamellae. The potential energy stored by the proton gradient and electrical charge is then used to drive the energetically unfavourable conversion of ADP and inorganic phosphate (P_i) to ATP and water.

The manganese-protein complex associated with light reaction II is exposed to the interior of the thylakoid. Consequently, the oxidation of water during light reaction II leads to release of hydrogen ions (protons) into the inner thylakoid space. Furthermore, it is likely that photoreaction II entails the transfer of electrons across the lamella toward its outer face, so that when plastoquinone molecules are reduced, they can receive protons from the outside of the thylakoid. When these reduced plastoquinone molecules are oxidized, giving up electrons to the cytochrome-iron-sulfur complex, protons are released inside the thylakoid. Because the lamella is impermeable to them, the release of protons inside the thylakoid by oxidation of both water and plastoquinone leads to a higher concentration of protons inside the thylakoid than outside it. In other words, a proton gradient is established across the lamella. Since protons are positively charged, the movement of protons across the thylakoid lamella during both light reactions results in the establishment of an electrical charge across the lamella.

An enzyme complex located partly in and on the lamellae catalyzes the reaction in which ATP is formed from ADP and inorganic phosphate. The reverse of this reaction is catalyzed by an enzyme called ATP-ase; hence, the enzyme complex is sometimes called an ATP-ase complex. It is also called the coupling factor. It consists of hydrophilic polypeptides (F_1), which project from the outer surface of the lamellae, and hydrophobic polypeptides (F_0), which are embedded inside the lamellae. F_0 forms a channel that permits protons to flow through the lamellar membrane to F_1. The enzymes in F_1 then catalyze ATP formation, using both the proton supply and the lamellar transmembrane charge.

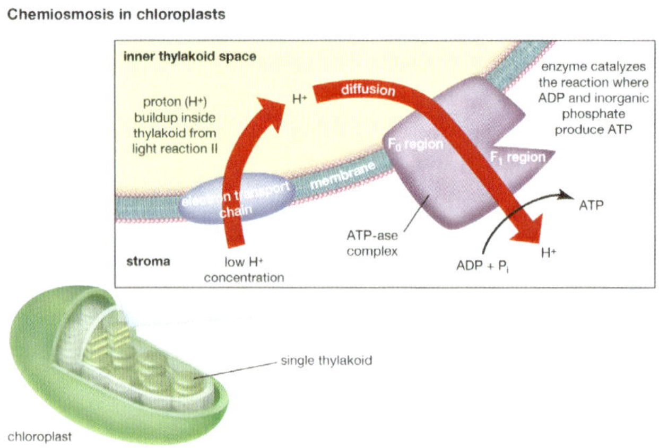

chemiosmosis in chloroplasts Chemiosmosis in chloroplasts that results in the donation of a proton for the production of adenosine triphosphate (ATP) in plants.

In summary, the use of light energy for ATP formation occurs indirectly: a proton gradient and electrical charge—built up in or across the lamellae as a consequence of electron flow in the light reactions—provide the energy to drive the synthesis of ATP from ADP and P_i.

Molecular Biology of Photosynthesis

Oxygenic photosynthesis occurs in a certain type of prokaryotic cells called cyanobacteria and eukaryotic plant cells (algae and higher plants). In eukaryotic plant cells, which contain chloroplasts and a nucleus, the genetic information needed for the reproduction of the photosynthetic apparatus is contained partly in the chloroplast chromosome and partly in chromosomes of the nucleus. For example, the carboxylation enzyme ribulose 1,5-bisphosphate carboxylase is a large protein molecule comprising a complex of eight large polypeptide subunits and eight small polypeptide subunits. The gene for the large subunits is located in the chloroplast chromosome, whereas the gene for the small subunits is in the nucleus. Transcription of the DNA of the nuclear gene yields messenger RNA (mRNA) that encodes the information for the synthesis of the small polypeptides. During this synthesis, which occurs on the cytosolic ribosomes, some extra amino acid residues are added to form a recognition leader on the end of the polypeptide chain. This leader is recognized by special receptor sites on the outer chloroplast membrane; these receptor sites then allow the polypeptide to penetrate the membrane and enter the chloroplast. The leader is removed, and the small subunits combine with the large subunits, which have been synthesized on chloroplast ribosomes according to mRNA transcribed from the chloroplast DNA. The expression of nuclear genes that code for proteins needed in the chloroplasts appears to be under control of events in the chloroplasts in some cases; for example, the synthesis of some nuclear-encoded chloroplast enzymes may occur only when light is absorbed by chloroplasts.

Plant Nutrition

Plants obtain food in two different ways. Autotrophic plants can make their own food from inorganic raw materials, such as carbon dioxide and water, through photosynthesis in the presence of sunlight. Green plants are included in this group. Some plants, however, are heterotrophic: they are totally parasitic and lacking in chlorophyll. These plants, referred to as holo-parasitic plants, are unable to synthesize organic carbon and draw all of their nutrients from the host plant.

Plants may also enlist the help of microbial partners in nutrient acquisition. Particular species of bacteria and fungi have evolved along with certain plants to create a mutualistic symbiotic relationship with roots. This improves the nutrition of both the plant and the microbe. The formation of nodules in legume plants and mycorrhization can be considered among the nutritional adaptations of plants. However, these are not the only type of adaptations that we may find; many plants have other adaptations that allow them to thrive under specific conditions.

Nutritional Requirements

Plants are unique organisms that can absorb nutrients and water through their root system, as well as carbon dioxide from the atmosphere. Soil quality and climate are the major determinants of plant distribution and growth. The combination of soil nutrients, water, and carbon dioxide, along with sunlight, allows plants to grow.

Chemical Composition of Plants

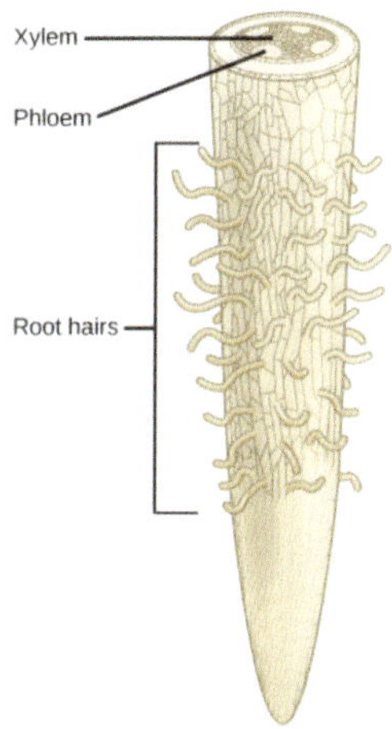

Figure: Water is absorbed through the root hairs and moves up the xylem to the leaves

Since plants require nutrients in the form of elements such as carbon and potassium, it is important to understand the chemical composition of plants. The majority of volume in a plant cell is water; it typically comprises 80 to 90 percent of the plant's total weight. Soil is the water source for land plants, and can be an abundant source of water, even if it appears dry. Plant roots absorb water from the soil through root hairs and transport it up to the leaves through the xylem. As water vapor is lost from the leaves, the process of transpiration and the polarity of water molecules (which enables them to form hydrogen bonds) draws more water from the roots up through the plant to the leaves. Plants need water to support cell structure, for metabolic functions, to carry nutrients, and for photosynthesis.

Plant cells need essential substances, collectively called nutrients, to sustain life. Plant nutrients may be composed of either organic or inorganic compounds. An organic compound is a chemical compound that contains carbon, such as carbon dioxide obtained from the atmosphere. Carbon that was obtained from atmospheric CO_2 composes the majority of the dry mass within most plants. An inorganic compound does not contain carbon and is not part of, or produced by, a living organism. Inorganic substances, which form the majority of the soil solution, are commonly called minerals: those required by plants include nitrogen (N) and potassium (K) for structure and regulation.

Essential Nutrients

Plants require only light, water and about 20 elements to support all their biochemical needs: these 20 elements are called essential nutrients (Table). For an element to be regarded as essential, three criteria are required: 1) a plant cannot complete its life cycle without the element; 2) no other element can perform the function of the element; and 3) the element is directly involved in plant nutrition.

Table. Essential Elements for Plant Growth

Macronutrients	Micronutrients
Carbon (C)	Iron (Fe)
Hydrogen (H)	Manganese (Mn)
Oxygen (O)	Boron (B)
Nitrogen (N)	Molybdenum (Mo)
Phosphorus (P)	Copper (Cu)
Potassium (K)	Zinc (Zn)
Calcium (Ca)	Chlorine (Cl)
Magnesium (Mg)	Nickel (Ni)
Sulfur (S)	Cobalt (Co)
	Sodium (Na)
	Silicon (Si)

Macronutrients and Micronutrients

The essential elements can be divided into two groups: macronutrients and micronutrients. Nutrients that plants require in larger amounts are called macronutrients. About half of the essential elements are considered macronutrients: carbon, hydrogen, oxygen, nitrogen, phosphorus, potassium, calcium, magnesium and sulfur. The first of these macronutrients, carbon (C), is required to form carbohydrates, proteins, nucleic acids, and many other compounds; it is therefore present in all macromolecules. On average, the dry weight (excluding water) of a cell is 50 percent carbon. As shown in figure, carbon is a key part of plant biomolecules.

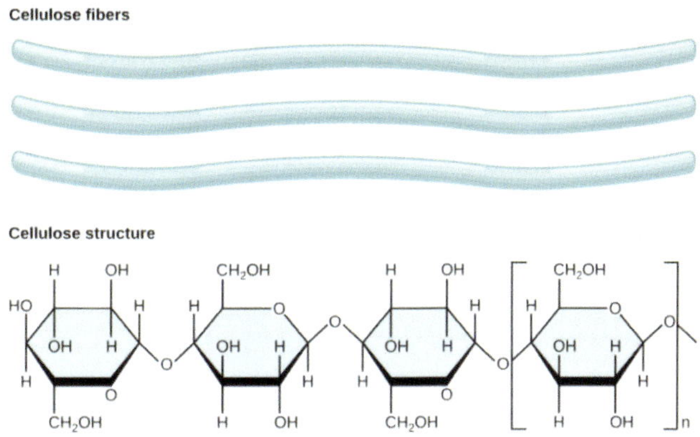

Figure. Cellulose, the main structural component of the plant cell wall, makes up over thirty percent of plant matter. It is the most abundant organic compound on earth.

The next most abundant element in plant cells is nitrogen (N); it is part of proteins and nucleic acids. Nitrogen is also used in the synthesis of some vitamins. Hydrogen and oxygen are macronutrients that are part of many organic compounds, and also form water. Oxygen is necessary for cellular respiration; plants use oxygen to store energy in the form of ATP. Phosphorus (P), another macromolecule, is necessary to synthesize nucleic acids and phospholipids. As part of ATP, phosphorus enables food energy to be converted into chemical energy through oxidative phosphorylation. Likewise, light energy is converted into chemical energy during photophosphorylation in photosynthesis, and into chemical energy to be extracted during respiration. Sulfur is part of certain amino acids, such as cysteine and methionine, and is present in several coenzymes. Sulfur also plays a role in photosynthesis as part of the electron transport chain, where hydrogen gradients play a key role in the conversion of light energy into ATP. Potassium (K) is important because of its role in regulating stomatal opening and closing. As the openings for gas exchange, stomata help maintain a healthy water balance; a potassium ion pump supports this process.

Magnesium (Mg) and calcium (Ca) are also important macronutrients. The role of calcium is twofold: to regulate nutrient transport, and to support many enzyme functions. Magnesium is important to the photosynthetic process. These minerals, along with the micronutrients, which are described below, also contribute to the plant's ionic balance.

In addition to macronutrients, organisms require various elements in small amounts. These micronutrients, or trace elements, are present in very small quantities. They include boron (B), chlorine (Cl), manganese (Mn), iron (Fe), zinc (Zn), copper (Cu), molybdenum (Mo), nickel (Ni), silicon (Si), and sodium (Na).

Figure: Nutrient deficiency is evident in the symptoms these plants show. This (a) grape tomato suffers from blossom end rot caused by calcium deficiency. The yellowing in this (b) Frangula alnus results from magnesium deficiency. Inadequate magnesium also leads to (c) intervenal chlorosis, seen here in a sweetgum leaf. This (d) palm is affected by potassium deficiency.

Deficiencies in any of these nutrients—particularly the macronutrients—can adversely affect plant growth. Depending on the specific nutrient, a lack can cause stunted growth, slow growth, or chlorosis (yellowing of the leaves). Extreme deficiencies may result in leaves showing signs of cell death.

Autotrophic Plants

Nitrogen Fixation: Root and Bacteria Interactions

Nitrogen is an important macronutrient because it is part of nucleic acids and proteins. Atmospheric nitrogen, which is the diatomic molecule N_2, or dinitrogen, is the largest pool of nitrogen in terrestrial ecosystems. However, plants cannot take advantage of this nitrogen because they do not have the necessary enzymes to convert it into biologically useful forms. However, nitrogen can be "fixed," which means that it can be converted to ammonia (NH_3) through biological, physical, or chemical processes. Biological nitrogen fixation (BNF) is the conversion of atmospheric nitrogen (N_2) into ammonia (NH_3), exclusively carried out by prokaryotes such as soil bacteria or cyanobacteria. Biological processes contribute 65 percent of the nitrogen used in agriculture. The following equation represents the process:

$$N_2 + 16\ ATP + 8e^- + 8H^+ \longrightarrow 2NH_3 + 16\ ADP + 16Pi + H_2$$

The most important source of BNF is the symbiotic interaction between soil bacteria and legume plants, including many crops important to humans. The NH3 resulting from fixation can be transported into plant tissue and incorporated into amino acids, which are then made into plant proteins. Some legume seeds, such as soybeans and peanuts, contain high levels of protein, and serve among the most important agricultural sources of protein in the world.

(a) (b) (c)

Figure: Some common edible legumes—like (a) peanuts, (b) beans, and (c) chickpeas—are able to interact symbiotically with soil bacteria that fix nitrogen

Soil bacteria, collectively called rhizobia, symbiotically interact with legume roots to form specialized structures called nodules, in which nitrogen fixation takes place. This process entails the reduction of atmospheric nitrogen to ammonia, by means of the enzyme nitrogenase. Therefore, using rhizobia is a natural and environmentally friendly way to fertilize plants, as opposed to chemical fertilization that uses a nonrenewable resource, such as natural gas. Through symbiotic nitrogen fixation, the plant benefits from using an endless source of nitrogen from the atmosphere. The process simultaneously contributes to soil fertility because the plant root system leaves behind some of the biologically available nitrogen. As in any symbiosis, both organisms benefit from the interaction: the plant obtains ammonia, and bacteria obtain carbon compounds generated through photosynthesis, as well as a protected niche in which to grow.

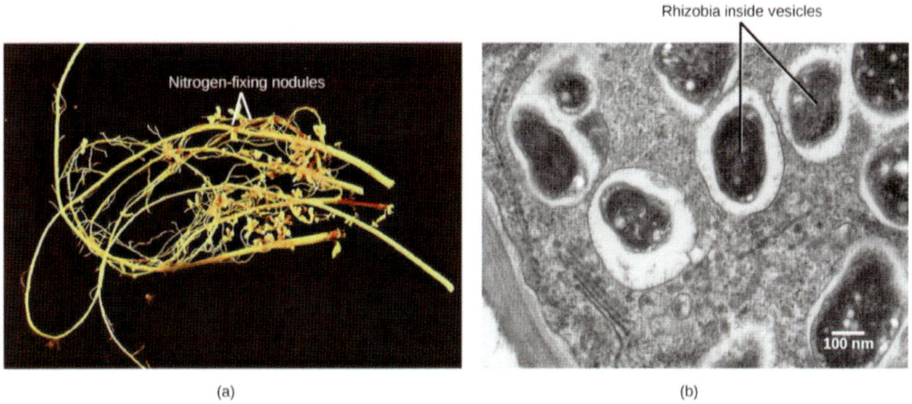

Figure 5. Soybean roots contain (a) nitrogen-fixing nodules. Cells within the nodules are infected with *Bradyrhyzobium japonicum*, a rhizobia or "root-loving" bacterium. The bacteria are encased in (b) vesicles inside the cell, as can be seen in this transmission electron micrograph.

Mycorrhizae: Symbiotic Relationship between Fungi and Roots

A nutrient depletion zone can develop when there is rapid soil solution uptake, low nutrient concentration, low diffusion rate, or low soil moisture. These conditions are very common; therefore, most plants rely on fungi to facilitate the uptake of minerals from the soil. Fungi form symbiotic associations called mycorrhizae with plant roots, in which the fungi actually are integrated into the physical structure of the root. The fungi colonize the living root tissue during active plant growth.

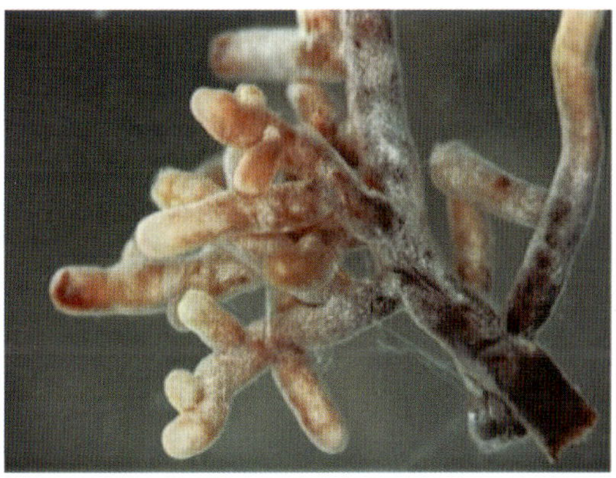

Figure. Root tips proliferate in the presence of mycorrhizal infection, which appears as off-white fuzz in this image

Through mycorrhization, the plant obtains mainly phosphate and other minerals, such as zinc and copper, from the soil. The fungus obtains nutrients, such as sugars, from the plant root. Mycorrhizae help increase the surface area of the plant root system because hyphae, which are narrow, can spread beyond the nutrient depletion zone. Hyphae can grow into small soil pores that allow access to phosphorus that would otherwise be unavailable to the plant. The beneficial effect on the plant is best observed in poor soils. The benefit to fungi is that they can obtain up to 20 percent of the total carbon accessed by plants. Mycorrhizae functions as a physical barrier to pathogens. It also provides an induction of generalized host defense mechanisms, and sometimes involves production of antibiotic compounds by the fungi.

There are two types of mycorrhizae: ectomycorrhizae and endomycorrhizae. Ectomycorrhizae form an extensive dense sheath around the roots, called a mantle. Hyphae from the fungi extend from the mantle into the soil, which increases the surface area for water and mineral absorption. This type of mycorrhizae is found in forest trees, especially conifers, birches, and oaks. Endomycorrhizae, also called arbuscular mycorrhizae, do not form a dense sheath over the root. Instead, the fungal mycelium is embedded within the root tissue. Endomycorrhizae are found in the roots of more than 80 percent of terrestrial plants.

Heterotrophic Plants

Some plants cannot produce their own food and must obtain their nutrition from outside sources—these plants are heterotrophic. This may occur with plants that are parasitic or saprophytic. Some plants are mutualistic symbionts, epiphytes, or insectivorous.

Plant Parasites

A parasitic plant depends on its host for survival. Some parasitic plants have no leaves. An example of this is the dodder, which has a weak, cylindrical stem that coils around the host and forms suckers. From these suckers, cells invade the host stem and grow to connect with the vascular bundles of the host. The parasitic plant obtains water and nutrients through these connections. The plant is a total parasite (a holoparasite) because it is completely dependent on its host. Other parasitic plants (hemiparasites) are fully photosynthetic and only use the host for water and minerals. There are about 4,100 species of parasitic plants.

Saprophytes

A saprophyte is a plant that does not have chlorophyll and gets its food from dead matter, similar to bacteria and fungi (note that fungi are often called saprophytes, which is incorrect, because fungi are not plants). Plants like these use enzymes to convert organic food materials into simpler forms from which they can absorb nutrients. Most saprophytes do not directly digest dead matter: instead, they parasitize fungi that digest dead matter, or are mycorrhizal, ultimately obtaining photosynthate from a fungus that derived photosynthate from its host. Saprophytic plants are uncommon; only a few species are described.

(a) (b)

Figure: (a) The dodder is a holoparasite that penetrates the host's vascular tissue and diverts nutrients for its own growth. The dodder has no chlorophyll and cannot produce its own food. (b) Saprophytes, like this Dutchmen's pipe (*Monotropa hypopitys*), obtain their food from dead matter and do not have chlorophyll

Symbionts

A symbiont is a plant in a symbiotic relationship, with special adaptations such as mycorrhizae or nodule formation. Fungi also form symbiotic associations with cyanobacteria and green algae (called lichens). Lichens can sometimes be seen as colorful growths on the surface of rocks and trees. The algal partner (phycobiont) makes food autotrophically, some of which it shares with the fungus; the fungal partner (mycobiont) absorbs water and minerals from the environment, which are made available to the green alga. If one partner was separated from the other, they would both die.

Epiphytes

An epiphyte is a plant that grows on other plants, but is not dependent upon the other plant for nutrition. Epiphytes have two types of roots: clinging aerial roots, which absorb nutrients from humus that accumulates in the crevices of trees; and aerial roots, which absorb moisture from the atmosphere.

Figure: (a) Lichens, which often have symbiotic relationships with other plants, can sometimes be found growing on trees. (b) These epiphyte plants grow in the main greenhouse of the *Jardin des Plantes* in Paris.

Insectivorous Plants

Figure: A Venus flytrap has specialized leaves to trap insects.

An insectivorous plant has specialized leaves to attract and digest insects. The Venus flytrap is popularly known for its insectivorous mode of nutrition, and has leaves that work as traps.

The minerals it obtains from prey compensate for those lacking in the boggy (low pH) soil of its native North Carolina coastal plains. There are three sensitive hairs in the center of each half of each leaf. The edges of each leaf are covered with long spines. Nectar secreted by the plant attracts flies to the leaf. When a fly touches the sensory hairs, the leaf immediately closes. Next, fluids and enzymes break down the prey and minerals are absorbed by the leaf. Since this plant is popular in the horticultural trade, it is threatened in its original habitat.

Plant Lipids

Lipids are small hydrophobic or amphiphilic molecules with a backbone of carbons. There are eight different classes of lipids; fatty acids, sterols, prenols, sphingolipids, glycerolipids, glycerophospholipids, saccharolipids and polyketides. Fatty acids consist of a hydrophilic carboxyl group with a hydrophobic tail, with different chain length. Fatty acids can be saturated or unsaturated and have different functional groups in addition to the carboxyl group. Further, they are often part of more complex lipids, including wax esters and glycerolipids. The sterols are based on a completely different structure, where the core is consisting of four rings, and are often functioning as hormones. Sterols also play an important role in membranes, where they affect the fluidity. The prenol lipids are all synthesized from different number of isoprene units, and includes carotenoids and other antioxidants. Sphingolipids always have a sphingoid base, consisting of the amino acid serine together with a fatty acid, but many different functional groups and side chains occur in this complex lipid class. Sphingolipids are constituents of membranes, but also involved in signaling. The glycerolipids includes all lipids containing a glycerol molecule. The most well known are the triacylglycerols, consisting of a glycerol with three fatty acid chains, which are used for energy storage in the cells. Glycerophospholipids, or just phospholipids, have a polar phosphate head group with two hydrophobic tails, attached via a glycerol molecule. Phospholipids are the main constituents of cell membranes. Saccharolipids are compound with fatty acids linked directly to a sugar backbone. The polyketides are structurally diverse but are always derived from condensation of ketoacyl subunits. These molecules often have antibiotic effects.

In plants, there are three main functions for lipids – energy storage, signaling and protection. In addition to this, amphiphilic lipids are the main constituents of membranes in the plant cell, as well as in every other eukaryotic organism. Lipids are also important for photosynthesis, both as constituents of the thylakoid membranes and by interacting with the photosynthetic protein complexes Photosystem I and II.

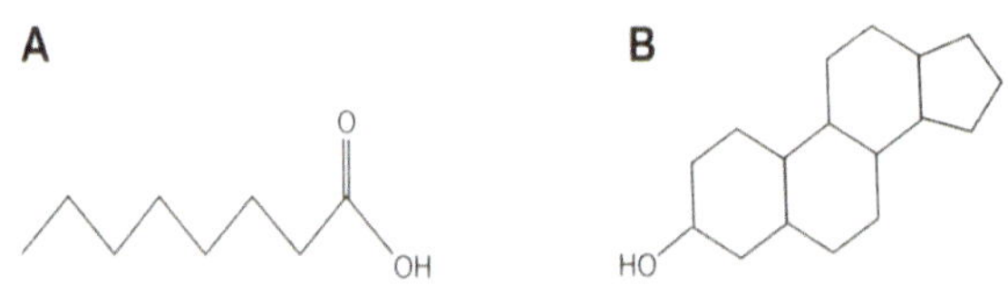

Figure. Two lipids, belonging to different classes. (A) Fatty acid (Octanoic acid) (B) Sterol.

The energy storage in plants mainly occurs in the form of starch, lipids or proteins, and which form is preferred varies between different species and tissues. Lipids are most often used for storage in the seeds, while starch is common in roots and other underground tissues. Lipids as energy storage usually mean triacylglycerols, which are stored in oil bodies inside the cell. However, waxes are sometimes used for this purpose, for example in jojoba seeds. Further, lipids play an important role in plant signaling; many hormones are lipids, e. g. the phytosterols (derived from sterol) and the jasmonates. These hormones are involved in different pathways, including defense response, programmed cell death and germination. Several lipids function as important second messengers in intracellular signaling, for example phosphatidic acid in the Phospholipase C and D signaling pathways and fatty acids in auxin signaling. Except for the involvement of lipids in defense signaling, more direct protective roles are also present; lipids are the main constituents in many defense barriers in plants, including the cuticle and wax layers of leafs, fruit and flowers. Due to the hydrophobic nature of lipid molecules, they are not able to move freely in the hydrophilic protoplasm or through the cell wall. Instead they need to be transported by other molecules. Some of these transporters have been identified, for example the protein Trigalactosyldiacylglycerol 4, which transfers lipids from the endoplasmic reticulum to the plastids. However, there are still several key transporters that have not been found. This includes the transporter protein of lipid molecules to the cuticle through the cell wall.

Non-Specific Lipid Transfer Proteins

The plant non-specific lipid transfer proteins (nsLTPs) are extensively studied, but yet their *in vivo* function is not elucidated. The nsLTPs were first discovered in potato tubers almost 40 years ago, although they were then known as phospholipid transfer proteins and thought to be involved in intracellular lipid trafficking. The ability to transfer phospholipids between membranes was shown *in vitro* in several studies and this hypothesis was widely accepted for many years, even after some nsLTPs were shown to be extracellular. The name phospholipid transfer protein was exchanged to lipid transfer protein, and later non-specific lipid transfer protein, when the proteins were shown to be able to bind and transfer different lipids, including fatty acids and fatty acyl-Coenzyme A. Similarly as the name, the hypotheses about the nsLTPs´ functions have come and gone during time. Outplayed is their suggested role in intracellular lipid trafficking, replaced by theories about pathogen defense, cuticle formation and pollen tube adhesion, among others.

The nsLTPs belong to the plant specific prolamin super family, which is characterized by conserved cysteine residues, low molecular mass and a high content of α-helices. Except for the nsLTPs, the prolamin superfamily consists of the α-amylase inhibitors, the prolamins, the 2S albumins, the indolines and the hybrid proline-rich proteins. Many of the proteins in this superfamily, including the nsLTPs, are known to cause food allergy in. In the Mediterranean area it is common with severe allergy to food derived from the rosaceae family (including almonds, apples and peach) and nsLTPs have been shown to be the allergens responsible for this. Later nsLTPs have also been identified as allergens in e.g. oranges, tomato, hazelnut, olive tree pollen and cabbage. Due to the similarity in structure between nsLTPs from different plants, there is a high rate of cross-reactivity.

The nsLTPs have been found in many different land plants, including Arabidopsis, rice, wheat, pine, peach and corn. The protein family is very large, with more than 150 members in wheat, 49 in Arabidopsis and 52 in rice. Lipid transfer proteins are found not only in plants, but in mammals as well. However, these proteins are not structurally or sequentially related to the lipid transfer proteins found in plants.

The proteins have traditionally been divided into two different types, according to size. The LTPs belonging to Type 1 was first discovered, and they have a size about 9 kDa. Several years later the smaller (usually 7 kDa) Type 2 was found in barley and wheat. The two original types have persisted through time, but more recently several additional types have been found. It is not yet clear whether the different types differ in function or lipid preferences.

Characteristics and Structure

A typical nsLTP consists of approximately 100 amino acids, of which 20-30 constitute the N-terminal signal peptide that directs the protein to the secretory pathway. Accordingly, several nsLTPs have been found in the cell wall, but some have been reported from intracellular organelles as well. Among the amino acids in the mature protein there are always eight cysteine residues conserved in a specific pattern, known as an eight cysteine motif (8CM). The 8CM in nsLTPs have the general form C-X_n-C-X_n-CC-X_n-CXC-X_n-C-X_n-C, where X is any amino acid and n any number of amino acids. The eight cysteines are involved in four intramolecular disulphide bridges that stabilize the tertiary structure of the protein. Thus, the nsLTPs are very stable to heat and different denaturing agents. Further, the nsLTP proteins are soluble and usually have a high pI due to many charged amino acids. There are few conserved amino acids except for the cysteines, especially when comparing proteins from different types. Another frequently occurring feature among the genes encoding nsLTPs is an intron located after the last cysteine in the 8CM.

MANSSAVLVLLVVTLLVATLLKETQAACGSSYDLAACLPAAESDIQPSAQCCTQLSSYLA
TDTPEECLCQTAYSPFFQSGADIQFAIKIPQKCNLSYRAGIQCNGNTIPGGQ

Figure. The amino acid sequence of a typical nsLTP (LTP1d from physcomitrella patens). The part with grey background is the N-terminal signal sequence directing the protein to the secretory pathway. The eight cysteines are shown pale grey. The arrow indicates the position of the intron in the corresponding gene.

The members of one type of nsLTPs, namely Type G, possess a C-terminal signal sequence in addition to the N-terminal one. This signal sequence leads to a posttranslational modification where a glycosylphosphatidylinositol(GPI)-anchor is added to the C-terminal end of the protein. The GPI-anchor attaches the protein on the extracellular side of the plasma membrane, by integration of the phosphatidylinositol into the membrane. The protein can be released from the anchor through cleavage by phospholipases. Further, many GPI-anchored proteins are occasionally taken into the cell by endocytosis. GPI-anchors are not unique for nsLTPs, but found on proteins from many different families in most eukaryotes. One example is the human protein thymocyte differentiation antigen-1, which is modulating signals for e.g. apoptosis, cellular adhesion and cell proliferation. Further, GPI-anchors also play an important role in prion diseases; prion proteins lacking the GPI-anchor that usually attach them to the cell surface of the host´s cells are less malign than the GPI-anchored ones.

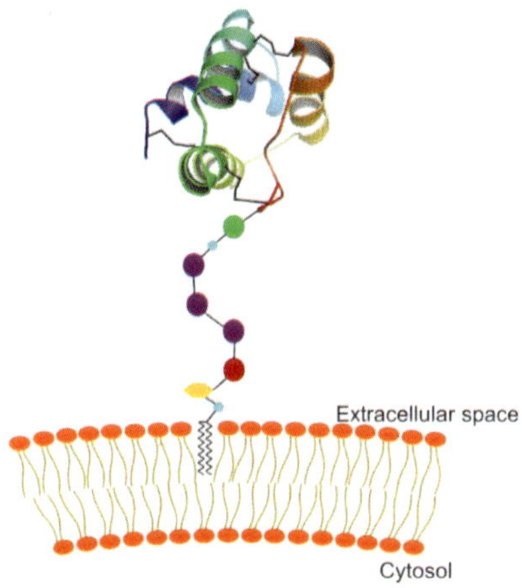

Figure: An nsLTP attached to the plasma membrane through a GPI-anchor. The anchor consists of one ethanolamine (green), three mannose molecules (purple), one glucosamine (red) and one inositol (yellow).

There is a phosphate molecule (cyan) between the ethanolamine and the mannoses, and between the inositol and the lipid tails. The inositol with the phosphate and the lipid tails together constitutes the phosphatidylinositol.

The crystal structure has been determined for several nsLTPs, among them one from peach, one from rice, one from Arabidopsis and one from maize. The structural studies have determined that the folding of nsLTP proteins occurs in two different ways, namely the Type 1 and the Type 2 fold. In the Type 1 fold, which is only found in the proteins belonging to Type 1, the disulphide bridges are formed by the cysteine residues 1-6, 2-3, 4-7 and 5-8. In the Type 2 fold, which is present in all types but Type 1, the fifth and the sixth cysteines have switched partners and the bridges are formed between the cysteine residues 1-5, 2-3, 4-7 and 6-8 instead. However, both variants give rise to similar tertiary structures with four or five α-helices and a central cavity. The interior of the cavity is lined by hydrophobic amino acids, making it suitable for lipid binding. In Type 1 the cavity is tunnel-like, with two entrances of different size, while there is only one opening in nsLTPs with the Type 2 fold. The cavity of Type 2 folded nsLTPs is also more flexible than that of Type 1, resulting in lower lipid specificity. For example, nsLTPs with the Type 1 fold are not able to bind sterols, in opposite to those with the Type 2 fold.

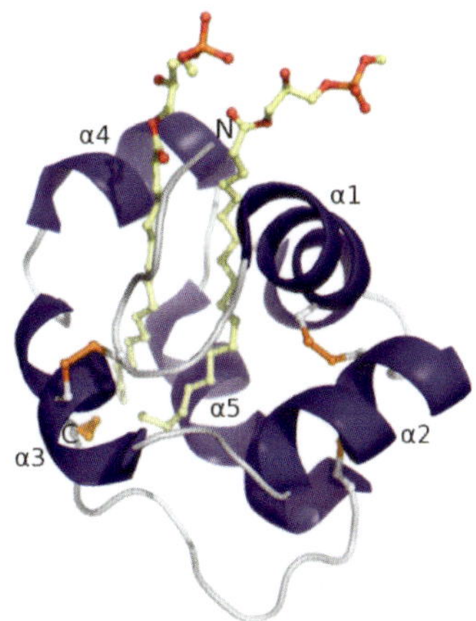

Figure. The crystal structure of the nsLTPd DIR1 from Arabidopsis, with the Type 2 fold, in complex with two lysophosphatidyl choline molecules (in yellow). The helices of the protein are shown in blue and the disulphide bonds in orange.

Lipid Binding

The nsLTPs have the ability to transfer lipids between monolayers *in vitro*. However, for most nsLTPs the lipid binding ability has not been examined, but they are considered as nsLTPs due to sequence similarity to known sequences. In cases where the lipid binding ability has been studied, most of the nsLTPs have been shown to be non-specific in terms of lipid binding ability. Consequently, they are able to bind various different lipids, including phospholipids, fatty acids and acyl-CoA. However, some lipids seem to be bound more efficiently than others. Using fluorescence labeled fatty acids, an nsLTP from maize showed highest affinity for those with chain lengths of 16-19 carbons, while hydroxyl groups or polyunsaturations reduced the affinity. Similar results are reported for a wheat nsLTP, where the method steady-state tyrosine fluorescence was used; the affinity is higher with chain lengths of 14-18 carbons compared to shorter chains and reduced by hydroxyl groups, but not unsaturations. Further, an nsLTP from ginkgo has been shown to have a low affinity for saturated and trans-unsaturated fatty acids, but high when they were cis-unsaturated. There are also some studies showing that the nsLTPs are able to bind two lipid molecules at the same time. Further, the volume of the tunnel has been shown to increase when a lipid is bound, wherefore it is impossible to predict the binding ability of an nsLTP by looking at the size of the tunnel. It has also been shown that conformational changes, where some amino acids are reoriented, occur within the protein upon lipid binding to better accommodate hydrophobic or hydrophilic side chains of the lipids. There is no clear difference in lipid preferences between nsLTPs from the different types, except for some bulky lipids that are too large for proteins with the more rigid Type 1 fold.

Expression

The typical nsLTP is highly expressed in epidermal tissues of the aerial part of the plant, including leaves, shoot meristems and flower. Others are found in almost all tissues, but with different expression patterns during different stages of plant development. For example, NtLTP1 from tobacco is mainly expressed in epidermis in adult tissue, but in young tissues the expression was much broader and found in roots and hypocotyl as well. Some nsLTPs are however specifically expressed in one or a few tissues, including root, anther, trichomes or seed. In many cases the expression of the nsLTPs are upregulated by different biotic and abiotic stresses, including abscisic acid and salicylic acid in strawberry and bacterial infection in barley. This have been shown for several additional nsLTPs from different species, including mungbean, bromegrass, sunflower, rice and Arabidopsis. Further, nsLTPs are often found to be differentially expressed in various mutants. One example is the double mutant of the RNA binding proteins TSN1 and TSN2 (Tudor-SN) in Arabidopsis, where six nsLTPs are downregulated during normal conditions, salt stress or both.

Proposed Functions

There are many different hypotheses regarding the functions of nsLTPs. One of the more well-established is a role in transport of lipids to the cuticle, through the cell wall. This has indirectly been shown for some Type G nsLTPs in Arabidopsis, where knockout/knockdown mutants have a lower wax load in the stem compared to wild-type plants. Their exact role in this lipid deposition, or the mechanism behind it, is however not clarified.

Another nsLTP from Arabidopsis, named DIR1, have been shown to be essential for long distance signaling in systemic acquired resistance (SAR) in Arabidopsis. The knockout mutant of DIR1 was shown to lack the ability to develop SAR in response to infection by *Pseudomonas syringae* and *Peronospora parasitica*. In agreement to this, several nsLTPs have been shown to be upregulated when exposed to biotic stresses, including the bacterium *Xanthomonas campestris* in cauliflower, the fungus *Puccinia melanocephala* in sugarcane and a tobacco mosaic virus in pepper. Further, some nsLTP have been shown to have antibiotic, antiviral or antifungal properties *in vitro*, supporting a role in defense against microorganisms. This is partly in agreement with the defense signaling hypothesis as well as the cuticle hypothesis, since one of the main functions of the cuticle is to protect the plants from pathogens.

Some nsLTPs have been found forming complexes with other proteins. One example is in tomato, where an nsLTP was shown to be necessary for the function of polygalacturonase, a protein involved in fruit ripening. In a similar manner, another nsLTP from Arabidopsis have been shown to bind calmodulin *in vitro*. Further, other nsLTPs seem to be involved in sexual reproduction, specifically in pollen tube growth and pollen tube adhesion, possibly by forming an adhesive matrix together with pectin. There is also evidence of nsLTPs required for normal pollen development, where plants lacking functional nsLTPs showed collapsed pollen grains or irregularities in the sporopollenin coat. NsLTPs have also been found necessary for the symbiotic relationship between cowpea, barrel clover and their respectively symbiotic partners.

The widely shifting hypotheses regarding the function of nsLTPs indicate a large functional divergence. However, there seems not to be a clear relationship between proteins of the same type and the function. Further, some of the proposed functions do not obviously involve any lipids, which would be expected by a lipid transfer protein. Clearly, further studies are required to fully understand the function and complexity of these elusive proteins.

Carbon Assimilation

Carbon fixation is a process that involves fixing a carbon into place. Just like you take that piece of paper and tack it to the bulletin board so that it won't float around or end up on the floor, plants want to take carbon dioxide (CO_2) out of the air and tack it to carbohydrates so it's not floating around all over the place. The process of taking inorganic carbon (usually CO_2) and tacking it to an organic molecule (usually a carbohydrate) is known as carbon fixation. Carbon fixation is also known as carbon assimilation.

The light reactions result in the formation of the high-energy compounds ATP and NADPH. While these compounds can be used to drive metabolic processes, one additional critical reaction must occur: the fixation of carbon dioxide. Without CO_2 fixation, the respiratory processes necessary to generate energy at night would result in an irreversible conversion of carbon compounds to CO_2. In addition, carbon fixation is required to provide energy to non-photosynthetic tissues within the plant, and to supply the raw material required for growth of the plant.

Animals lack the ability to perform net carbon dioxide fixation; reactions such as the pyruvate carboxylase reaction only result in temporary increases in the number of carbon atoms in the compound. Plants, however, have the capability of using the energy obtained from light to drive the permanent incorporation of carbon dioxide into structural molecules. The process for fixing and assimilating carbon is called the Calvin cycle in honor of the 1961 Nobel Laureate Melvin Calvin.

Overview of the Calvin Cycle

The Calvin cycle begins with the five-carbon carbohydrate ribulose-1,5- bisphosphate.

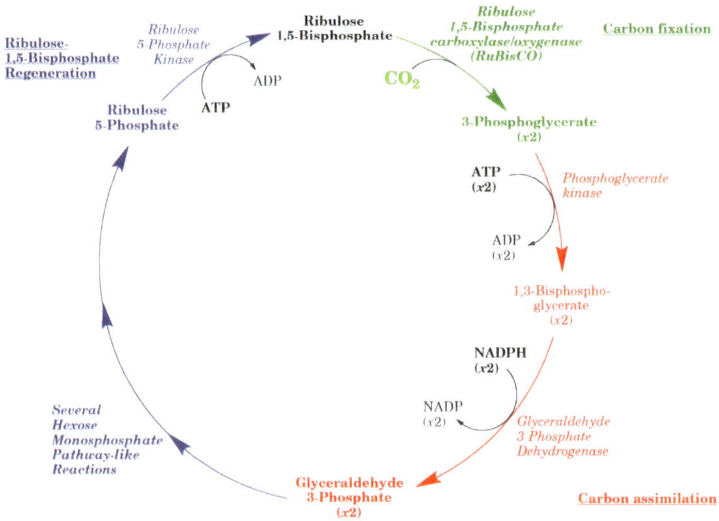

Ribulose-1,5-bisphosphate acts as the carbon dioxide acceptor; following carbon dioxide addition, the resulting six-carbon compound is cleaved into two three-carbon 3-phosphoglycerate molecules. The 3-phosphoglycerate is then converted to glyceraldehyde-3-phosphate using energy obtained from the light reactions of photosynthesis. The glyceraldehyde-3-phosphate has several possible fates; however, the continuation of the carbon fixation process requires regenerating the ribulose-1,5-bisphosphate starting material. The cycle thus has three phases: carbon fixation, glyceraldehyde-3-phosphate formation and ribulose-1,5- bisphosphate regeneration.

Rubisco

The rate-limiting enzyme in the Calvin cycle is ribulose 1,5-bisphosphate carboxylase/oxygenase (better known as rubisco). In most plants, rubisco is a complex of 8 large (53 kDa) and 8 small (14 kDa) subunits ($\alpha_8\beta_8$). The large subunit is coded by a chloroplast gene, while the small subunit gene is located in the nucleus. In photosynthetic bacteria, rubisco is usually a dimer of proteins homologous to the plant large subunit.

RuBisCo is an intensively studied protein, because it is one of the most important enzymes in existence. RuBisCo is present at a concentration of ~250 mg/ml in the chloroplast stroma, comprising ~50% of the total chloroplast protein. Plants production of RuBisCo exceeds that of any other enzyme, and probably any other protein, on earth. Current estimates suggest that the biosphere contains ~10 kg of rubisco per human (this works out to about 0.02 moles of the enzyme per person). This tremendous amount of protein is necessary because the catalytic efficiency of rubisco is fairly low: it has a k_{cat} of 1 to 3 sec^{-1}, and a rather low affinity for carbon dioxide. It is also necessary because RuBisCo catalyzes the wasteful oxygenase side reaction fixing sufficient carbon to support growth requires large amounts of this enzyme.

The RuBisCo reaction is moderately complex. The general reaction mechanism is shown below.

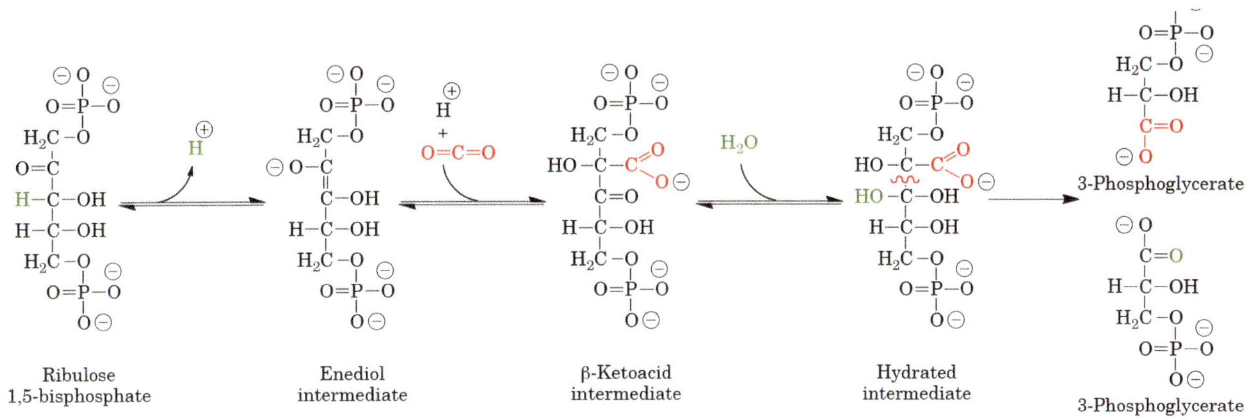

The enzyme has a basic residue that abstracts a proton from the 3-position carbon of the ribulose-1,5-bisphosphate substrate, producing the enzyme-bound ene-diol intermediate. The ene-diol acts as the carbon dioxide acceptor, with the carbon dioxide molecule forming a bond to the 2-position carbon. The six-carbon β -ketoacid intermediate is then attacked by a water molecule, resulting in cleavage of the bond between the former 2-position and 3-position carbons of the ribulose-1,5- bisphosphate molecule. This results in the release of two 3-phosphoglycerate molecules (one of which, as released, contains a negatively charged carbon ion which then obtains a proton from the aqueous solvent).

The overall reaction is therefore:

Ribulose 1, 5-bisphosphate + CO_2 + $H_2O \rightarrow$ 2x 3-phosphoglycerate + 2 H^+

The rubisco reaction thus yields two triose phosphate molecules from the single pentose bisphosphate. Note that ATP is not required for the rubisco reaction, a property that differs from most other carboxylase reactions; in this case, the reaction has a large negative ΔG° because of the release of the two triose phosphates.

Carbon Assimilation Reactions

The carbon-fixing rubisco reaction thus results in the release of 2 three-carbon 3- phosphoglycerate molecules. The 3-phosphoglycerate molecules are converted to 1,3- bisphosphoglycerate by phosphoglycerate kinase; this reaction requires ATP as a phosphate donor. The 1,3-bisphosphoglycerate is then converted to glyceraldehyde- 3-phosphate by glyceraldehyde-3-phosphate dehydrogenase.

These two reactions use chloroplast isozymes of the glycolytic enzymes phosphoglycerate kinase and glyceraldehyde-3-phosphate dehydrogenase (note that the glyceraldehyde-3-phosphate dehydrogenase uses NADPH as a substrate rather than NADH). Because two 3-phosphoglycerate molecules are released by rubisco, two ATP and two NADPH are required to convert all of the triose phosphate to glyceraldehyde-3-phosphate. The ATP and NADPH required are produced by the light reactions, which therefore drive the entire process.

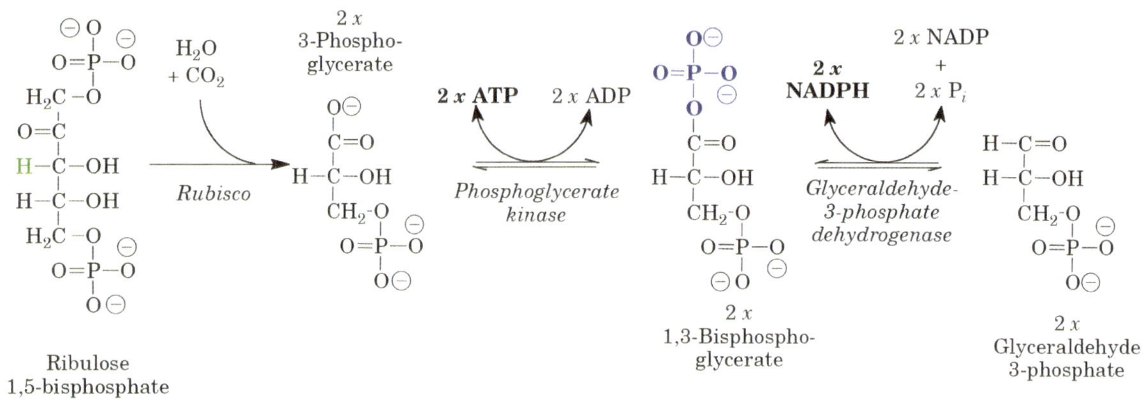

The phosphoglycerate kinase and glyceraldehyde-3-phosphate dehydrogenase reactions are reversible. In chloroplasts exposed to light and carbon dioxide, these reactions result in glyceraldehyde-3-phosphate formation due to the high concentrations of ATP, NADPH, and 3-phosphoglycerate.

Regeneration of ribulose-1,5-bisphosphate

Glyceraldehyde-3-phosphate can be converted to glucose without further input of energy. However, unless the ribulose 1,5-bisphosphate is regenerated the rubisco reaction will be unable to continue.

The carbon assimilation process begins with three ribulose 1,5-bisphosphate (containing a total of 15 carbons); it ends with six glyceraldehyde-3-phosphate (containing a total of 18 carbons), a net gain of three carbons from carbon dioxide. Assuming that five triose phosphates are used to regenerate the starting ribulose 1,5-bisphosphate, the remaining triose phosphate can then be diverted to other processes.

The regeneration process uses enzymes related to those of the hexose monophosphate pathway. Three-carbon units in the form of glyceraldehyde-3- phosphate or dihydroxyacetone phosphate enter the pathway in a total of five places. The result of the pathway is the formation of three ribulose 1,5-bisphosphate molecules from the five trioses.

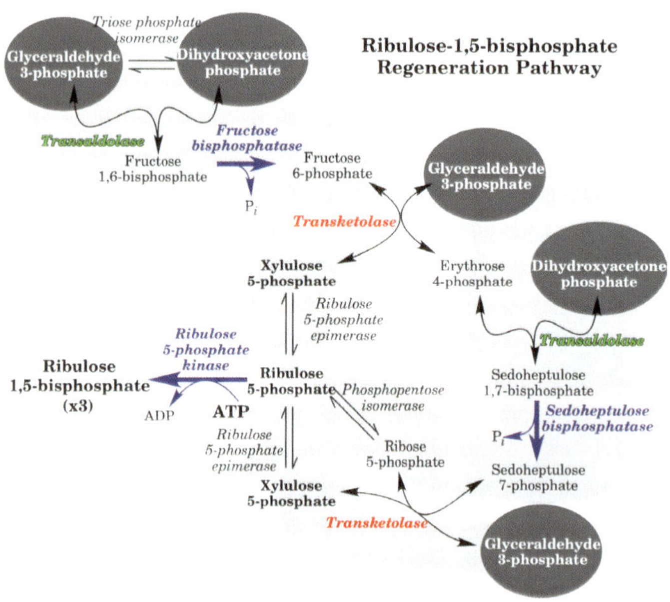

Three irreversible steps, fructose bisphosphatase, sedoheptulose bisphosphatase, and ribulose-5-phosphate kinase make the regeneration pathway irreversible. Note that the regeneration pathway only uses five trioses; this means that the sixth triose formed can be used for other purposes. Animals lack rubisco, sedoheptulose bisphosphatase, and ribulose-5-phosphate kinase, and therefore are incapable of the carbon assimilation reactions (in addition to lacking the requisite, non-metabolic, energy source).

Glyceraldehyde-3-phosphate and dihydroxyacetone phosphate are effectively interchangeable due to the action of the reversible glycolytic enzyme triose phosphate isomerase. Glyceraldehyde-3-phosphate and dihydroxyacetone phosphate can be combined in an aldol condensation reaction by transaldolase (an isozyme of the glycolytic aldolase) to produce fructose-1,6-bisphosphate. The fructose-1,6- bisphosphate is then converted to fructose-6-phosphate by fructose bisphosphatase, the first irreversible enzyme in the pathway.

The fructose-6-phosphate then acts as a substrate for transketolase. Transketolase is a thiamin pyrophosphate-dependent enzyme that catalyzes twocarbon transfer reactions. In this case, it transfers two carbons from fructose-6- phosphate to glyceraldehyde-3-phosphate, yielding the four-carbon erythrose-4- phosphate, and the five-carbon xylulose-5-phosphate. Ribulose-5-phosphate epimerase converts the xylulose-5-phosphate to the first of the ribulose-5-phosphate molecules produced by the regeneration pathway.

The erythrose-4-phosphate, an aldotetrose, acts as a substrate for another transaldolase reaction, condensing with dihydroxyacetone phosphate to form the seven-carbon compound sedoheptulose-1,7-bisphosphate. Sedoheptulose bisphosphatase catalyzes the second irreversible reaction of the regeneration pathway, releasing sedoheptulose-7-phosphate.

The sedoheptulose-7-phosphate acts as a substrate for a second transketolase reaction, combining with the final glyceraldehyde-3-phosphate to yield two pentose phosphates: ribose-5-phosphate, and a second xylulose-5-phosphate. Both of these pentose phosphates are then converted to ribulose-5-phosphate molecules.

The pathway produces a total of three ribulose-5-phosphate from the five triose phosphates. Ribulose-5-phosphate kinase catalyzes the final irreversible step of the regeneration pathway, the ATP-dependent phosphorylation of ribulose-5-phosphate to produce the rubisco substrate, ribulose-1,5-bisphosphate.

The net reaction of running the Calvin cycle to produce one excess triose phosphate

(in the form of glyceraldehyde-3-phosphate is:

6 NADPH + 3 CO_2 + 9 ATP + 6 H_2O $\rightarrow$ glyceraldehyde-3-phosphate + 9 ADP + 8 Pi + 6 NADP

The light reactions produce roughly three ATP for every two NADPH; thus, *in principle*, one complete oxygen-evolving light reaction results in fixation of one molecule of carbon dioxide. In practice, however, some ATP and NADPH are used for other purposes, and therefore the light reactions and carbon assimilation reactions are not always stoichiometrically coupled.

Transketolase Reaction

The enzyme transketolase contains a thiamin pyrophosphate cofactor. As in most enzymes that use this cofactor, in transketolase, the thiamin pyrophosphate cofactor becomes transiently covalently modified during the reaction. In transketolase the two-carbon unit is removed from the ketose (in the example below, from xylulose-5-phosphate) and given to the aldose acceptor (in the example, to ribose-5-phosphate). The reaction generates an aldehyde in the remaining portion of the donor molecule, and maintains the ketone originally present in the transferred two-carbon unit.

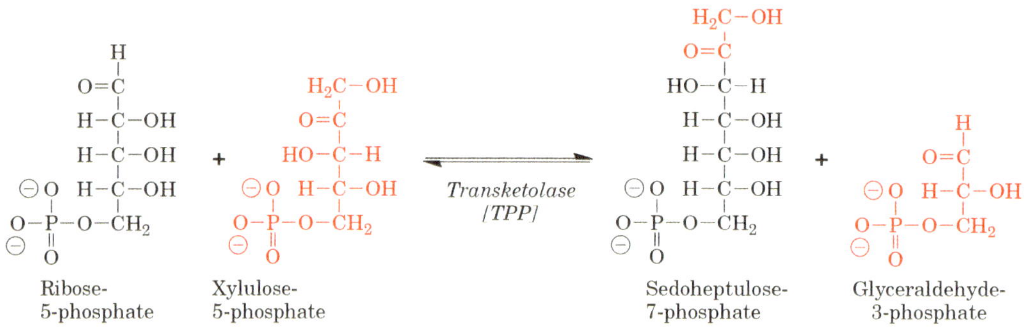

Utilization of Assimilated Carbon

The excess trioses produced have several possible fates. These fates must involve release of phosphate from the carbohydrate, otherwise phosphate becomes limiting. (Light allows carbon fixation, but does not increase the amount of phosphate present.)

1) Starch production: Starch is an energy storage form of glucose. The starch produced and stored within the chloroplast acts as an energy source to maintain metabolism during the night.

Starch is a mixture of amylose, a linear α-1,4-glucose polymer and amylopectin, an α-1,4-glucose polymer with α-1,6 branches. Amylopectin differs from glycogen only in having fewer branch points.

The diagram at right shows the structural features of the glucose polymers.

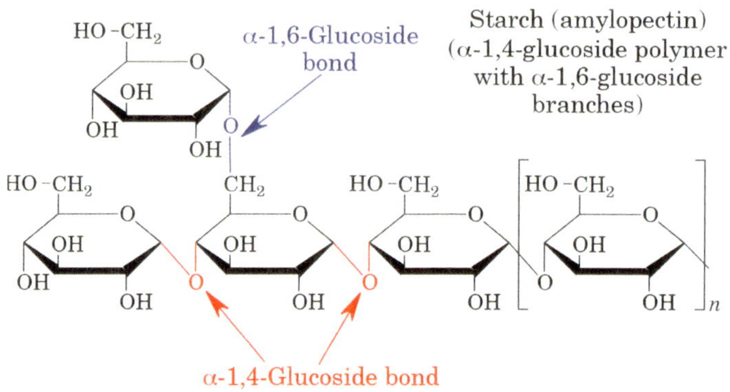

The starch synthetic system is generally similar to glycogen synthesis in animals. However, the starch synthase uses ADP-glucose rather than UDP-glucose as the glucose donor.

Production of starch is regulated primarily by ADP-glucose pyrophosphorylase, which acts as the control point and rate-limiting step for starch synthesis. The enzyme uses ATP and glucose-1-phosphate as substrates, yielding ADP-glucose and pyrophosphate as products. The reaction is reversible, with pyrophosphate hydrolysis acting as the driving force. The ADP-glucose produced acts as a substrate for the starch synthase that adds the glucose unit to the growing starch chain. ADPglucose pyrophosphorylase is allosterically regulated; in most plants it is stimulated by 3-phosphoglycerate (an indicator that the cell is actively fixing carbon) and is inhibited by phosphate (which is elevated when ATP levels are low).

The synthesis of starch from triose phosphates is summarized in the chart below.

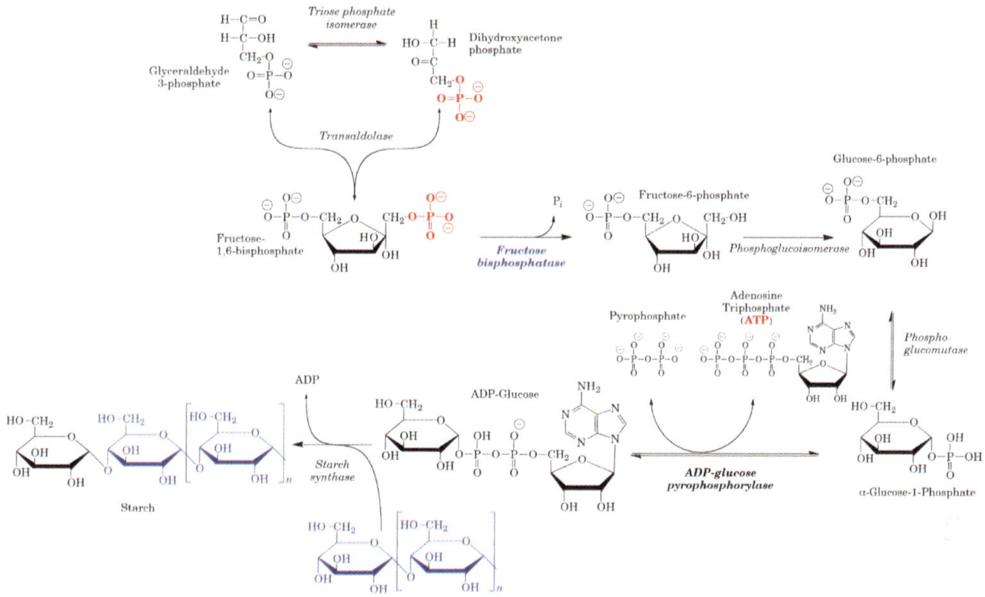

2) Triose Export: Utilization of the fixed carbon requires translocation of the triose phosphate molecules synthesized from the chloroplast to the cytoplasm. This process is performed by a phosphate/triose phosphate antiport located in the chloroplast inner membrane. Dihydroxyacetone phosphate produced in the carbon assimilation reactions leaves the chloroplast in exchange for phosphate (which is required for the synthesis of additional triose phosphates). The cytoplasmic phosphate is released from the carbohydrate during both glycolysis and sucrose synthesis.

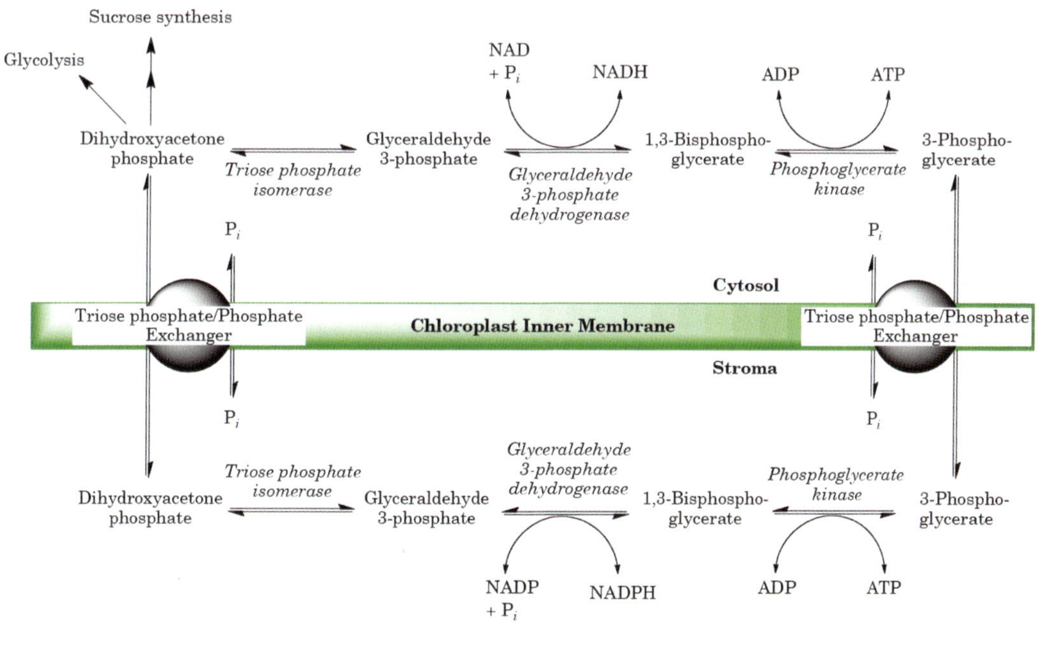

The triose transporter can also be used as part of a shuttle system for moving ATP and reducing equivalents generated during the light reactions out of the chloroplast. The ATP and reduced nicotinamide cofactors can then be used to support metabolism in the other parts of the cell. To allow this, dihydroxyacetone phosphate leaves the chloroplast in exchange for phosphate. However, the dihydroxyacetone phosphate is converted to 3-phosphoglycerate using the glycolytic enzymes triose phosphate isomerase, glyceraldehyde-3-phosphate dehydrogenase, and phosphoglycerate kinase. These reactions generate NADH and ATP in the cytoplasm. The 3-phosphoglycerate produced then re-enters the chloroplast using the phosphate/triose phosphate antiport, and is reconverted to dihydroxyacetone phosphate using the enzymes of the carbon assimilation pathway.

The shuttle system can therefore be used to move ATP and reducing equivalents, without using carbon compounds; alternatively, the dihydroxyacetone phosphate can leave the chloroplast, with only phosphate returning.

3) Sucrose synthesis: Sucrose is a non-reducing (and therefore chemically less reactive) disaccharide of fructose and glucose used to transport energy from photosynthetic apparatus to non-photosynthetic cells. Production of sucrose occurs in the cytoplasm.

As with starch production, sucrose synthesis is regulated by availability of triose phosphates. The relative levels of starch and sucrose production are tightly regulated to prevent the photosynthetic cell from depleting its energy stores.

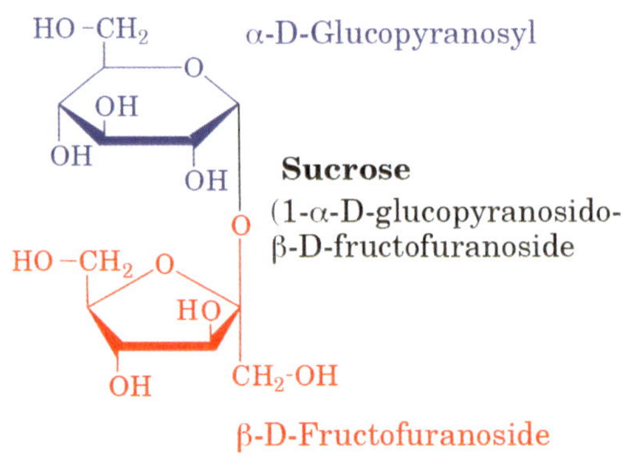

4) Glycolysis: Plants, like all eukaryotic cells, contain mitochondria; some of the triose phosphates generated during photosynthesis (or at night during starch breakdown) are used in the cytoplasm and mitochondria for generation of energy and biosynthetic intermediates. The glycolytic and TCA cycle pathways of plants are similar to those of animals.

Regulation of Carbon Assimilation

Unless ATP and NADPH generated by the light reactions is available, the rubisco reaction and the other carbon assimilation reactions merely waste energy to no useful purpose. Therefore, several of the Calvin cycle enzymes are tightly regulated to prevent the cycle in the absence of light. (In other words the so-called ."dark reactions." do not occur in the dark.)

The regulation of rubisco occurs at several steps. The free rubisco enzyme has a high affinity for ribulose-1,5-bisphosphate but is inactive. In order to become activated, carbon dioxide must bind to a critical lysine residue (Lys201 of the large subunit); the presence of ribulose-1,5-bisphosphate in the binding site prevents this from happening. Removal of the ribulose-1,5-bisphosphate requires an ATPhydrolysis dependent enzyme, rubisco activase, which catalyzes the release of ribulose-1,5-bisphosphate and the covalent modification of Lys^{201}.

The carbamate derivative of rubisco requires Mg^{2+} for activity (it is thought that the cation is required to stabilize the charges that develop on some of the reaction intermediates). The concentration of Mg^{2+} in the stroma increases during the light reactions, because the thylakoid lumen releases Mg^{2+} when the internal proton concentration is high (*i.e.* when protons are pumped into the lumen during the light reactions). Proton pumping also increases pH of the stroma to ~8; rubisco is most active at this pH.

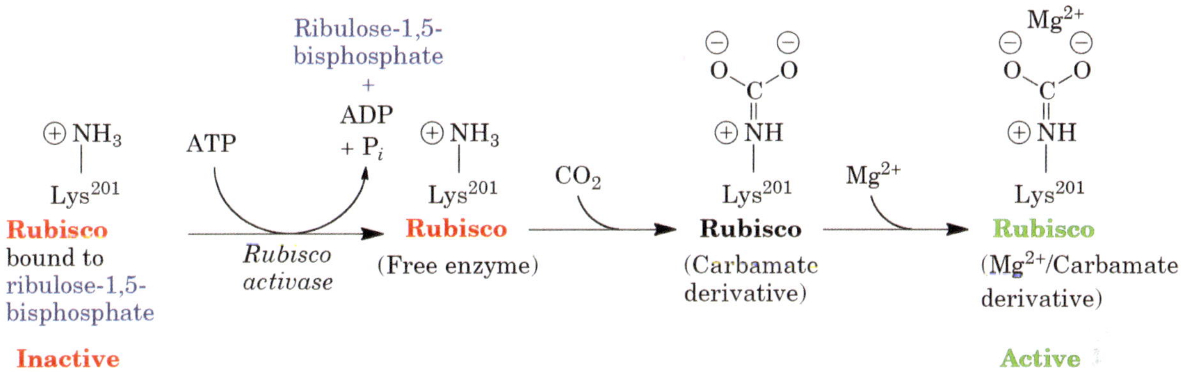

Although the carbamate derivative of rubisco is potentially active, it is also regulated by an endogenous inhibitor. 2-Carboxyarabinitol 1-phosphate is related to the β -ketoacid intermediate of the rubisco reaction, and binds to rubisco with high affinity. 2-Carboxyarabinitol 1-phosphate is synthesized in the dark and broken down in response to light by other enzymes.

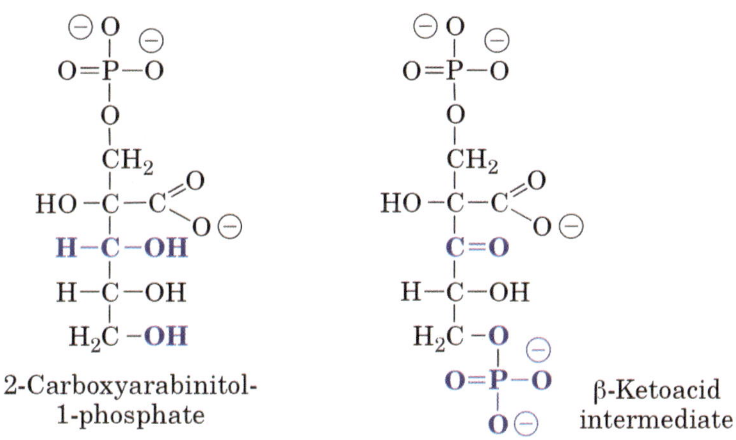

Thus, rubisco is inactive:

1. When bound to an inhibitor,
2. When bound to its ribulose-1,5-bisphosphate substrate if the enzyme is not covalently modified by carbon dioxide, and
3. In the absence of Mg^{2+}.

Rubisco is active: 1) when Lys^{201} is carbamoylated, 2) when bound to Mg^{2+}, 3) when 2-Carboxyarabinitol 1-phosphate has been degraded by light-dependent enzymes, and 4) under the conditions of elevated pH that occur when the chloroplast is exposed to light.

Fructose bisphosphatase and sedoheptulose bisphosphatase, two of the irreversible enzymes of the regeneration pathway are also stimulated by elevated pH and elevated Mg^{2+} concentration. In addition, these enzymes contain a pair of cysteine residues that are critical for activity. In the absence of light, the enzymes become oxidized, and form a disulfide bond between the cysteine side-chains. Illumination of the chloroplast results in reduction of the soluble ferredoxin (a protein found in the electron transport chain involved in NADPH production). Reduced ferredoxin can donate electrons to thioredoxin (a process catalyzed by thioredoxin reductase). The reduced thioredoxin then reduces the disulfide bonds of fructose bisphosphatase and sedoheptulose bisphosphatase to free sulfhydryl groups in order to activate these enzymes. These enzymes are therefore activated indirectly by light.

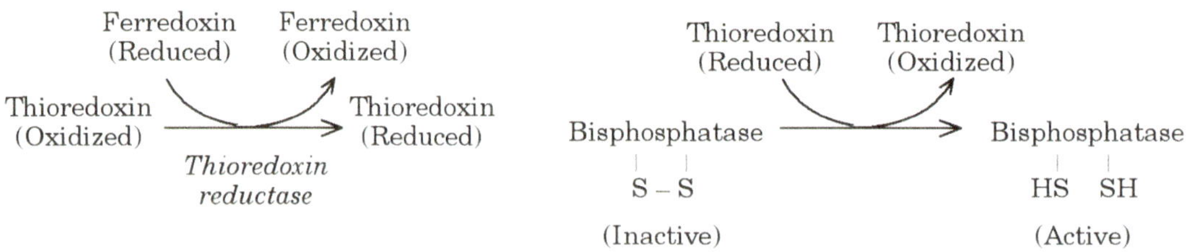

Photorespiration

Rubisco, in addition to its critically important carboxylase activity, is capable of using oxygen as a substrate in place of carbon dioxide. This results in oxygenase activity (hence the "o" in the word "rubisco"). The product of the oxygenase activity must then be converted back to Calvin cycle intermediates by the poorly named "photorespiration pathway".

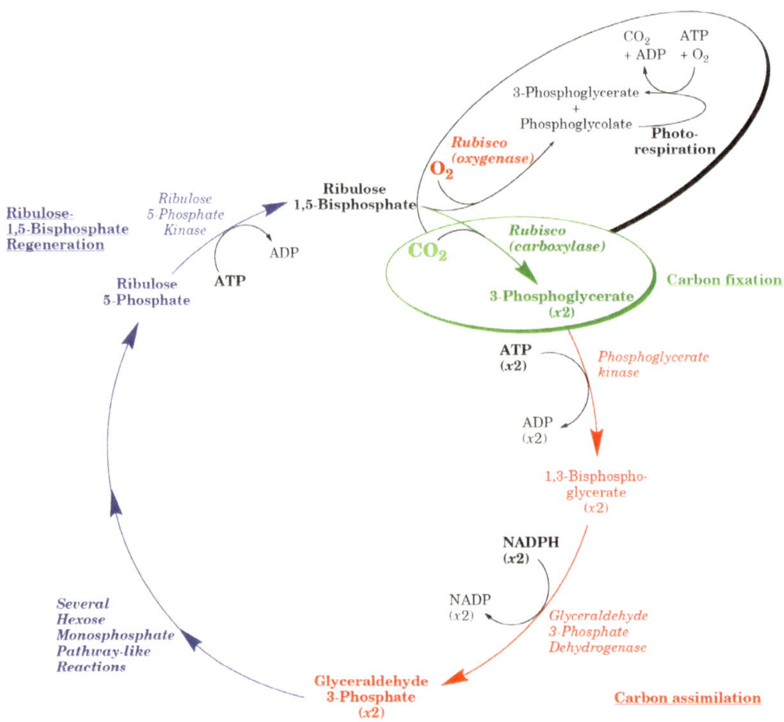

The rubisco oxygenase activity results in the formation of a molecule of the two carbon compound phosphoglycolate and a molecule of 3-phosphoglycerate.

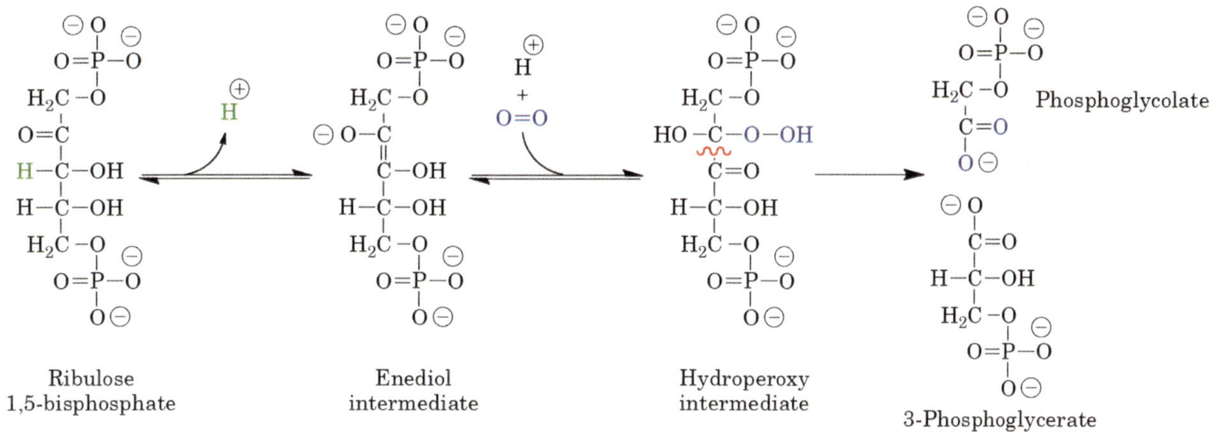

The phosphoglycolate produced must be converted to 3-phosphoglycerate by the photorespiration pathway. The process is called photorespiration because it results in oxygen consumption and carbon dioxide production; in effect it uses energy to run the rubisco reaction in reverse. Photorespiration uses several different compartments, and requires glycine as a pseudo-coenzyme (the glycine carbon effectively comes from phosphoglycolate, but the amino group nitrogen is lost). In the overall process, two molecules of phosphoglycolate are converted to one molecule of 3-phosphoglycerate, with loss of nitrogen as ammonia, loss of carbon dioxide, and the utilization of a molecule of ATP.

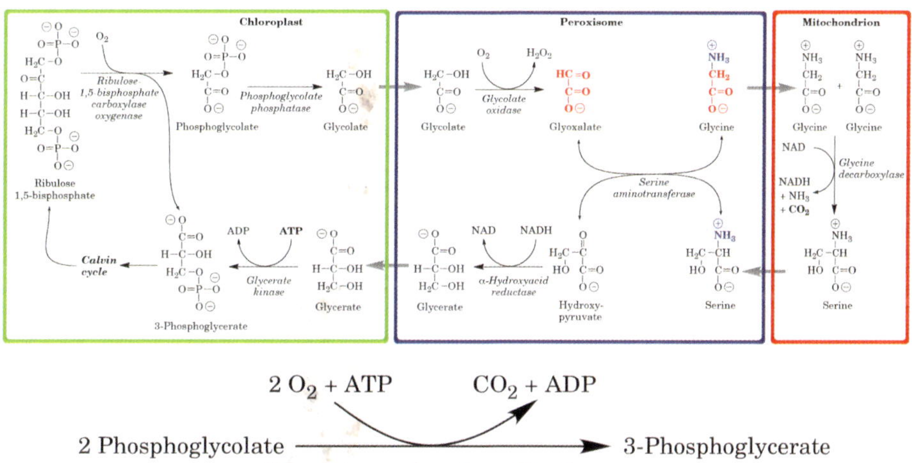

Photorespiration Pathway

Photorespiration is a remarkably wasteful process; both conversion of phosphoglycolate to 3-phosphoglycerate and regeneration of ribulose-1,5-phosphate require ATP. The energy lost in this process has been estimated to decrease plant growth by as much as 50%.

The purpose of the oxygenase activity of rubisco is not known. However, no organisms have evolved a rubisco that lacks this oxygenase activity. It is possible that the oxygenase may act as protection from damaging photooxidation; light in absence of CO_2 appears to result in photooxidation damage to the cell. Alternatively, the limited specificity of the enzyme may be a vestige of the historical time period when free oxygen was rare.

The mechanistic reason for the oxygenase activity is clear. The rubisco K_m for CO2 is ~ $9 \mu M$, a value that increases with temperature. The rubisco Km for O2 is ~ $350 \mu M$. Atmospheric partial pressures of oxygen and carbon dioxide result in dissolved concentrations in water for carbon dioxide of ~ $11 \mu M$ and for oxygen of ~ $250 \mu M$. Under normal conditions, the ratio of carbon dioxide to oxygen utilization is roughly 3 to 1. At higher temperatures, the Km for carbon dioxide rises, and the result is that the oxygenase activity becomes an even larger fraction of the total activity of rubisco.

Plants that use the pathways discussed above are called C3 plants, to differentiate them from the rarer C4 types that use a modified pathway intended to decrease photorespiration. The C4 plants temporarily fix carbon dioxide using phosphoenolpyruvate carboxylase (an enzyme that does not react with oxygen) in mesophyll cells. The phosphoenolpyruvate carboxylase reaction converts phosphoenolpyruvate to oxaloacetate. The oxaloacetate is then converted to malate by malate dehydrogenase. The malate moves to the bundle-sheath cells by passing through plasmodesmata (structures that function like gap junctions), where malic enzyme releases the temporarily fixed carbon dioxide. The result is a much higher carbon dioxide concentration in the local environment near the rubisco.

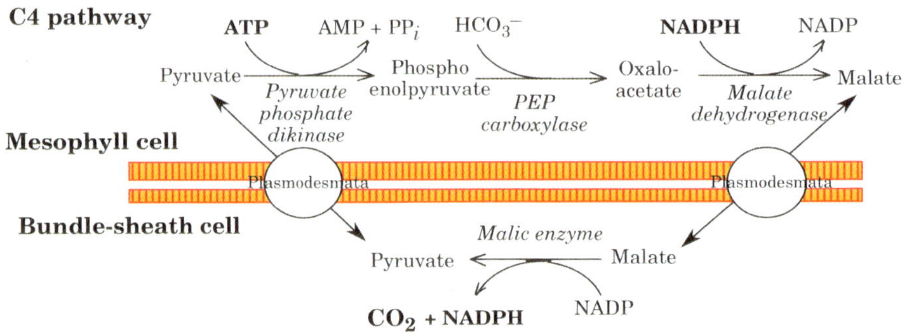

The pyruvate must be converted back to phosphoenolpyruvate using an unusual enzyme, pyruvate phosphate dikinase, which uses ATP to phosphorylate both the pyruvate and an inorganic phosphate (the pyrophosphate produced is then hydrolyzed to help drive the reaction). The C4 pathway thus requires the additional energy of two ATP equivalents per carbon fixed. However, the reduction in wasted energy due to lower oxygenase activity (and therefore lower photorespiration) allows C4 plants to outgrow C3 plants under conditions of high temperature with intense sunlight.

C3 plants perform both photorespiration and the normal carbon-fixing Calvin cycle. C4 plants also perform both photorespiration and an identical carbon-fixing Calvin cycle. However, by adding an additional (energy-dependent) pathway, and by creating a spatial separation between the outside of the plant and the carbon fixation machinery, C4 plants perform much less oxygenase activity than do C3 plants, and therefore are more energy efficient under some conditions.

In C3 plants, most photosynthesis occurs in the mesophyll cells. These cells are exposed to atmospheric oxygen and carbon dioxide. As a result, rubisco can use either carbon dioxide or oxygen as substrate, and therefore performs both carboxylase and oxygenase reactions. At high temperatures, rubisco has reduced affinity for carbon dioxide. Because the affinity for oxygen does not change significantly, the reduced affinity for carbon dioxide means that the oxygenase activity occurs more frequently at high temperatures.

In C4 plants, the mesophyll cell fixes carbon dioxide temporarily by converting the three-carbon phosphoenolpyruvate to four-carbon oxaloacetate (C4 = four-carbon). This carbon is then transported to the bundle-sheath cell further inside the plant, where carbon dioxide is released.

Because carbon dioxide concentration inside the bundle-sheath cell is higher, and because oxygen concentration is lower than in the mesophyll cell, the rubisco performs far fewer oxygenase reactions and more carboxylase reactions. This becomes important at high temperatures.

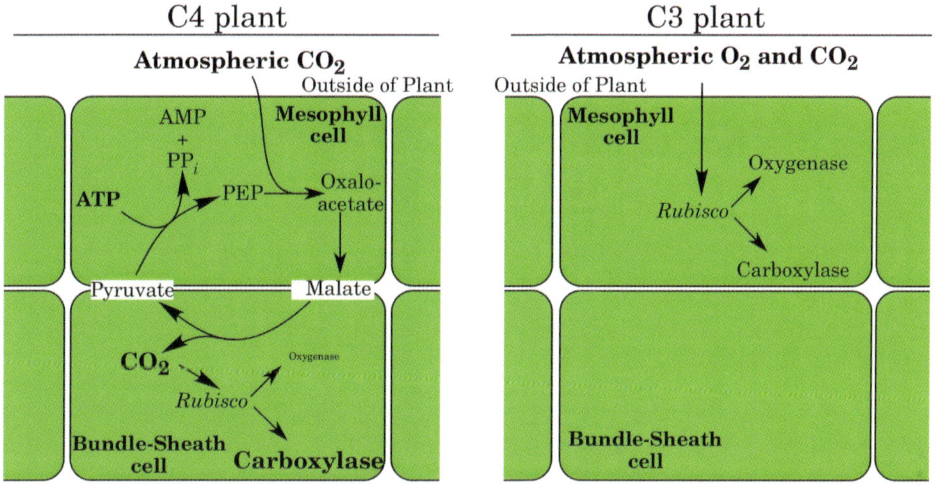

If the rubisco performs only carboxylase activity C4 plants require more energy per carbon fixed. However, if the C3 plant oxygenase/carboxylase ratio is high (e.g., at high temperatures), the energy saved by not performing the oxygenase pathway in C4 plants means that C4 plants may grow faster than C3 plants.

Catabolism in Plants

Plant cells contain low levels of protein in comparison with animal cells mainly because of the high amount of carbohydrate (cellulose and others) that compose most of a plant's structure. However, the importance of proteins and amino acids, the building blocks for proteins, cannot be overlooked. Besides their role as protein constituents, amino acids are also involved in a plethora of cellular reactions and therefore they influence a number of physiological processes such as plant growth and development, intracellular pH control, generation of metabolic energy or redox power, and resistance to both abiotic and biotic stress. Furthermore, a role for amino acids during signaling in plants has recently been discusse. As a consequence, one can expect that the regulation of amino acid catabolism involves a wide set of both general and specific regulators and shows significant differences among plant species, tissues, and developmental stages. It should be noted that the importance of protein metabolism has been revisited in recent years. The function of specific amino acids and their degradation has been extensively investigated in different plant organs. However, our current knowledge concerning amino acid catabolism in general remains rather fragmented. In contrast, the regulation of amino acid biosynthesis has received considerable attention in the past, mainly in the context of biotechnological approaches aiming at increasing the concentration of essential amino acids in crop plants.

Biological Role of Protein Catabolism

The flux through amino acid catabolic pathways can be expected to change massively throughout the life cycle of a plant. During germination, which initially occurs in the absence of light, seed storage proteins are degraded to provide amino acids for the biosynthesis of the proteins required by the growing plant. In addition, the energy demand of the young seedling has to be covered by amino acid oxidation and degradation of other storage compounds, such as fatty acids and starch, until the photosynthetic apparatus is fully functional.

In growing photosynthetically active cells, amino acid biosynthesis is up-regulated to provide substrates for the highly active protein synthesis. Protein turnover and amino acid degradation are less important during this stage. In nonmeristematic tissue, amino acids for protein biosynthesis can mainly be provided by protein turnover, at least in the absence of stress. However, it also has to be considered that amino acids have several functions in addition to their role as a protein constituent, which require tightly controlled steady state levels. In fact, some amino acids (e.g., serine, proline, and leucine) have been shown to act as signaling molecules themselves and others are precursors for the synthesis of phytohormones or other secondary metabolites with signaling function. In addition, cysteine is very reactive and therefore toxic if it is allowed to accumulate above a certain level.

During senescence, nutrients are reallocated from the source leaves to sink tissues such as developing seeds, and therefore protein and amino acid degradation is especially high. Similarly, in conditions leading to scarcity of amino acids in plants, proteins constitute reservoirs of amino acids that catabolic programs, such as proteasomemediated degradation and autophagy, mobilize. Amino acids are subsequently recycled and allocated for the synthesis of specific proteins required under nutrient limitation. Furthermore, during carbon starvation or the normal life cycle, proteins are degraded, and the complete oxidation of their amino acids produces the energy required to fuel the particular needs of certain organs (e.g. stressed leaves or roots). Hence, the accurate sensing of amino acid levels seems to be a key point for the efficient regulation of protein and amino acid synthesis and catabolism, as well as for the control of energy production. In this context, regulation of amino acid content, fluxes, and transport through the plant are critical for plant adaptation to carbon and nitrogen status, development, and defense. The molecular mechanisms underlying the regulation of plant amino acid catabolism are largely unknown but can be expected to be very complex.

Autophagy and Protein Degradation

Large-scale nutrient recycling observed during several stages of a plant's life cycle (e.g. seed production, developmental and stress-induced senescence) is achieved by a process called autophagy. Interestingly, autophagy has also been shown to be relevant for the vegetative phase of plant development, as it provides energy for growth during the night. Cytosolic constituents and whole organelles are enclosed by autophagosomes and subsequently delivered to the vacuoles for degradation. The breakdown of proteins into amino acids or peptides is performed by different classes of proteases. Studies on Arabidopsis thaliana reported the predominant involvement of cysteine and serine proteases but also a role for the proteasome in the degradation of carbonylated proteins, which are accumulated during senescence. Notably, the genes coding for the major proteases involved in senescence-associated protein degradation are largely unknown.

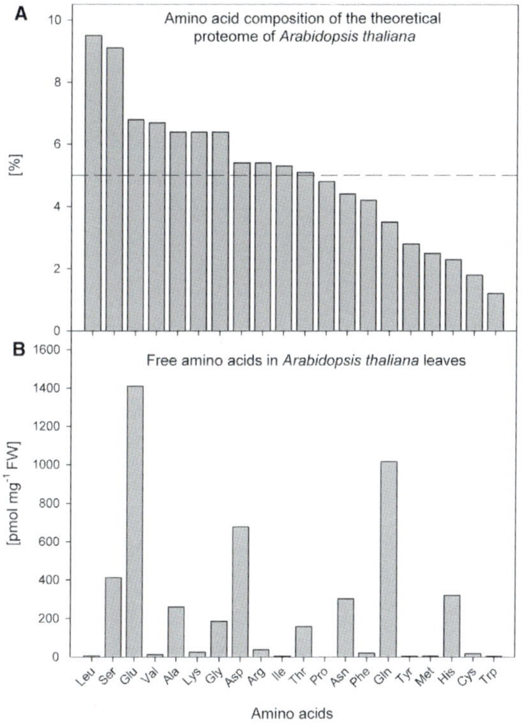

Figure. Protein-Bound and Free Amino Acids in Arabidopsis thaliana

(A) Frequencies of amino acids in the theoretical proteome were calculated based on all protein sequences defined by the Arabidopsis TAIR10 protein database. (B) Concentrations of free amino acids are from Watanabe et al. (Arabidopsis Col0 cultivated on soil in a growth chamber under short-day light condition [8-h day, 140–160 mmol m^{-2} s^{-1} , 20°C; 16-h night, 16°C] for 14 days, transferred to a greenhouse for an additional growth period under short-day conditions [8-h day, 20°C; 16-h night, 16°C]. Material: base of the leaf; leaf harvested 4 h after onset of light).

Pool Sizes of Protein-Bound and Free Amino Acids in Plants

Considering the manifold functions of the individual amino acids, their pool sizes are of critical importance. Pool sizes of free amino acids not only depend on the ratio of amino acid biosynthesis and degradation but also on the biosynthesis and the breakdown of proteins. To understand more how this actually happens, we first asked: what is the interdependence of the pools of free amino acids on one hand and the pools of the protein-bound amino acids on the other? To answer this question, we need to consider that, if the total protein content of the leaf accounts for about 2% of the fresh weight, each individual protein-bound amino acid accounts, on average, for 0.1% of the leaf fresh weight. Considering a molecular mass of 100 Da for an average amino acid, this corresponds to pool sizes of the protein-bound amino acids in the range of 10 nmol/mg fresh weight. In fact, pool sizes of the 20 amino acids only vary by less than a factor of 10. Within the theoretical proteome of Arabidopsis thaliana, leucine represents the most abundant amino acid (9.2%), while tryptophan is least abundant (1.2%;). Interestingly, the proportions of protein-bound amino acids of the theoretical proteome nicely correlate with the average proportions determined for plant-based foodstuff.

By contrast, pools sizes of the free amino acids are much smaller and highly diverse. In expanding Arabidopsis leaves, concentrations of free amino acids vary between 1000 pmol/mg fresh weight (glutamate). These large differences reflect the various functional roles of the individual amino acids. In leaves, glutamate, glutamine, aspartate, and asparagine are the primary products of nitrogen assimilation and therefore pools are large, especially in the light. Furthermore, many amino acids represent precursors for the biosynthesis of other nitrogenous compounds such as nucleotides, phytohormones, or secondary metabolites. In addition, the serine content can be high under conditions of increased photorespiration. Finally, pools of all amino acids are much induced during stress. Proline is known to significantly increase during the stress response in several plants and considered to represent a compatible osmolyte. Moreover, branched-chain amino acids are also much induced during various stresses. A meta-study on the metabolic response of Arabidopsis to abiotic stresses revealed that lysine and threonine are much induced under several stress situations. Furthermore, virtually all pools of free amino acids increase substantially during leaf senescence.

In summary, pools of free amino acids are on average 100- to 1000-fold smaller than the corresponding pools of proteinbound amino acids. At the same time, they may change dynamically and substantially in response to either environmental factors or developmental stages. Pool sizes of free amino acids are thus of great importance, not only because they are required for protein biosynthesis but also due to the numerous additional functions of amino acids for other metabolic pathways and in the frame of signal transduction processes. Precisely adjusted pool sizes of free amino acids are reached by the regulated interplay of amino acid and protein biosynthesis on one side and protein and amino acid degradation on the other.

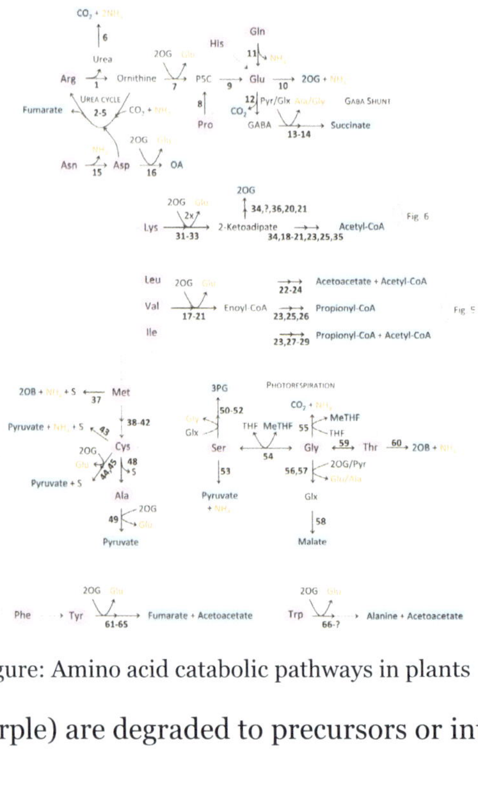

Figure: Amino acid catabolic pathways in plants

Amino acids (highlighted in purple) are degraded to precursors or intermediates of the TCA cycle (highlighted in blue). The nitrogen-containing products are given in orange letters. Enzymatic steps are numbered from 1 to 66, and enzyme names can be found in table together with some further information. Dashed gray arrows indicate synthesis pathways of amino acids that require other amino acids as a precursor. Complex pathways involving several reaction steps are shown in detail in figures (indicated by the gray boxes). 2OB, 2-oxobutyrate; 2OG, 2-oxoglutarate; 3PG, 3-phosphoglycerate; GABA, g-aminobutyric acid; Glx, glyoxylate; MeTHF, methylene tetrahydrofolate; OA, oxaloacetate; P5C, 1-pyrroline-5-carboxylate; Pyr, pyruvate; S, sulfur-containing product; THF, tetrahydrofolate.

Pathways of Amino Acid Catabolism in Plants

Complete degradation pathways for all 20 proteinogenic amino acids have been described in animals, and there are several alternative pathways present in prokaryotes. In plants, corresponding knowledge so far is limited. Notwithstanding, several reaction steps involved in plant amino acid degradation have been described. Others can be postulated based on predicted enzymes identified by sequence homology in enzymes of other groups of organisms. Interestingly, some plant-specific variations are also apparent and are discussed below. Figure shows an overview of our current knowledge of amino acid catabolic pathways in land plants. Enzymes are numbered from 1 to 66, and information such as Arabidopsis accession numbers, subcellular localization, experimental evidence, and expression characteristics is provided in the table below.

Step 1: Nitrogen is Removed as Ammonium and Transferred to Storage Compounds

Nitrogen can be removed from amino acids by deamination reactions producing 2-oxoacids and ammonium. The mitochondria matrix enzyme glutamate dehydrogenase (reaction 10) catalyzes the oxidative deamination of glutamate to 2-oxoglutarate and free ammonium transferring electrons to NAD+ or NADP+ . Serine and threonine can also be directly deaminated by dehydratases (reactions 53 and 60), and the amino groups of methionine and cysteine are released as ammonium during the reactions performed by methionine-g-lyase (reaction 37) and cysteine desulfhydrase (reaction 43). In contrast to the reaction performed by glutamate dehydrogenase, these reactions are not linked to NADH production. Free ammonium is also produced during oxidative decarboxylation of glycine (reaction 55). The amino groups of the remaining amino acids, which cannot be directly deaminated, are transferred to 2-oxoglutarate producing glutamate . Several aminotransferases with different substrate specificities have been described (reactions 7, 13, 16, 17, 49, 50, 56, 57, 61, 66). However, the enzymes catalyzing transamination of cysteine (reaction 44) and a-aminoadipate (reaction 33) remain to be identified. Since the glutamate dehydrogenase shunt enables the netto release of carbon skeletons from several amino acids, it acts as a branch point between carbon and nitrogen metabolism. As such, it is particularly relevant for the maintenance of respiration during carbohydrate limitation.

If required for further metabolism, ammonium can be re-assimilated by glutamine synthase and incorporated into amino acids with high nitrogen content. In particular, asparagine, glutamine, and arginine are used as nitrogen storage and transport compounds. In sink tissues, the amide groups of asparagine and glutamine and the guanidine group of arginine are hydrolyzed by asparaginase (reaction 15), glutaminase (reaction 11), and arginase (reaction 1), respectively, to remobilize the nitrogen .

Step 2: Carbohydrates are Remobilized to Sink Tissues or Oxidized During Respiration

The carbon skeletons of amino acids are generally converted to precursors or intermediates of the tricarboxylic acid (TCA) cycle. Thus, they contribute to mitochondrial metabolism and ATP production. In addition, the oxidation of some amino acids (leucine, isoleucine, valine, lysine, and proline) directly feeds electrons into the mitochondrial electron transport chain. In order to estimate the maximal energy yield that can be achieved from the oxidation of the individual amino acids, we counted all oxidation steps. NAD-dependent reactions can lead to the translocation of 10 protons from the mitochondrial matrix into the intermembrane space if electrons enter the respiratory chain at complex I (four protons) and are transferred to oxygen via complex III (four protons) and complex IV (two protons). Accordingly, when electrons are fed into the respiratory chain at the level of ubiquinone, the maximal number of translocated protons is six. This is the case for the FAD-dependent enzymes succinate dehydrogenase and proline dehydrogenase (reaction 8), as well as for isovaleryl-CoA dehydrogenase (reaction 19) and D-2-hydroxyglutarate dehydrogenase (reaction 36), which transfer electrons to ubiquinone via the electron-transfer flavoprotein/electron-transfer flavoprotein oxidoreductase system (reactions 20 and 21). ATP production was then estimated assuming that three protons are required for the synthesis of each ATP molecule plus one additional proton for the ADP/ATP antiport across the inner mitochondrial membrane. The number of ATPs resulting directly from substrate-level phosphorylation in the TCA cycle was added. Energy yield ranges from 2.5 ATP for glycine up to 34 ATP for tyrosine, which is comparable to the oxidation of glucose as a substrate. Our results show that ATP production from branched-chain amino acids (BCAA) and lysine catabolism is particularly high. Remarkably, carbohydrate starvation induces these pathways exactly, and mutants of enzymes associated show an accelerated senescence phenotype under extended darkness . The physiological relevance of amino acids as alternative respiratory substrates has also been demonstrated during less severe stress conditions experienced by most plants at some stage during their life cycle, such as drought or short light periods.

Aspartate, Alanine, Asparagine and Glutamine

Catabolic pathways for some amino acids can be very short. For instance, transamination of aspartate produces oxaloacetate (reaction 16), and alanine is directly converted to pyruvate by alanine aminotransferases (reaction 49). Also, asparagine and glutamine are metabolized to aspartate and glutamate, respectively (reactions 15 and 11). Others are rather more complicated and are discussed in the following sections.

Arginine

Arginine is used for nitrogen storage and transport in many plants. Thus, mobilization of nitrogen from source tissues re quires arginine degradation, and the enzymes involved are induced during senescence and germination. Arginine is hydrolyzed in the mitochondria by arginase (reaction 1) producing urea and ornithine. Further hydrolysis of urea catalyzed by cytosolic urease (reaction 6) yields ammonia and carbamate, which rapidly decays non-enzymatically forming a second molecule of ammonia and carbon dioxide. Urease is the only nickel-containing enzyme known in plants and requires three accessory proteins for activation. Complete degradation of arginine proceeds by transamination of ornithine to 1-pyrroline-5-carboxylate (P5C; reaction 7) and subsequent oxidation to glutamate via glutamic semialdehyde by NAD(P)+ dependent 1-pyrroline-5-carboxylate dehydrogenase (reaction 9) in the mitochondrial matrix . Alternatively, for nitrogen recycling, ornithine can be transferred to the plastids and re-enter arginine synthesis (reactions 2–5; Slocum, 2005). These reactions are equivalent to the urea cycle, which eliminates excess nitrogen in animals. Indeed, urea strongly accumulates during leaf senescence and is used for long-distance nitrogen transport in the phloem, indicating a similar function in plants. After decarboxylation, arginine and ornithine are also metabolized to polyamines such as putrescine, spermidine, and spermine.

Proline

Proline acts as an osmolyte and a chemical chaperone and is therefore accumulated by plants under various stress conditions. Its catabolism takes place in the mitochondria and proceeds via two oxidation steps to glutamate. Proline dehydrogenase (reaction 8) converts proline to P5C, which is also produced by transamination of ornithine (see above), and transfers electrons to ubiquinone. Oxidation of P5C to glutamate is then catalyzed by 1-pyrroline-5-carboxylate dehydrogenase (reaction 9).

Glutamate

The breakdown of glutamate can be achieved by oxidative deamination to 2-oxoglutarate (reaction 10). In addition, decarboxylation of glutamate by cytosolic glutamate decarboxylases (reaction 12) produces g-aminobutyrate (GABA), which is transaminated to succinic semialdehyde (reaction 13) and oxidized to succinate by succinate semialdehyde dehydrogenase (reaction 14) in the mitochondria. The so-called GABA shunt bypasses two steps of the TCA cycle and carries a high metabolic flux in illuminated leaves, whereas the classical TCA cycle is prevalent during heterotrophic plant metabolism at night. The physiological significance of GABA, which accumulates up to millimolar concentrations in certain tissues and under stress conditions, remains largely unclear. Several functions have been suggested, such as the regulation of cytosolic pH and osmolarity, defense against insects, and signaling. Recent evidence from metabolome and transcriptome studies also revealed a close association of GABA with central metabolism under carbon limitation, indicating that it might play a role in coordinating carbon–nitrogen balance and even mediate a starvation response in plant cells.

Pathway	No.	Description	Accession numbers	Localization
Arg/Pro	1	Arginase	AT4G08870, AT4G08900	mi
	2	Carbamoyl-phosphate synthase	AT1G29900, AT3G27740	pl
	3	Ornithine carbamoyltransferase	AT1G75330	pl
	4	Argininosuccinate synthase	AT4G24830	pl
	5	Argininosuccinate lyase	AT5G10920	pl
	6	Urease + accessory proteins	AT1G67550, AT2G35035, AT1G21840, AT2G34470	cy
	7	Ornithine aminotransferase	AT5G46180	mi
	8	Proline dehydrogenase	AT3G30775, AT5G38710	mi
	9	d-1-Pyrroline-5-carboxylate dehydrogenase	AT5G62530	mi

Glu	10	Glutamate dehydrogenase	AT5G18170,AT5G07440,AT3G03910	mi
Gln	11	Glutaminase	AT5G60540	cy
	12	Glutamate decarboxylase	AT5G17330, AT1G65960, AT2G02000, AT2G02010, AT3G17760	cy
	13	GABA transaminase	AT3G22200	mi
	14	Succinate semialdehyde dehydrogenase	AT1G79440	mi
Asn/Asp	15	Asparaginase	AT5G08100, AT3G16150	cy, pe
	16	Aspartate aminotransferase	AT2G30970, AT4G31990, AT2G30970, AT2G22250, AT1G62960, AT4G31990, AT5G11520, AT5G19550, AT2G22250, AT1G62800, AT1G62800	pl, mi, pe, cy
BCAA	17	Branched-chain amino acid transaminase	AT1G10060, AT1G10070, AT3G49680, AT3G19710, AT5G65780, AT1G50110, AT3G05190, AT5G27410, AT1G50090	mi, pl, cy
	18	Branched-chain a-keto acid dehydrogenase complex	AT5G09300, AT3G13450, AT3G06850, AT3G17240, AT1G48030, AT1G21400, AT1G55510	mi
	19	Isovaleryl-CoA-dehydrogenase	AT3G45300	mi
	20	Electron transfer flavoprotein (ETF)	AT1G50940, AT5G43430	mi
	21	ETF:ubiquinoneoxidoreductase	AT2G43400	mi
	22	Methylcrotonyl-CoA carboxylase	AT1G03090, AT4G34030	mi
	23	Enoyl-CoA hydratase	AT1G76150, AT4G29010, AT4G16800,AT4G16210	pe, mi
	24	Hydroxmethylglutaryl-CoA lyase	AT2G26800	mi
	25	3-Hydroxyacyl-CoA dehydrogenase	AT3G15290, AT4G29010, AT3G06860	pe
	26	3-Ketoacyl-CoA thiolase	AT2G33150,AT5G48880,AT1G04710	pe, mi
	27	Hydroxyacyl-CoA hydrolase	AT5G65940, AT1G06550, AT2G30650, AT2G30660, AT3G60510, AT4G31810, AT4G13360	pe, mi
	28	3-Hydroxyisobutyrate dehydrogenase	AT4G20930, AT4G29120	mi
	29	Methylmalonate-semialdehyde dehydrogenase	AT2G14170	mi
	30	Acyl-CoA oxidase	AT4G16760, AT5G65110, AT1G06290, AT3G51840, AT2G35690, AT1G06310	pe

Lys	31	Lysine-ketoglutarate reductase/ saccharopine dehydrogenase	AT4G33150	cy
	32	Aldehyde dehydrogenase 7B4	AT1G54100	cy
	34	2-Oxoglutarate dehydrogenase complex	AT5G65750, AT3G55410, AT3G17240, AT4G26910, AT5G55070, AT1G48030	mi
	35	Acetyl-CoA acetyltransferase	AT5G48230, AT5G47720, AT5G48230	cy
	36	D-2-Hydroxyglutarate dehydrogenase	AT4G36400	mi
Met	37	Methionine g-lyase	AT1G64660	cy
	38	S-Adenosylmethionine synthase	AT1G02500, AT4G01850, AT2G36880, AT3G17390	cy, nu
	39	S-Adenosylmethionine-dependent methyltransferase	AT5G38780, AT5G38100, AT5G37970, AT5G37990	cy
	40	Adenosylhomocysteinase	AT4G13940, AT3G23810	cy
	41	Cystathionine b-synthase homolog	AT4G14880, AT3G22460	cy
	42	Cystathionine g-lyase homolog	AT3G57050, AT3G01120	pl
	43	Cysteine desulfhydrase	AT5G28030, AT1G48420, AT3G26115	cy, mi
	45	Sulfurtransferase	AT1G79230, AT1G16460, AT4G01050, AT5G66040, AT5G66170, AT3G25480, AT1G09280, AT2G40760, AT1G17850, AT2G42220, AT3G08920, AT5G19370, AT4G24750, AT5G55130, AT4G27700, AT4G35770, AT2G17850, AT2G21045	cy, mi, pl, nu, er
Cys	46	Sulfur dioxygenase	AT1G53580	mi
	47	Sulfite oxidase	AT3G01910	pe
	48	Cysteine desulfurase	AT5G65720, AT1G08490, AT3G62130	mi, pl, cy
Ala	49	Alanine aminotransferase	AT1G17290, AT1G72330	mi
Gly/Ser/ Thr	50	Serine-glyoxylate aminotransferase	AT2G13360	pe
	51	Glycerate dehydrogenase	AT1G79870, AT1G68010, AT1G12550	cy, mi, pe
	52	D-Glycerate kinase	AT1G80380	cy
	53	Serine dehydratase	AT4G11640	cy
	54	Serine hydroxymethyltransferase	AT4G37930, AT4G32520, AT4G13930, AT5G26780, AT1G22020, AT1G36370, AT4G13890	mi, pl, cy
	55	Glycine cleavage system	AT4G33010, AT1G32470, AT2G35370, AT2G35120, AT2G26080, AT1G11860, AT3G17240, AT1G48030	mi
	56	Alanine-glyoxylate aminotransferase	AT2G38400, AT4G39660, AT3G08860	mi
	57	Glutamate-glyoxylate aminotransferase	AT1G23310, AT1G70580	pe

	58	Malate synthase	AT5G03860	pe
	59	Threonine aldolase	AT1G08630, AT3G04520	cy
	60	Threonine dehydratase	AT3G10050	pl
Aromatic	61	Tyrosine aminotransferase	AT4G23600, AT5G53970, AT5G36160	cy
	62	4-Hydroxyphenylpyruvate dioxygenase	AT1G06570	cy
	63	Homogentisate 1,2-dioxygenase	AT5G54080	cy
	64	Maleylacetoacetate isomerase	AT2G02390	cy
	65	Fumarylacetoacetase	AT1G12050, AT3G16700, AT4G15940	cy, mi
	66	Tryptophan aminotransferase	AT1G70560, AT1G23320, AT4G24670, AT1G34060, AT1G34040	cy, va

Table. List of Enzymes Involved in Amino Acid Catabolism.

Subcellular localizations were taken from SUBA3 or, if available, from experimental evidence in the literature. cy, cytosol; er, endoplasmic reticulum; mi, mitochondrion; nu, nucleus; pe, peroxisome; pl, plastid; va, vacuole.

Histidine

Histidine is metabolized to glutamate by four enzymatic steps in animals(histidine ammonia-lyase, urocanatehydratase, imidazole propionase, and formimidoylglutamase). However, to our knowledge, this pathway has not yet been investigated at all in plants.

Leucine, Isoleucine and Valine

BCAA catabolism is rather complex and has not been completely unraveled in plants. The initial steps in the degradation pathways of leucine, isoleucine, and valine are identical. First, branched-chain 2-oxoacids are produced by transamination (reaction 17). Seven different isoforms of BCAA transaminase have been identified in Arabidopsis so far, localized in different compartments. They also catalyze the final step in BCAA synthesis, and the mitochondrial isoform BCAT2 has been shown to be especially relevant for degradation. The oxidative decarboxylation of a-ketoacids producing acyl-CoAs is catalyzed by the branched-chain a-ketoacid dehydrogenase complex (reaction 18), which is functionally and structurally similar to pyruvate dehydrogenase and 2-oxoglutarate dehydrogenase and shares the same E3 subunit. Subsequently, isovaleryl-CoA dehydrogenase (reaction 19) oxidizes isovaleryl-CoA derived from leucine catabolism, but probably also 2-methylbutanoyl-CoA and 2-methylpropanolyCoA from isoleucine and valine degradation, transferring the electrons into the mitochondrial electron transport chain at the level of ubiquinone via electron transfer flavoprotein (ETF, reaction 20) and ETF: ubiquinone oxidoreductase (ETFQO, reaction 21). Leucine catabolism proceeds with a well-characterized carboxylation step catalyzed by biotin-containing methylcrotonyl-CoA carboxylase (reaction 22) to produce 3-methyl-glutaconyl-CoA.

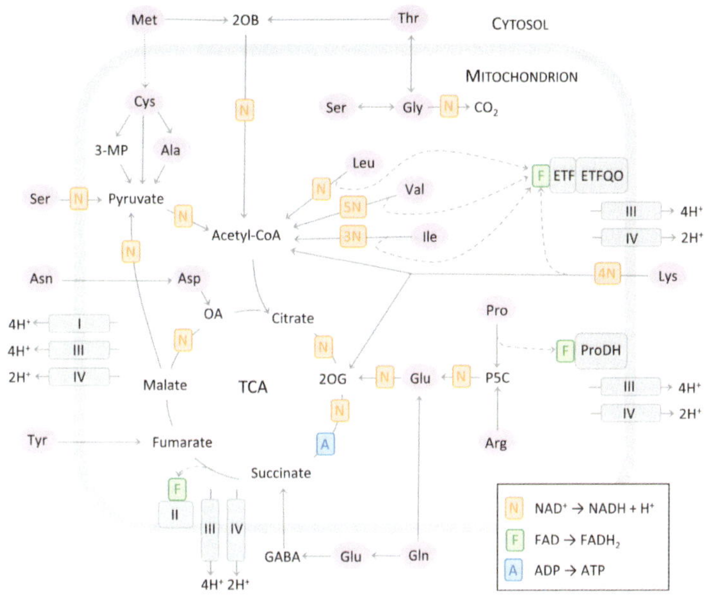

Figure. Subcellular localization and energy yield of amino acid catabolic pathways

Amino acids (highlighted in purple) are degraded to precursors or intermediates of the TCA cycle. Most reaction steps of the catabolic pathways occur either in the mitochondria or cytosol. Additional isoforms with different localizations have been omitted from this scheme for the sake of clarity but can be found in table. Note that for threonine dehydratase, which converts Thr to 2OB, only a plastid localized isoform has been described so far (plastid not shown in the figure). In order to estimate the amount of ATP that can be produced from the degradation of the individual amino acids, oxidation steps are marked with an orange "N" for NAD-dependent dehydrogenases, and with a green "F" for FADdependent dehydrogenases. The blue "A" indicates ATP production via substrate-level phosphorylation. The number of protons translocated across the inner mitochondrial membrane by respiratory chain complexes I, III, and IV is also shown. A calculation of the possible energy yield from complete oxidation of the individual amino acids can be found in table. 2OB, 2-oxobutyrate; 2OG, 2-oxoglutarate; 3MP, 3-mercaptopyruvate; ETF, electron-transfer flavoprotein; ETFQO, electron-transfer flavoprotein:ubiquinone oxidoreductase; GABA, g-aminobutyric acid; OA, oxaloacetate; P5C, 1-pyrroline-5-carboxylate; ProDH, proline dehydrogenase; TCA, tricarboxylic acid cycle.

The subsequent steps can be inferred from homology to the animal system, but the respective enzymes and their physiological functions have not been analyzed in plants to date. 3- Methyl-glutaconyl-CoA as well as the enoyl-CoAs derived from isoleucine and valine catabolism are probably converted to hydroxyacyl-CoAs by enoyl-CoA hydratase (reaction 23). Afterward, the three pathways split. 3-HydroxymethylglutarylCoA lyase (reaction 24) produces acetoacetate and acetylCoA in the final reaction of leucine degradation. The conversion of acetoacetate to acetyl-CoA is catalyzed by acetoacetyl-CoA synthase and acetyl-CoA acetyltransferase (reaction 35) in animals; however, for the first reaction, no plant homolog has been identified so far. Breakdown of 2-methyl-3- hydroxybutyryl-CoA derived from isoleucine catabolism to acetyl-CoA and propionyl-CoA requires two enzymatic steps: oxidation by 3-hydroxyacyl-CoA dehydrogenase (reaction 25) and thiolytic cleavage by 3-ketoacyl-CoA thiolase (reaction 26). 3-Hydroxyisobutyryl-CoA from valine oxidation is probably hydrolyzed to 3-hydroxyisobutyrate (reaction 27) followed by an oxidation step (reaction 28). Finally, methylmalonate-semialdehyde dehydrogenase (reaction 29) catalyzes the oxidative decarboxylation of methylmalonate semialdehyde yielding propionyl-CoA.

Breakdown of propionyl-CoA leads to acetyl-CoA in plants and thus requires a different set of enzymes than the propionylCoA carboxylase pathway, which converts propionyl-CoA to succinyl-CoA in animals. The enzymes involved in this pathway have not been identified so far. Interestingly, by using radioactive labeling, a sequence of reactions completely analogous to the conversion of isobutyryl-CoA to propionyl-CoA has been demonstrated for propionyl catabolism in peanut (Arachis hypogaea) mitochondria. Thus, it seems plausible that this final part of the pathway could be catalyzed by the same set of enzymes (reactions 23 and 27–29), except isovaleryl-CoA dehydrogenase, which does not use propionyl-CoA as a substrate and therefore has to be replaced by a different acyl-CoA dehydrogenase (reaction 30). It is still a matter of debate whether the entire BCAA degradation pathway is localized in the mitochondria. Some reaction steps are also required for peroxisomal fatty acid b oxidation, and localization of the involved enzyme isoforms is unclear.

Amino acid	**NADH**	**FADH2**	**Direct ATP**	**P ATPa**
Tyrosine	11	3	2	34
Leucine	10	4	1	32
Lysine	10	3	2	31.5
Isoleucine	9	3	2	29
Proline	8	3	1	25.5
Arginine	8	2	1	24
Valine	8	2	1	24
Glutamate, glutamine	7	2	2	22.5
Methionine, threonine	6	1	1	17.5
Aspartate, asparagine	4	1	1	12.5
Alanine,serine, cysteine	4	1	1	12.5
Glycine	1	0	0	2.5
Glucose	10	2	4	32

Table . Possible Energy Yield from Complete Oxidation of Amino Acids. NADH/FADH2, number of NAD/FAD molecules reduced during the pathway; direct ATP, ATP produced by substrate-level phosphorylation in the TCA cycle. a Assuming four protons required per ATP.

Lysine

A complete degradation pathway for lysine has not yet been identified in plants. However, the initial steps are already well characterized. Conversion of lysine to saccharopine and subsequently to 2-aminoadipic-6-semialdehyde is catalyzed by two functionally independent domains in the bifunctional cytosolic enzyme lysine-ketoglutarate reductase (LKR)/saccharopine dehydrogenase (SDH) (reaction 31). This enzyme is regulated at the posttranslational level via calcium and phosphorylation and, together with the committed step in the synthesis pathway, controls lysine homeostasis. Lysine can also be converted to L-pipecolate, a regulator of inducible plant immunity. Since L-pipecolate is catabolized by oxidation to $L-\Delta^1$ -pipeidine-6-carboxylate, which spontaneously isomerizes to 2-aminoadipic-6-semialdehyde, this route has been suggested as a possible bypass of the LKR/SDH reaction during lysine degradation.

The next steps in the catabolic pathway are still highly speculative. Labeling studies in different crop species identified glutamate as a product of lysine degradation. This finding led to the conclusion that lysine is probably converted to acetyl-CoA as in animals, then further metabolized to 2-oxoglutarate in the TCA cycle and transaminated to glutamate. Interestingly, an alternative route, which directly produces 2-oxoglutarate via D-2-hydroxyglutarate dehydrogenase (reaction 36), has recently been demonstrated in plants.

The conversion of 2-aminoadipic-6-semialdehyde to acetyl-CoA probably requires eight reaction steps. First, cytosolic aldehyde dehydrogenase 7B4 (reaction 32) catalyzes the oxidation to 2-aminoadipate, which is transaminated to 2-ketoadipate (reaction 33) and then oxidatively decarboxylated to glutaryl-CoA (reaction 34). Since this reaction is equivalent to the oxidative decarboxylation of 2-oxoglutarate in the TCA cycle, it might be catalyzed by the 2-oxoglutarate dehydrogenase complex. A specific glutaryl-CoA dehydrogenase has not been identified in plants. However, isovaleryl-CoA dehydrogenase has been shown to be involved in lysine catabolism and could catalyze the oxidation of glutaryl-CoA to glutaconyl-CoA (reaction 19). In animals, glutaconyl-CoA metabolism then proceeds via crotonyl-CoA, 3-hydroxybutyryl-CoA, and acetoacetyl-CoA (reactions 23, 25, and 35). Interestingly, feeding experiments with isovaleryl-CoA dehydrogenase-deficient Arabidopsis mutants recently provided evidence for a slight modification of this reaction sequence in plants. 2-Oxopentanoate $\left(= \alpha - \text{ketovalerate}\right)$ accumulated strongly indicating that the terminal carboxyl group of 2-ketoadipate is probably removed prior to the oxidation step catalyzed by isovaleryl-CoA dehydrogenase (Figure, left branch). However, enzymes catalyzing the decarboxylation of either glutaconylCoA or 2-ketoadipate have not yet been identified, and experimental evidence proving the functional significance for this part of the pathway in plants is still missing.

D-2-Hydroxyglutarate has been identified as an intermediate of lysine catabolism in plants, but the route of its production remains elusive. One possibility to generate D-2-hydroxyglutarate from glutaryl-CoA would be a reaction sequence analogous to the conversion of succinyl-CoA to malate in the TCA cycle. Alternatively, D-2-hydroxyglutarate could be produced from glutaconyl-CoA by an unknown enzyme activity. Only the final reaction step in this alternative lysine catabolic pathway, oxidation of D-2- hydroxyglutarate to 2-oxoglutarate by D-2-hydroxyglutarate dehydrogenase (reaction 36), is well characterized. Electrons are transferred to the mitochondrial respiratory chain via the ETF/ETFQO complex (reactions 20 and 21).

Glycine, Serine and Threonine

The aspartate family pathway synthesizes, through several different metabolic branches, lysine, threonine, methionine, and isoleucine. Furthermore, methionine and threonine are precursors of isoleucine synthesis, and cysteine is produced from serine. In plants, it is clear that this pathway is of high importance from a nutritional perspective given that it culminates with the synthesis of the four essential amino acids described above. However, these reactions are metabolically expensive and are not likely to play a role during amino acid catabolism. In contrast, threonine, glycine, and serine can be interconverted easily, so that the reactions shown (reactions 54 and 59) can be relevant for synthesis of the product amino acid as well as degradation of the substrate amino acid.

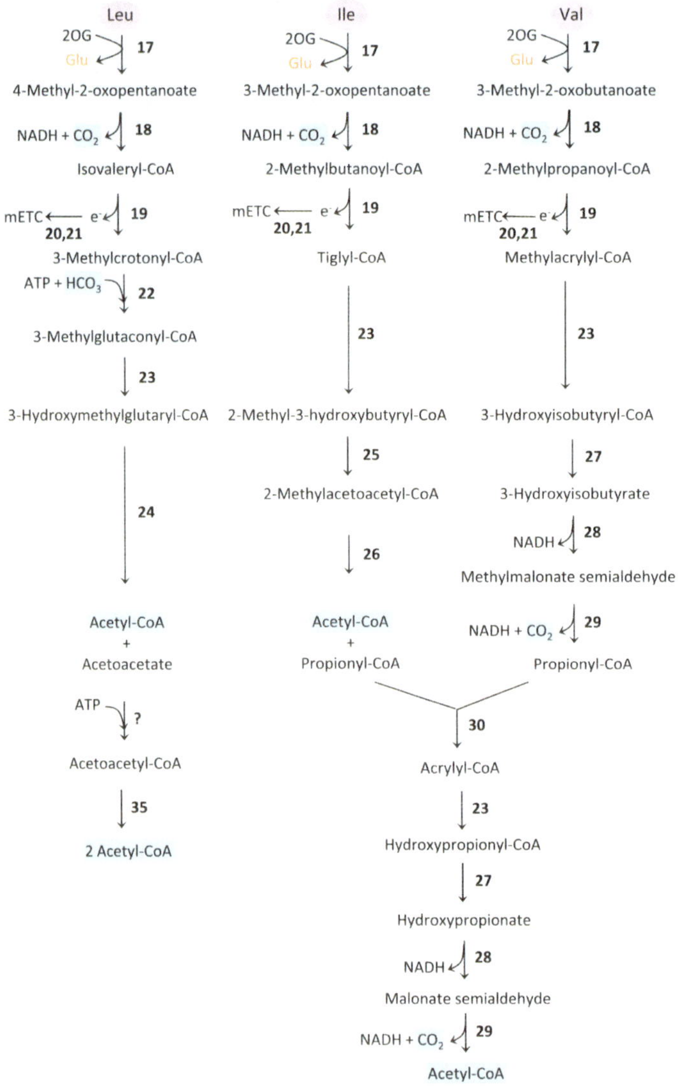

Figure: Branched-Chain Amino Acid Catabolism

The degradation of leucine, isoleucine, and valine finally leads to acetyl-CoA and CO2 (marked in blue). The amino groups are transferred to glutamate (orange letters). Enzymes catalyzing the initial steps of the three pathways have already been characterized in plants (indicated by the light blue boxes). Subsequent reactions (shown in gray boxes) have been deduced from homology and indirect experimental evidence, and further confirmation is still required. 2OG, 2-oxoglutarate; e, electrons; mETC, mitochondrial electron transport chain.

The complete degradation of glycine to carbon dioxide and ammonium is catalyzed by the mitochondrial multi-enzyme system glycine decarboxylase (reaction 55). A methyl group is transferred to tetrahydrofolate and thus contributes to C1 metabolism. It is still a matter of debate whether the four enzymes involved in the reaction sequence known as P, H, L, and T protein form a stable complex. The glycine cleavage system is required for photorespiration in C3 plants and closely interacts with serine hydroxymethyltransferase (reaction 54), which methylates glycine to serine using methylene tetrahydrofolate as a cofactor. Due to their role in photorespiration, these enzymes are strongly induced by light and can become the most abundant mitochondrial matrix proteins during the day. The photorespiratory cycle proceeds via transamination of serine to hydroxypyruvate catalyzed by serine-glyoxylate aminotransferase (reaction 50) and NADH-dependent reduction to glycerate (reaction 51) in the peroxisomes. Phosphorylation by glycerate kinase (reaction 52) in the plastids then generates 3-phosphoglycerate, which can reenter the Calvin–Benson cycle or be reduced to glyceraldehyde 3-phosphate and used for glycolytic degradation or carbohydrate synthesis. The alternative route of serine degradation via pyruvate requires serine dehydratase activity, shown to be present in the cytosolic enzyme serine racemase. Glycine can also be transaminated by alanine-glyoxylate aminotransferase (reaction 56) or glutamate-glyoxylate aminotransferase (reaction 57), and the resulting glyoxylate can be acetylated in the peroxisomes to produce malate. However, whether this route is relevant for glycine catabolism has not been tested.

Threonine aldolase (reaction 59) catalyzes the reversible conversion of threonine to glycine plus acetaldehyde. The strong (50-fold) accumulation of threonine in threonine aldolase 1 knockout mutants demonstrates its physiological role in threonine catabolism. The alternative reaction, deamination of threonine to 2- oxobutyrate (reaction 60), is part of the isoleucine synthesis pathway and therefore regulated by feedback inhibition. A threonine dehydrogenase, which is involved in threonine catabolism in animals and prokaryotes, has not yet been identified in plants.

Cysteine

The carbon skeleton of cysteine is converted to pyruvate via removal of the amino and sulfhydryl groups. So far, three different routes have been described for the catalysis of these reactions (Figure). Cytosolic cysteine desulfhydrase (reaction 43) deaminates cysteine producing ammonia, pyruvate, and hydrogen sulfide. Knockout of this enzyme leads to an accumulation of cysteine indicating a function in cysteine homeostasis. In addition, alterations in autophagosome formation and stomatal opening were detected, which could be reversed by exogenous supplementation with hydrogen sulfide (H2S). Thus, cysteine desulfhydrase also regulates production of the signaling molecule H2S. A D-cysteine-specific desulfhydrase is localized in the mitochondria and has also been implicated in cysteine catabolism. All isoforms of this enzyme are induced during senescence, suggesting a role during nutrient mobilization. Reduced sulfur for the synthesis of iron–sulfur clusters, biotin, and thiamin is further produced by cysteine desulfurases, also known as NifS-like proteins (reaction 48). Different isoforms are present in the mitochondria, cytosol, and plastids. However, since cysteine desulfurases exclusively transfer the sulfur to iron–sulfur cluster scaffold proteins, they are most likely not relevant for cysteine catabolism. An additional mitochondrial pathway catalyzing complete degradation of cysteine begins with the transamination to 3-mercaptopyruvate by a presently unknown aminotransferase . A sulfurtransferase (reaction 45) then transfers the sulfhydryl group to glutathione, and the resulting glutathione persulfide is oxidized to sulfite by the sulfurdioxygenase ETHE1 (reaction 46). Sulfite is either converted to thiosulfate by addition of a persulfide group (reaction 45) or transferred to the peroxisomes and oxidized to sulfate by sulfite oxidase (reaction 47). ETHE1 is highly induced by carbohydrate starvation, and mutants show premature senescence under light-limiting growth conditions indicating that this pathway is functionally significant for the use of amino acids as an energy source.

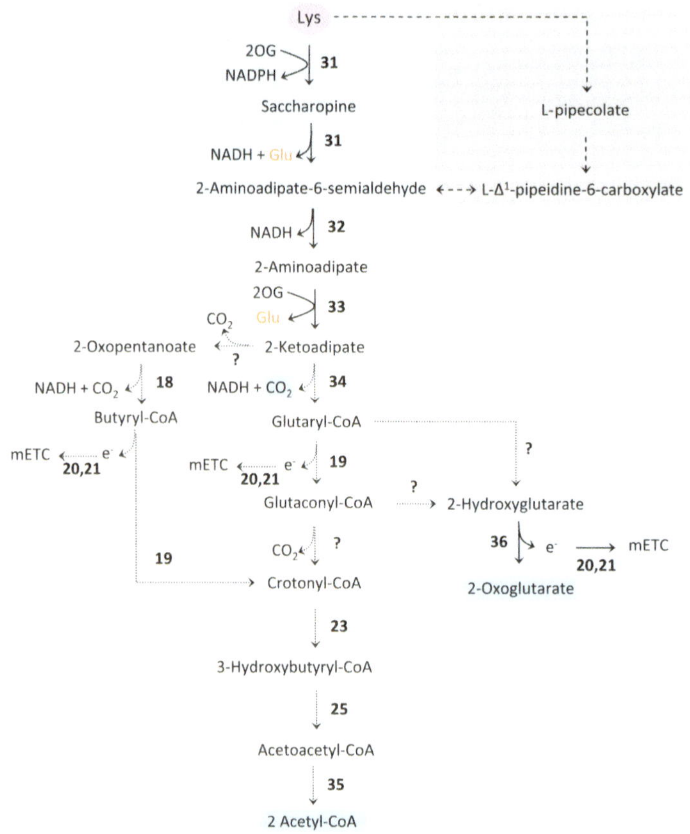

Figure: Lysine Catabolism

Lysine can be catabolized to either acetyl-CoA or 2-oxoglutarate (marked in blue), and both amino groups are transferred to glutamate (orange letters). Individual reaction steps have already been extensively characterized in plants (indicated by the light blue boxes). However, the major part of the pathway (shown in the light gray box) still requires experimental confirmation, and some steps are highly speculative (dotted arrows). A potential bypass of the reaction catalyzed by the bifunctional enzyme lysine-ketoglutarate reductase/saccharopine dehydrogenase (reaction 31) via L-pipecolate is also indicated (dark gray box). 2OG, 2-oxoglutarate; e, electrons; mETC, mitochondrial electron transport chain.

Methionine

A significant amount of methionine is converted to S-adenosylmethionine (SAM) by S-adenosylmethionine synthase. SAM serves as a methyl donor in transmethylation reactions and is a precursor for the synthesis of ethylene, biotin, and polyamines. Methylation of methionine results in S-methylmethionine (SMM), a possible storage and transport form of methionine. Both metabolites can be recycled back to methionine. Methionine feeding leads to an increase in SAM and SMM levels, and a reduction in the expression level of S-adenosylmethionine synthase significantly increases methionine levels, showing that this pathway is also relevant for methionine homeostasis. However, it is doubtful whether the reverse transsulfuration pathway (reactions 39–42), which converts SAM to cysteine during methionine catabolism in animals, also takes place in plants. The first two steps, transfer of the methyl group to an acceptor (reaction 39) and cleavage of the adenosyl group (reaction 40), are required for the recycling of SAM to methionine in the methyl cycle and clearly occur in plants. Cystathionine-b-synthase (reaction 41) and cystathionine-g-lyase (reaction 42) subsequently convert homocysteine to cysteine in animals, and homologs of both genes are present in the Arabidopsis genome. However, earlier studies found only limited capacity to metabolize methionine to cysteine in Spinacia oleracea (spinach) and Lemna paucicostata (duckweed).

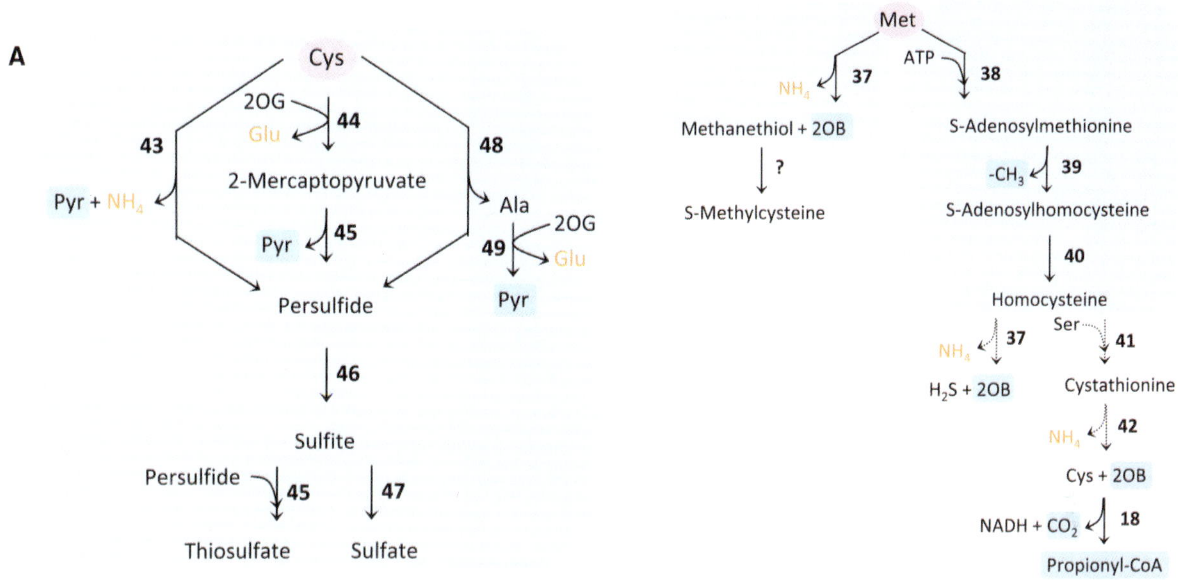

Figure: Sulfur Amino Acid Catabolism.

(A) Cysteine catabolism. Three different routes have been described that convert cysteine to pyruvate (marked in blue). The amino group is either released as ammonium or transferred to glutamate (orange letters). The thiol group is oxidized to thiosulfate or sulfate.

(B) Methionine catabolism. Methionine can be converted to methanethiol and 2-oxobutyrate or to homocysteine via S-adenosylmethionine, and the respective enzymes are known in plants (indicated by the light blue boxes). The subsequent steps shown in gray boxes are still rather speculative. Products of the carbon skeleton are highlighted in blue and the nitrogen-containing products are given in orange letters.

An alternative pathway for methionine degradation that is not present in animals has been identified in plants,. Methionine-g-lyase (reaction 37) converts methionine to 2-oxobutyrate, ammonium, and methanethiol in the cytosol. 2-Oxobutyrate can be used as a precursor for isoleucine synthesis or degraded via oxidation to acetyl-CoA. The further fate of methanethiol is not yet entirely clear. Part of it is probably emitted as a volatile compound, and the remaining is converted to S-methylcysteine, which could play a role as a storage compound. However, given that these results have been obtained by feeding an Arabidopsis cell suspension culture with 2 mM methionine in the presence of sucrose, they might rather be a detoxification reaction of excess methionine. Recombinant methionine-g-lyase also has high activity with homocysteine as a substrate, indicating that complete degradation of methionine could proceed alternatively via this route. A physiological role of methionine-g-lyase for the remobilization of sulfur under sulfate limitation and in seed production has been shown in Arabidopsis.

Tyrosine

Dixon and Edwards demonstrated that the complete degradation pathway of tyrosine, known from the mammalian system, is also functionally present in Arabidopsis. Transamination of tyrosine (reaction 61) produces 4-hydroxyphenylpyruvate, which is converted to homogentisate by 4-hydroxyphenylpyruvate dioxygenase (reaction 62). In plants, homogentisate is the precursor of tocopherols such as vitamin E and plastoquinones. Catabolism proceeds via oxidative ring cleavage catalyzed by homogentisate-1,2-dioxygenase (reaction 63). The product maleylacetoacetate is isomerized to fumarylacetoacetate by glutathione-S-transferase zeta 1 (reaction 64), which has maleylacetoacetate isomerase activity. The final step in tyrosine degradation is the hydrolysis of fumarylacetoacetate to fumarate and acetoacetate (reaction 65). Thus, an association of tyrosine and fumarate metabolism inside and outside the mitochondria should be considered.

Phenylalanine

Phenylalanine is hydroxylated to tyrosine prior to degradation in animals. However, the catabolic pathway for phenylalanine in plants remains largely elusive since no phenylalanine hydroxylase homolog has been found in the Arabidopsis genome. Furthermore, evidence from feeding experiments with different species of higher plants indicates that phenylalanine is not degraded to any great extent . However, it is important to mention that a unique folate-dependent phenylalanine hydroxylase localized in chloroplasts has been identified in non-flowering plants.

Tryptophan

Studies on tryptophan metabolism in plants have focused on its role as a precursor for the synthesis of diverse secondary metabolites such as auxin, phytoalexins, glucosinolates, and alkaloids . Although research efforts on herbicides have allowed significant advances in the understanding of the synthesis of aromatic amino acids, to our knowledge, the degradation pathway of tryptophan has not been studied in plants so far.

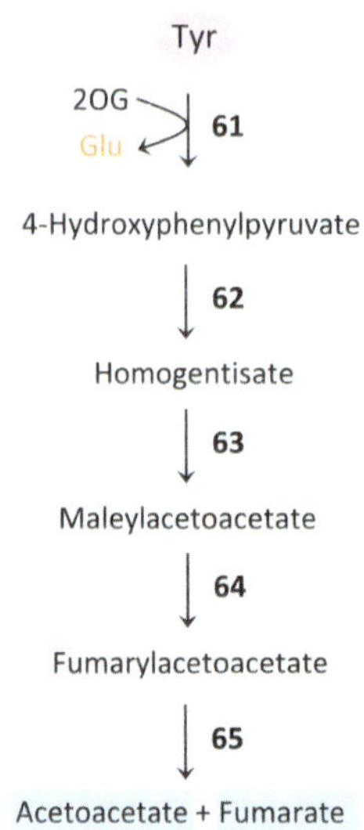

Figure. Tyrosine Catabolism. The complete pathway oxidizing tyrosine to acetoacetate and fumarate (marked in blue) has been fully characterized in plants. The amino group is transferred to glutamate (orange letters). For enzyme names and further information, please see Table 1 and Supplemental Table 1. 2OG, 2-oxoglutarate.

Subcellular Compartmentalization of Amino Acid Catabolic Pathways

Amino acid synthesis takes place mainly in the chloroplasts, where reducing power as well as ATP can be provided directly. In contrast, most reactions of the catabolic pathways are localized in the mitochondria or the cytosol. From a metabolic perspective, this localization makes sense, since on the one hand the carbohydrate skeletons are converted to substrates of the TCA cycle. On the other hand, amino acid catabolism is highly relevant for the senescence process, and chloroplasts are degraded very early in this process, whereas mitochondria remain fully functional until the end. For some steps (e.g. reactions 25 and 30 from BCAA oxidation or reactions 57 and 58 from glycine metabolism), transfer of the intermediates into the peroxisomes might be necessary, since the enzymes involved are also required for peroxisomal pathways such as fatty acid b-oxidation and the glyoxylate cycle. However, dual localization of individual enzymes or isoforms is also likely and has to be tested experimentally.

Regulation of Amino Acid Catabolism

Local amino acid concentrations depend on the synthesis and degradation rates of proteins as well as amino acids and on transport processes. Protein synthesis and degradation are obviously not suitable to control the level of individual amino acids. These processes rather modify the total pool size, which is especially relevant when bulk degradation takes place (e.g. during germination and senescence) leading to a massive increase in free amino acid concentrations. However, protein synthesis is energetically expensive and complex. Accordingly, cells must sense intracellular amino acid pools to couple their abundance to their usage. Amino acid transport is relevant for the distribution of nitrogen within the plant and is regulated mainly at the transcriptional level. However, a potential regulatory protein, GDU1, has also been described recently. To maintain normal metabolic function, homeostasis of individual amino acids has to be tightly regulated due to their specific functions in signaling, stress response, and the production of secondary metabolites. Also, during senescence or carbohydrate starvation, steady state levels of amino acids change dramatically with a relative increase in branched-chain and aromatic amino acids, which can be used most efficiently as alternative respiratory substrates. In this regard, synthesis and degradation rates of individual amino acids are probably the most relevant parameters for adjusting their concentrations.

Less and Galili concluded from bioinformatics analysis of a large microarray dataset that free amino acid concentrations are regulated by a combination of transcriptional and posttranslational control. Amino acid biosynthesis is mainly modulated via allosteric feedback inhibition of individual enzymes by the end product, and therefore transcript levels are relatively stable. In contrast, transcription of catabolic enzymes changes considerably in response to different stress conditions. Interestingly, genes of the BCAA degradation pathway also show an oscillating diurnal transcription pattern with induction during the night, when they are required for energy metabolism. Some details are already available on transcription factors involved in the induction of proline catabolic enzymes during darkness and stress relief . The first enzyme of lysine catabolism (LKR/SDH, reaction 31) is a clear exception to this regulation pattern, since it is extensively regulated via post-translational modifications.

Labeling experiments using 13CO2 revealed that de novo biosynthesis rates of amino acids largely differ. While alanine was labeled rapidly (>70% after 1 h labeling pulse), phenylalanine, glycine, and serine were only moderately and several others (e.g., glutamate, threonine, lysine, leucine, isoleucine, proline, and asparagine) were very slowly ($<10\%$ after 1 h) (labeled. It can be anticipated that labeling rates for individual amino acids in addition to their biosynthesis rates depend on their pool size, their frequencies within the proteome, and their turnover rates, as well as cellular protein biosynthesis and degradation rates.

Molecular Plant

It is also important to highlight that a growing body of evidence suggests that there is not always a strict correlation between transcript levels and changes in the abundance of the encoded proteins. This apparent discordance is not only due to limitations in the analysis procedures for each component but can also be explained by dynamic rates of protein turnover , which makes understanding the regulation of protein and amino acid catabolism a rather exciting research field. Recent research into plant metabolism has often included the usage of modeling and correlationbased approaches. Since protein and amino acid degradation pathways are involved in a complex metabolic network, integration of the results obtained using different postgenomic techniques will be important for this research field in the future. Notably, when revisiting amino acid catabolism, we realized that a large-scale comparison of relevant transcript abundance (encoding specific enzymes and metabolite transporters) with metabolic fluxes has yet to be made. Given the recognized importance of amino acids, this is rather surprising and highlights that further examination of the catabolic enzymes should provide significant insights into an integrated overview of the metabolic connections of the catabolic pathways of amino acids in higher plants.

References

- David Oakley Hall; K. K. Rao; Institute of Biology (1999). Photosynthesis. Cambridge University Press. ISBN 978-0-521-64497-6. Retrieved 3 November 2011
- Role-of-phytochemistry-in-systematic-botany-taxonomy-34721: biologydiscussion.com, Retrieved 10 June 2018
- Fuks, Bruno; Fabrice Homblé (8 July 1996). "Mechanism of Proton Permeation through Chloroplast Lipid Membranes" (PDF). Plant Physiology. 112 (2): 759–766. doi:10.1104/pp.112.2.759. PMC 158000. PMID 8883387. Retrieved 25 August 2013
- López-García, Purificación; Eme, Laura; Moreira, David (2017). "Symbiosis in eukaryotic evolution". Journal of Theoretical Biology. 434: 20–33. doi:10.1016/j.jtbi.2017.02.031
- Photosynthesis, science: britannica.com, Retrieved 16 March 2018
- Jones, Daniel (2003) [1917], Peter Roach, James Hartmann and Jane Setter, eds., English Pronouncing Dictionary, Cambridge: Cambridge University Press, ISBN 3-12-539683-2
- Keeling, Patrick J. (2004). "Diversity and evolutionary history of plastids and their hosts". American Journal of Botany. 91 (10): 1481–93. doi:10.3732/ajb.91.10.1481. PMID 21652304
- Photosynthesis-Carbon-Assimilation-Processes: rosehulman.edu, Retrieved 14 May 2018
- Milo, Ron; Philips, Rob. "Cell Biology by the Numbers: How large are chloroplasts?". book.bionumbers.org. Retrieved 7 February 2017

Chapter 4

Plant Genetics and Breeding

Genes, variation and heredity in plants are studied under a branch of botany called plant genetics. A significant area of plant genetics is plant breeding, which involves modifying the traits of plants to obtain desired yields. This chapter has been written to provide an understanding of plant genetics and breeding through the elucidation of topics related to plant DNA, RNA, doubled haploidy and transformation vector.

Plant Genetics

In biology, a gene is a unit of heredity transferred from a parent to offspring. Regions or groups of genes make up a plant DNA structure and are what determine some characteristics of the offspring. More simply put, a gene contains a particular set of instructions.

In horticulture, a plant's genes influence the development of the plant, as they are parts of its chromosomes and are inherited through sexual propagation.

Plant genes or genetics are what researchers aim to manipulate in their quest for genetically modified organisms (GMOs).

Gregor Mendel was the founding father of the genetic laws of inheritance. In his research he studied the inherited traits of pisum sativum (peas), such as their shapes, sizes, and colors. Likewise, Barbara McClintock discovered the jumping genes while she was studying maize.

Just like any other multi-cellular organism, the fundamental principles of genetics is the same in the inheritances of plants. For example, just like in animals, plants also have somatic mutations regularly. However, it is possible that the mutations can contribute to the germ line as flowers develop at the ends of branches composed of somatic cells.

These mutant branches are known as sports and can also be cultivated if they are economically desirable. Polyploidy is also more common and so plant genetics are a bit different from those of animals. Polyploid are those having more than two paired (homologous) sets of chromosomes.

There are two ways by which genes can be transformed: the gene gun method and the agrobacterium method.

The gene gun method is especially useful in transforming monocot species like corn and rice and is also known as biolistics.

The agrobacterium method has been successfully practiced in dicots, i.e. broadleaf plants, such as soybeans and tomatoes, for many years. It is also effective in monocots like grasses, including corn and rice. This method is also preferred over the gene gun method as it is easier to monitor due to a greater frequency of single-site insertions of foreign DNA.

DNA

DNA is the hereditary or genetic material, present in all cells, that carries information for the structure and function of living things.

In the plant kingdom, DNA, or deoxyribonucleic acid, is contained within the membrane-bound cell structures of the nucleus, mitochondria, and chloroplasts. DNA has several properties that are unique among chemical molecules.

It is universal to all living organisms, having the same structure and function in each. It is capable of reproducing itself in a process known as self-replication. This property allows cell division, and thus continuity, growth, and repair.

It carries in its structure the genetic code, or set of instructions, for cellular development and maintenance. Finally, it undergoes changes in chemical structure, from both environmental and internal causes, called mutations, which contribute to evolution, diversity, and disease.

Chemical Structure

DNA is a simple molecule, consisting of four nucleotides. Each nucleotide has a five-carbon sugar (deoxyribose), a phosphate, and one of four possible nitrogenous bases: the double-ringed purines of adenine (A) and guanine (G) and the single-ringed pyrimidines of thymine (T) and cytosine (C).

Most of the properties of DNA relate to the unique bonds that form among the nucleotides: The sugarphosphate components align themselves linearly, while the nitrogen rings bond perpendicularly.

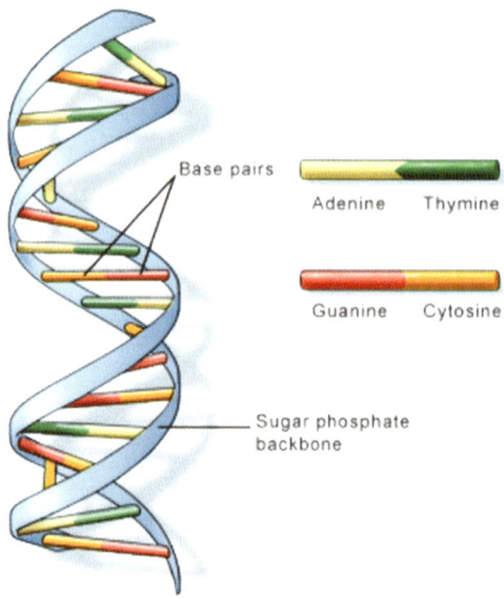

The nitrogen rings further bond in a very specific fashion:A always pairs with T, and Galways pairs with C. The DNAmolecule thus appears as a ladder, the sides being sugar-phosphate; the rungs, the A-T and G-C pairs.

Further bonding and folding produces a structure shaped like a spiral ladder, known as a double helix. This double helix is compactly packaged in ropelike structures known as chromosomes, which are visible under a light microscope before and during cell division.

During the daily life of the cell, the DNA appears as an indistinguishable dark mass call chromatin (an inclusive term referring to DNA and the proteins that bind to it, located in the nuclei of eukaryotic cells).

Watson and Crick Model

In the mid-1800's Gregor Mendel, an Austrian monk, postulated that genetic material existed. He discovered the laws of heredity using the pea and other plants in his garden to study the inheritance of such traits as flower color.

Nearly seventy years passed before scientists James Watson and Francis Crick, in 1953, proposed the double helix as the most plausible model for each of the unique properties of the molecule. Their model was verified by X-ray diffraction techniques soon afterward.

Several researchers, working at Columbia University and elsewhere in the United States, had led the way prior to Watson and Crick by discovering the chemical composition of this genetic material and the nitrogenous base pairing: The amount of adenine always equaled that of thymine and likewise with guanine and cytosine.

The Watson and Crick model also suggested that the two sides, or strands, of DNA run in opposite directions. That is, the phosphate sugar of one side points upward, while the other strand points downward.

This property is known as antiparallel bonding. The Watson and Crick model could easily explain how DNA replicates during cell division and how genetic information is encoded in its structure.

Self-Replication

Self-replication, which allows the continuity of generations and the growth and repair of individual organisms, occurs during cell division. DNA must be able to produce exact copies of itself. The molecule is uniquely designed for this: A series of enzyme-mediated steps allows the double helix to unwind or unzip, like a zipper, separating the two strands.

Next, nucleotides from digested food enter first the cell and then the nucleus. They bind to a corresponding nucleotide: A with T and G with C. The process continues until two new double-stranded molecules of DNA have formed, each a copy of the other and each going into the new cells that resulted from cell division.

Protein Synthesis

The information encoded in DNA allows for all the development and maintenance of the cell and the organism. The language of this code lies in a linear reading of adjacent nucleotides on each strand. Every three nucleotides specify or fit a particular amino acid, the individual units of proteins.

A second molecule, ribonucleic acid (RNA), copies the molecular structure of DNA and brings the information outside the nucleus into the surrounding cytoplasm of the cell,where the amino acids are assembled, in specified order, to produce a protein.

Post production modifications of these proteins, such as the addition of sugars, fats, or metals, allow a vast array of functional and structural diversity. Plant DNA codes for a variety of substances that are unique to plants. These products sustain not only the plants themselves but also entire ecological niches, as well as humankind.

Mitochondrial and Chloroplastic DNA

A second, independently functioning set of DNA exists in two organelles outside the cell's nucleus, the mitochondria and the chloroplast. It is in the mitochondria, the power sources of cells, where carbohydrates, fats, and proteins are broken down to their raw elements with the release of stored chemical bond energy in the form of heat (calories).

The second region in which DNA is housed outside the nucleus is in the chloroplast, a structure unique to plant cells. In chloroplasts, photosynthesis occurs, the process by which plants are able to transform carbon dioxide, water, and solar energy to produce sugars and, later, fats and proteins, with the release of oxygen. This critical process undertaken by plants sustains most life on earth.

Both mitochondrial and chloroplastic DNA replicate separately from nuclear DNA during cell division. It is postulated that these organelles once, billions of years ago, may have been independently living organisms that were incorporated into other cells to form the eukaryotic cells that make up nonbacterial life-forms such as fungi, protists, plants, and animals.

Plant Proteins

A large array of proteins that are unique to plants are encoded on plant DNA. A group that has received much attention are the so-called phytochemicals, substances with powerful health benefits. Well studied classes are few, including the flavonoids, phytosterols, carotenoids, indoles, coumarins, organosulfurs, terpenes, saponins, lignans, and isothiocyanates.

Each group contains specific proteins that are both antioxidants and anticarcinogens protecting animal cells from cancer-causing agents. The carotenoids, such as beta-carotene, found in orange and yellow fruits and vegetables, and lycopenes, found in tomatoes, appear to protect animals against heart disease and stroke as well as cancer.

The phytosterols, like those found in soybeans, are estrogen like compounds that mimic female hormones. These appear to protect female organs from cancers and also appear to lower cholesterol.

Plant Hormones

Large segments of plant DNA are devoted to coding for specialized plant hormones. Hormones are substances that are produced by one group of cells, circulate to another site, and affect the DNA of the target cells. In plants, these hormones control cell division, growth, and differentiation.

There are five well-described classes of plant hormones: the auxins, gibberellins, cytokinins, ethylene, and abscisic acid. Among the auxins' functions is allowing phototropism, the property that makes plants bend toward the light.

Produced in the roots, auxins travel to stems,making cells elongate on the dark side of plant tissue. Ethylene is a gaseous substance that ripens fruits and causes them to drop from the plant. Abscisic acid contributes to the aging and falling of leaves.

RNA

Ribonucleic acid or RNA is a polymer or chain of nucleotide units, each comprising a nitrogenous base (adenine, cytosine, guanine, or uracil), a five-carbon sugar (ribose), and a phosphate group. The sugar and phosphate groups form the polymer's backbone, while the nitrogenous bases extending from the backbone provide RNA's distinctive properties.

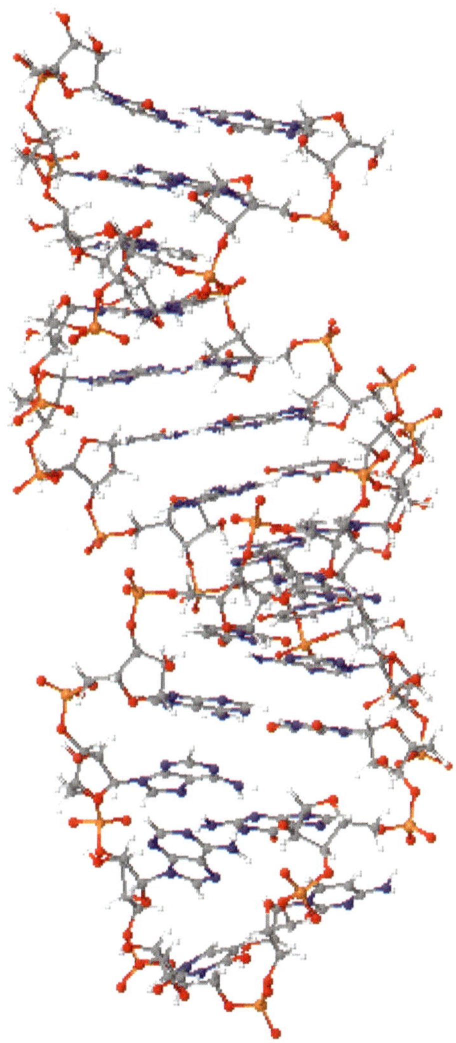

A hairpin loop from a pre-mRNA. Notice the single strand with its nitrogen-rich (blue) bases extending from its oxygen-rich (red) backbone.

In living cells, RNA in different configurations fulfills several important roles in the process of translating genetic information from deoxyribonucleic acid (DNA) into proteins. One type of RNA (messenger(m) RNA) acts as a messenger between DNA and the protein synthesis complexes known as ribosomes; a second type (ribosomal(r) RNA) forms vital portions of the structure of ribosomes; a third type (transfer(t) RNA) is an essential guide to deliver the appropriate protein building blocks, amino acids, to the ribosome; and other types of RNA, microRNAs (miRNAs) play a role in regulating gene expression, while small nuclear(sn) RNA helps with assuring that mRNA contains no nucleotide units that would lead to formation of a faulty protein. RNA also serves as a genetic blueprint for certain viruses, and some RNA molecules (called ribozymes) are also involved in the catalysis of biochemical reactions.

RNA is very similar to DNA, but differs in a few important structural details. RNA is usually single stranded, while DNA naturally seeks its stable form as a double stranded molecule. RNA nucleotides contain ribose while DNA nucleotides contain the closely related sugar deoxyribose. Furthermore, RNA uses the nucleotide uracil in its composition, instead of the thymine that is present in DNA. RNA is transcribed from DNA by enzymes called RNA polymerases and is generally further processed by other enzymes, some of them guided by non-coding RNAs.

Single-stranded RNA is similar to the protein polymer in its natural propensity to fold back and double up with itself in complex ways assuming a variety of biologically useful configurations.

The connectedness of living organisms can be seen in the ubiquitousness of RNA in living cells and in viruses throughout nature, and in the universal role of RNA in protein synthesis.

Chemical and Stereochemical Structure

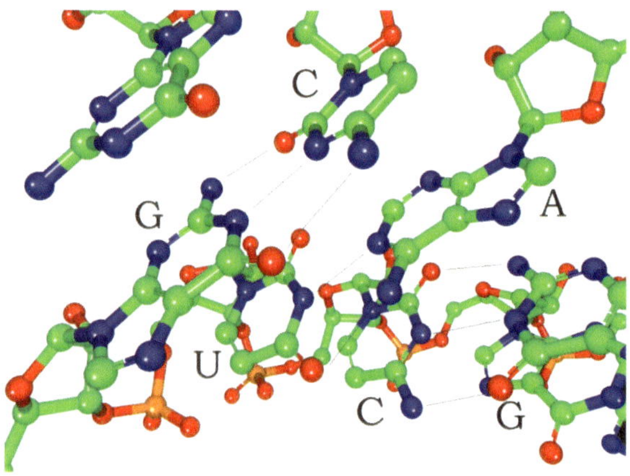

Base-pairing in a siRNA (small interfering RNA) segment, a double-stranded type of RNA. Hydrogen atoms are not shown.

RNA is a nucleic acid, a complex, high-molecular-weight macromolecule composed of nucleotide chains whose sequence of bases conveys genetic information.

A nucleotide is a chemical compound comprising three components: a nitrogen-containing base, a pentose (five-carbon) sugar, and one or more phosphate groups. The nitrogen-containing base of a nucleotide (also called the nucleobase) is typically a derivative of either purine or pyrimidine. The most common nucleotide bases are the purines adenine and guanine and the pyrimidines cytosine and thymine (or uracil in RNA).

Nucleic acids are polymers of repeating units (called monomers). Specifically, they often comprise long chains of nucleotide monomers connected by covalent chemical bonds. RNA molecules may comprise as few as 75 nucleotides or more than 5,000 nucleotides, while a DNA molecule may comprise more than 1,000,000 nucleotide units.

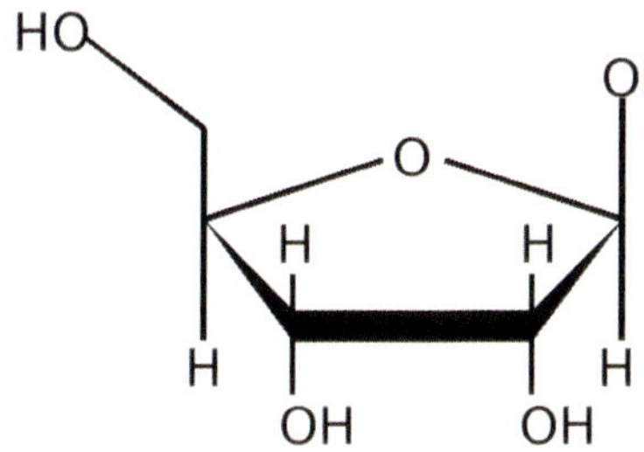

Ribose in acyclic form

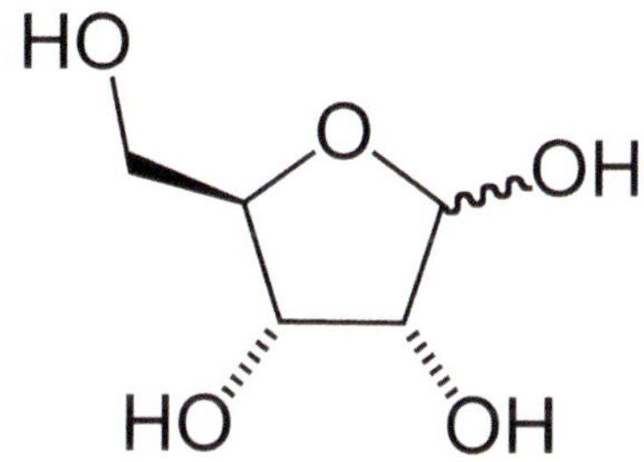

A conventional skeletal formula

In RNA, the sugar component, ribose is a water-soluable, pentose sugar (monosaccharide with five carbon atoms). Ribose has the chemical formula C_{5H10O5}.

Ribose is an aldopentose, which means a pentose sugar with an aldehyde functional group in position one. An aldehyde group comprises a carbon atom bonded to a hydrogen atom and double-bonded to an oxygen atom (chemical formula O=CH-). Ribose forms a five-member ring with four carbon atoms and one oxygen. Hydroxyl (-OH) groups are attached to three of the carbons. The fourth carbon in the ring (one of the carbon atoms adjacent to the oxygen) has attached to it the fifth carbon atom and a hydroxyl group.

The RNA polymer features a ribose and phosphate backbone with one of four different nucleotide bases—adenine, guanine, cytosine, and uracil—attached to each ribose-phosphate unit.

There are also numerous modified bases and sugars found in RNA that serve many different roles. Pseudouridine (Ψ), in which the linkage between uracil and ribose is changed from a C–N bond to a C–C bond, and ribothymidine (T), are found in various places (most notably in the TΨC loop of tRNA). Another notable modified base is hypoxanthine (a deaminated guanine base whose nucleoside is called inosine). Inosine plays a key role in the Wobble Hypothesis of the genetic code. There are nearly 100 other naturally occurring modified nucleosides, of which pseudouridine and nucleosides with 2'-O-methylribose are by far the most common. The specific roles of many of these modifications in RNA are not fully understood. However, it is notable that in ribosomal RNA, many of the post-translational modifications occur in highly functional regions, such as the peptidyl transferase center and the subunit interface, implying that they are important for normal function.

The most important structural feature of RNA that distinguishes it from DNA is the presence of a hydroxyl group at the 2'-position of the ribose sugar. The presence of this functional group enforces the C3'-endo sugar conformation (as opposed to the C2'-endo conformation of the deoxyribose sugar in DNA) that causes the helix to adopt the A-form geometry rather than the B-form most commonly observed in DNA. This results in a very deep and narrow major groove and a shallow and wide minor groove. A second consequence of the presence of the 2'-hydroxyl group is that in conformationally flexible regions of an RNA molecule (that is, not involved in formation of a double helix), it can chemically attack the adjacent phosphodiester bond to cleave the backbone.

Comparison with DNA

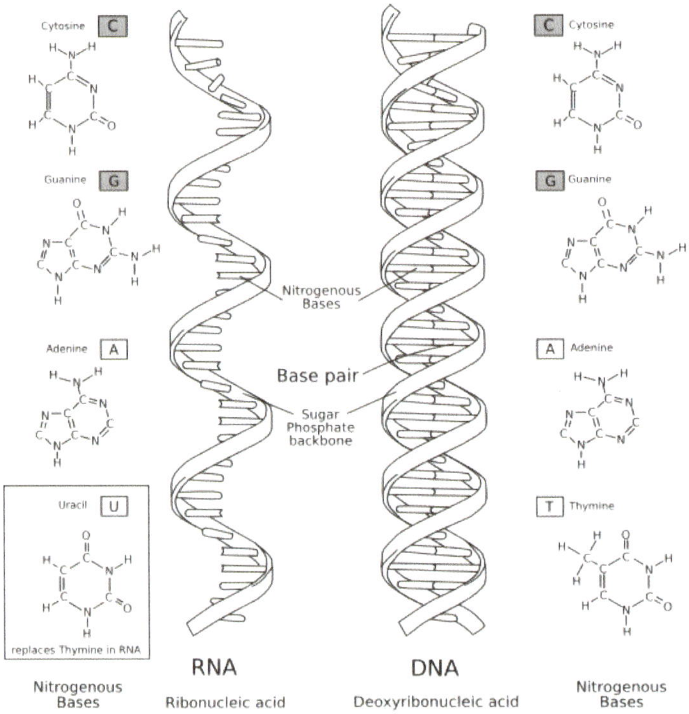

Left: An RNA strand, with its nitrogenous bases. Right: Double-stranded DNA.

The most common nucleic acids are deoxyribonucleic acid (DNA) and ribonucleic acid (RNA). The main role of DNA is the long-term storage of genetic information. DNA is often compared to a blueprint, since it contains instructions for constructing other components of the cell, such as proteins and RNA molecules. The DNA segments that carry genetic information are called *genes*, but other DNA sequences have structural purposes or are involved in regulating the expression of genetic information. RNA, also, may serve more than one purpose, but it is most commonly identified as the intermediate between the DNA blueprint and the actual workings of the cell, serving as the template for the synthesis of proteins from the genetic information stored in DNA.

RNA and DNA differ in three main ways:

First, unlike DNA which is double-stranded, RNA is intrinsically a single-stranded molecule in most of its biological roles and has a much shorter chain of nucleotides. (While RNA is usually single-stranded, the RNA molecule also quite commonly forms double-helical regions where a given strand has folded back on itself. Double-stranded RNA is found also in certain viruses.)

Secondly, while DNA contains *deoxyribose*, RNA contains *ribose*. There is no hydroxyl group attached to the pentose ring in the 2' position in DNA, whereas RNA has two hydroxyl groups. These hydroxyl groups make RNA less stable than DNA because it is more prone to hydrolysis. ("Deoxy" simply indicates that the sugar lacks an oxygen atom present in ribose, the parent compound.)

Thirdly, the complementary nucleotide to adenine is not thymine, as it is in DNA, but rather uracil, which is an unmethylated form of thymine.

Most biologically active RNAs, including tRNA, rRNA, snRNAs, and other non-coding RNAs (such as the signal recognition particle(SRP) RNAs), contain extensively base paired regions that have folded together to form double stranded helices. Structural analysis of these RNAs reveals that they are highly structured with tremendous variety with collections of short helices packed together into structures much more akin to proteins than to DNA, which is usually limited to long double-stranded helices. Through such a variety of structures, RNAs can achieve chemical catalysis, like enzymes. For instance, determination of the structure of the ribosome—an enzyme that catalyzes peptide bond formation—revealed that its active site is composed entirely of RNA.

Synthesis

Synthesis of RNA is usually catalyzed by an enzyme, RNA polymerase, using DNA as a template. Initiation of synthesis begins with the binding of the enzyme to a promoter sequence in the DNA (usually found "upstream" of a gene). The DNA double helix is unwound by the helicase activity of the enzyme. The enzyme then progresses along the template strand in the 3' -> 5' direction, synthesizing a complementary RNA molecule with elongation occurring in the 5' -> 3' direction. The DNA sequence also dictates where termination of RNA synthesis will occur.

There are also a number of RNA-dependent RNA polymerases as well that use RNA as their template for synthesis of a new strand of RNA. For instance, a number of RNA viruses (such as poliovirus) use this type of enzyme to replicate their genetic material. Also, it is known that RNA-dependent RNA polymerases are required for the RNA interference pathway in many organisms.

Biological Roles

RNA's great variety of possible structures and chemical properties permits it to perform a much greater diversity of roles than in the cell than DNA. Three principal types of RNA are involved in protein synthesis:

- Messenger RNA (mRNA) serves as the template for the synthesis of a protein. It carries information from DNA to the ribosome.
- Transfer RNA (tRNA) is a small chain of nucleotides that transfers a specific amino acid to a growing polypeptide chain at the ribosomal site of synthesis. It pairs the amino acid to the appropriate three-nucleotide codon on the mRNA molecule.
- Ribosomal RNA (rRNA) molecules are extremely abundant and make up at least 80 percent of the RNA molecules found in a typical eukaryotic cell. In the cytoplasm, usually three or four rRNA molecules combine with many proteins to perform a structural and essential catalytic role, as components of the ribosome.

RNA also may serve as a catalyst for reactions and as a genetic blueprint, rather than DNA, in various viruses. Some RNA, including tRNA and rRNA, is non-coding in that it is not translated into proteins.

Messenger RNA (mRNA)

Messenger RNA is RNA that carries information from DNA to the ribosome sites of protein synthesis in the cell. In eukaryotic cells, once mRNA has been transcribed from DNA, it is "processed" before being exported from the nucleus into the cytoplasm, where it is bound to ribosomes and translated into its corresponding protein form with the help of tRNA. In prokaryotic cells, which do not have nucleus and cytoplasm compartments, mRNA can bind to ribosomes while it is being transcribed from DNA. After a certain amount of time the message degrades into its component nucleotides, usually with the assistance of ribonucleases.

Non-coding RNA

RNA genes (also known as non-coding RNA or small RNA) are genes that encode RNA that is not translated into a protein. The most prominent examples of RNA genes are those coding for transfer RNA (tRNA) and ribosomal RNA (rRNA), both of which are involved in the process of translation. Two other groups of non-coding RNA are microRNAs (miRNA) which regulate the expression of genes through a process called RNA interference (RNAi), and small nuclear RNAs (snRNA), a diverse class that includes for example the RNAs that form spliceosomes that excise introns from pre-mRNA.

Transfer RNA (tRNA)

Transfer RNA is a small RNA chain of about 74-95 nucleotides that transfers a specific amino acid to a growing polypeptide chain at the ribosomal site of protein synthesis, during translation. It has sites for amino-acid attachment and an anticodon region for codon recognition that binds to a specific sequence on the messenger RNA chain through hydrogen bonding. It is a type of non-coding RNA.

Ribosomal RNA (rRNA)

Ribosomal RNA is the catalytic component of the ribosomes, the protein synthesis factories in the cell. Eukaryotic ribosomes contain four different rRNA molecules: 18S, 5.8S, 28S, and 5S rRNA. Three of the rRNA molecules are synthesized in the nucleolus, and one is synthesized elsewhere. rRNA molecules are extremely abundant and make up at least 80 percent of the RNA molecules found in a typical eukaryotic cell.

Catalytic RNA

Although RNA contains only four bases, in comparison to the twenty-odd amino acids commonly found in proteins, certain RNAs (called ribozymes) are still able to catalyze chemical reactions. These include cutting and ligating other RNA molecules, and also the catalysis of peptide bond formation in the ribosome.

Genetic Blueprint in Some Viruses

Some viruses contain either single-stranded or double-stranded RNA as their source of genetic information. Retroviruses, for example, store their genetic information as RNA, though they replicate in their hosts via a DNA intermediate. Once in the host's cell, the RNA strands undergo reverse transcription to DNA in the cytosol and are integrated into the host's genome. Human immunodeficiency virus (or HIV) is a retrovirus thought to cause acquired immune deficiency syndrome (AIDS), a condition in which the human immune system begins to fail, leading to life-threatening opportunistic infections.

Double-stranded RNA (dsRNA) is RNA with two complementary strands, similar to the DNA found in all cells. dsRNA forms the genetic material of some viruses called dsRNA viruses. In eukaryotes, long RNA such as viral RNA can trigger RNA interference, where short dsRNA molecules called siRNAs (small interfering RNAs) can cause enzymes to break down specific mRNAs or silence the expression of genes. siRNA can also increase the transcription of a gene, a process called RNA activation. siRNA is often confused with miRNA; siRNAs are double-stranded, whereas miRNAs are single-stranded.

RNA World Hypothesis

The RNA world hypothesis proposes that the earliest forms of life relied on RNA both to carry genetic information (like DNA does now) and to catalyze biochemical reactions like an enzyme. According to this hypothesis, descendants of these early lifeforms gradually integrated DNA and proteins into their metabolism.

In the 1980s, scientists discovered that certain RNA molecules (called *ribozymes*) may function as enzymes, whereas previously only proteins were believed to have catalytic ability. Many natural ribozymes catalyze either their own cleavage or the cleavage of other RNAs, but they have also been found to catalyze the aminotransferase activity of the ribosome.

The discovery of ribozymes provides a possible explanation for how early RNA molecules might have first catalyzed their own replication and developed a range of enzymatic activities. Known as the RNA world hypothesis, this explanation posits that RNA evolved before either DNA or proteins from free-floating nucleotides in the early "primordial soup." In their function as enzymes, RNA molecules might have begun to catalyze the synthesis of proteins, which are more versatile than RNA, from amino acid molecules. Next, DNA might have been formed by reverse transcription of RNA, with DNA eventually replacing RNA as the storage form of genetic material. Although there are remaining difficulties with the RNA world hypothesis, it remains as a possible key to understanding the origin and development of the multi-functional nature of nucleic acids, the interconnectedness of life, and its common origins.

RNA Secondary Structures

The functional form of single stranded RNA molecules, just like proteins, frequently requires a specific tertiary structure. The scaffold for this structure is provided by secondary structural elements, which arise through the formation of hydrogen bonds within the interfolded molecule. This leads to several recognizable "domains" of secondary structure like hairpin loops, bulges, and internal loops. The secondary structure of RNA molecules can be predicted computationally by calculating the minimum free energies (MFE) structure for all different combinations of hydrogen bondings and domains. There has been a significant amount of research directed at the RNA structure prediction problem.

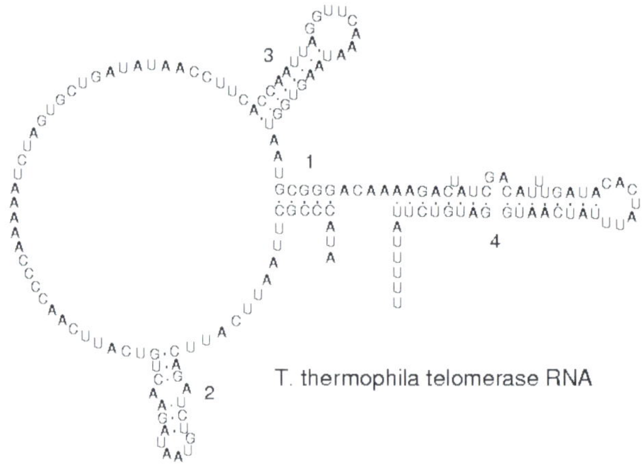

Secondary structure of an RNA from a telomerase.

Plant Breeding

Plant breeding is an application of genetic principles to produce plants that are more useful to humans.

This is accomplished by selecting plants found to be economically or aesthetically desirable, first by controlling the mating of selected individuals, and then by selecting certain individuals among the progeny. Such processes, repeated over many generations, can change the hereditary makeup and value of a plant population far beyond the natural limits of previously existing populations. It emphasizes the application of genetic principles to the improvement of plants; the biological factors underlying plant breeding are dealt with in the article heredity.

barley: development of drought-resistant varieties
Efforts to develop drought-resistant varieties of barley.

Plant breeding is an ancient activity, dating to the very beginnings of agriculture. Probably soon after the earliest domestications of cereal grains, humans began to recognize degrees of excellence among the plants in their fields and saved seed from the best for planting new crops. Such tentative selective methods were the forerunners of early plant-breeding procedures.

The results of early plant-breeding procedures were conspicuous. Most present-day varieties are so modified from their wild progenitors that they are unable to survive in nature. Indeed, in some cases, the cultivated forms are so strikingly different from existing wild relatives that it is difficult even to identify their ancestors. These remarkable transformations were accomplished by early plant breeders in a very short time from an evolutionary point of view, and the rate of change was probably greater than for any other evolutionary event.

Scientific plant breeding dates back hardly more than 50 years. The role of pollination and fertilization in the process of reproduction was not widely appreciated even 100 years ago, and it was not until the early part of the 20th century that the laws of genetic inheritance were recognized and a beginning was made toward applying them to the improvement of plants. One of the major facts that has emerged during the short history of scientific breeding is that an enormous wealth of genetic variability exists in the plants of the world and that only a start has been made in tapping its potential.

Goals

The plant breeder usually has in mind an ideal plant that combines a maximum number of desirable characteristics. These characteristics may include resistance to diseases and insects; tolerance to heat and frost; appropriate size, shape, and time to maturity; and many other general and specific traits that contribute to improved adaptation to the environment, ease in growing and handling, greater yield, and better quality. The breeder of fancy show plants must also consider aesthetic appeal. Thus the breeder can rarely focus attention on any one characteristic but must take into account the manifold traits that make the plant more useful in fulfilling the purpose for which it is grown.

Increase of Yield

One of the aims of virtually every breeding project is to increase yield. This can often be brought about by selecting obvious morphological variants. One example is the selection of dwarf, early maturing varieties of rice. These dwarf varieties are sturdy and give a greater yield of grain. Furthermore, their early maturity frees the land quickly, often allowing an additional planting of rice or other crop the same year.

Another way of increasing yield is to develop varieties resistant to diseases and insects. In many cases the development of resistant varieties has been the only practical method of pest control. Perhaps the most important feature of resistant varieties is the stabilizing effect they have on production and hence on steady food supplies. Varieties tolerant to drought, heat, or cold provide the same benefit.

Modifications of Range and Constitution

Another common goal of plant breeding is to extend the area of production of a crop species. A good example is the modification of grain sorghum since its introduction to the United States about 100 years ago. Of tropical origin, grain sorghum was originally confined to the southern Plains area and the Southwest, but earlier maturing varieties were developed until grain sorghum is now an important crop as far north as North Dakota.

Development of crop varieties suitable for mechanized agriculture has become a major goal of plant breeding in recent years. Uniformity of plant characters is very important in mechanized agriculture because field operations are much easier when the individuals of a variety are similar in time of germination, growth rate, size of fruit, and so on. Uniformity in maturity is, of course, essential when crops such as tomatoes and peas are harvested mechanically.

The nutritional quality of plants can be greatly improved by breeding. For example, it is possible to breed varieties of corn (maize) much higher in lysine than previously existing varieties. Breeding high-lysine maize varieties for those areas of the world where maize is the major source of this nutritionally essential amino acid has become a major goal in plant breeding.

In breeding ornamentals, attention is paid to such factors as longer blooming periods, improved keeping qualities of flowers, general thriftiness, and other features that contribute to usefulness and aesthetic appeal. Novelty itself is often a virtue in ornamentals, and the spectacular, even the bizarre, is often sought.

Methods of Plant Breeding

Mating Systems

Plant mating systems devolve about the type of pollination, or transferal of pollen from flower to flower. A flower is self-pollinated (a "selfer") if pollen is transferred to it from any flower of the same plant and cross-pollinated (an "outcrosser" or "outbreeder") if the pollen comes from a flower on a different plant. About half of the more important cultivated plants are naturally cross-pollinated, and their reproductive systems include various devices that encourage cross-pollination; *e.g.*, protandry (pollen shed before the ovules are mature, as in the carrot and walnut), dioecy (stamens and pistils borne on different plants, as in the date palm, asparagus, and hops), and genetically determined self-incompatibility (inability of pollen to grow on the stigma of the same plant, as in white clover, cabbage, and many other species).

Other plant species, including a high proportion of the most important cultivated plants such as wheat, barley, rice, peas, beans, and tomatoes, are predominantly self-pollinating. There are relatively few reproductive mechanisms that promote self-pollination; the most positive of which is failure of the flowers to open (cleistogamy), as in certain violets. In barley, wheat, and lettuce the pollen is shed before or just as the flowers open; and in the tomato pollination follows opening of the flower, but the stamens form a cone around the stigma. In such species there is always a risk of unwanted cross-pollination.

In controlled breeding procedures it is imperative that pollen from the desired male parent, and no other pollen, reaches the stigma of the female parent. When stamens and pistils occur in the same flower, the anthers must be removed from flowers selected as females before pollen is shed. This is usually done with forceps or scissors. Protection must also be provided from "foreign" pollen. The most common method is to cover the flower with a plastic or paper bag. When the stigma of the female parent becomes receptive, pollen from the desired male parent is transferred to it, often by breaking an anther over the stigma, and the protective bag is replaced. The production of certain hybrids is, therefore, tedious and expensive because it often requires a series of delicate, exacting, and properly timed hand operations. When male and female parts occur in separate flowers, as in corn (maize), controlled breeding is easier.

A cross-pollinated plant, which has two parents, each of which is likely to differ in many genes, produces a diverse population of plants hybrid (heterozygous) for many traits. A self-pollinated plant, which has only one parent, produces a more uniform population of plants pure breeding (homozygous) for many traits. Thus, in contrast to outbreeders, self-breeders are likely to be highly homozygous and hence true breeding for a specified trait.

Breeding Self-pollinated Species

The breeding methods that have proved successful with self-pollinated species are (1) mass selection; (2) pure-line selection; (3) hybridization, with the segregating generations handled by the pedigree method, the bulk method, or by the backcross method; and (4) development of hybrid varieties.

Mass Selection

In mass selection, seeds are collected from (usually a few dozen to a few hundred) desirable appearing individuals in a population, and the next generation is sown from the stock of mixed seed. This procedure, sometimes referred to as phenotypic selection, is based on how each individual looks. Mass selection has been used widely to improve old "land" varieties, varieties that have been passed down from one generation of farmers to the next over long periods.

An alternative approach that has no doubt been practiced for thousands of years is simply to eliminate undesirable types by destroying them in the field. The results are similar whether superior plants are saved or inferior plants are eliminated: seeds of the better plants become the planting stock for the next season.

A modern refinement of mass selection is to harvest the best plants separately and to grow and compare their progenies. The poorer progenies are destroyed and the seeds of the remainder are harvested. It should be noted that selection is now based not solely on the appearance of the parent plants but also on the appearance and performance of their progeny. Progeny selection is usually more effective than phenotypic selection when dealing with quantitative characters of low heritability. It should be noted, however, that progeny testing requires an extra generation; hence gain per cycle of selection must be double that of simple phenotypic selection to achieve the same rate of gain per unit time.

Mass selection, with or without progeny test, is perhaps the simplest and least expensive of plant-breeding procedures. It finds wide use in the breeding of certain forage species, which are not important enough economically to justify more detailed attention.

Pure-line Selection

Pure-line selection generally involves three more or less distinct steps: (1) numerous superior appearing plants are selected from a genetically variable population; (2) progenies of the individual plant selections are grown and evaluated by simple observation, frequently over a period of several years; and (3) when selection can no longer be made on the basis of observation alone, extensive trials are undertaken, involving careful measurements to determine whether the remaining selections are superior in yielding ability and other aspects of performance. Any progeny superior to an existing variety is then released as a new "pure-line" variety. Much of the success of this method during the early 1900s depended on the existence of genetically variable land varieties that were waiting to be exploited. They provided a rich source of superior pure-line varieties, some of which are still represented among commercial varieties. In recent years the pure-line method as outlined above has decreased in importance in the breeding of major cultivated species; however, the method is still widely used with the less important species that have not yet been heavily selected.

A variation of the pure-line selection method that dates back centuries is the selection of single-chance variants, mutations or "sports" in the original variety. A very large number of varieties that differ from the original strain in characteristics such as colour, lack of thorns or barbs, dwarfness, and disease resistance have originated in this fashion.

Hybridization

During the 20th century planned hybridization between carefully selected parents has become dominant in the breeding of self-pollinated species. The object of hybridization is to combine desirable genes found in two or more different varieties and to produce pure-breeding progeny superior in many respects to the parental types.

Genes, however, are always in the company of other genes in a collection called a genotype. The plant breeder's problem is largely one of efficiently managing the enormous numbers of genotypes that occur in the generations following hybridization. As an example of the power of hybridization in creating variability, a cross between hypothetical wheat varieties differing by only 21 genes is capable of producing more than 10,000,000,000 different genotypes in the second generation. At spacings normally used by farmers, more than 50,000,000 acres would be required to grow a population large enough to permit every genotype to occur in its expected frequency. While the great majority of these second generation genotypes are hybrid (heterozygous) for one or more traits, it is statistically possible that 2,097,152 different pure-breeding (homozygous) genotypes can occur, each potentially a new pure-line variety. These numbers illustrate the importance of efficient techniques in managing hybrid populations, for which purpose the pedigree procedure is most widely used.

Pedigree breeding starts with the crossing of two genotypes, each of which have one or more desirable characters lacked by the other. If the two original parents do not provide all of the desired characters, a third parent can be included by crossing it to one of the hybrid progeny of the first generation (F_1). In the pedigree method superior types are selected in successive generations, and a record is maintained of parent–progeny relationships.

The F_2 generation (progeny of the crossing of two F_1 individuals) affords the first opportunity for selection in pedigree programs. In this generation the emphasis is on the elimination of individuals carrying undesirable major genes. In the succeeding generations the hybrid condition gives way to pure breeding as a result of natural self-pollination, and families derived from different F_2 plants begin to display their unique character. Usually one or two superior plants are selected within each superior family in these generations. By the F_5 generation the pure-breeding condition (homozygosity) is extensive, and emphasis shifts almost entirely to selection between families. The pedigree record is useful in making these eliminations. At this stage each selected family is usually harvested in mass to obtain the larger amounts of seed needed to evaluate families for quantitative characters. This evaluation is usually carried out in plots grown under conditions that simulate commercial planting practice as closely as possible. When the number of families has been reduced to manageable proportions by visual selection, usually by the F_7 or F_8 generation, precise evaluation for performance and quality begins. The final evaluation of promising strains involves (1) observation, usually in a number of years and locations, to detect weaknesses that may not have appeared previously; (2) precise yield testing; and (3) quality testing. Many plant breeders test for five years at five representative locations before releasing a new variety for commercial production.

The bulk-population method of breeding differs from the pedigree method primarily in the handling of generations following hybridization. The F_2 generation is sown at normal commercial planting rates in a large plot. At maturity the crop is harvested in mass, and the seeds are used to establish the next generation in a similar plot. No record of ancestry is kept. During the period of bulk propagation natural selection tends to eliminate plants having poor survival value. Two types of artificial selection also are often applied: (1) destruction of plants that carry undesirable major genes and (2) mass techniques such as harvesting when only part of the seeds are mature to select for early maturing plants or the use of screens to select for increased seed size. Single plant selections are then made and evaluated in the same way as in the pedigree method of breeding. The chief advantage of the bulk population method is that it allows the breeder to handle very large numbers of individuals inexpensively.

Often an outstanding variety can be improved by transferring to it some specific desirable character that it lacks. This can be accomplished by first crossing a plant of the superior variety to a plant of the donor variety, which carries the trait in question, and then mating the progeny back to a plant having the genotype of the superior parent. This process is called backcrossing. After five or six backcrosses the progeny will be hybrid for the character being transferred but like the superior parent for all other genes. Selfing the last backcross generation, coupled with selection, will give some progeny pure breeding for the genes being transferred. The advantages of the backcross method are its rapidity, the small number of plants required, and the predictability of the outcome. A serious disadvantage is that the procedure diminishes the occurrence of chance combinations of genes, which sometimes leads to striking improvements in performance.

Hybrid Varieties

The development of hybrid varieties differs from hybridization in that no attempt is made to produce a pure-breeding population; only the F_1 hybrid plants are sought. The F_1 hybrid of crosses between different genotypes is often much more vigorous than its parents. This hybrid vigour, or heterosis, can be manifested in many ways, including increased rate of growth, greater uniformity, earlier flowering, and increased yield, the last being of greatest importance in agriculture.

By far the greatest development of hybrid varieties has been in corn (maize), primarily because its male flowers (tassels) and female flowers (incipient ears) are separate and easy to handle, thus proving economical for the production of hybrid seed. The production of hand-produced F_1 hybrid seed of other plants, including ornamental flowers, has been economical only because greenhouse growers and home gardeners have been willing to pay high prices for hybrid seed.

Recently, however, a built-in cellular system of pollination control has made hybrid varieties possible in a wide range of plants, including many that are self-pollinating, such as sorghums. This system, called cytoplasmic male sterility, or cytosterility, prevents normal maturation or function of the male sex organs (stamens) and results in defective pollen or none at all. It obviates the need for removing the stamens either by hand or by machine. Cytosterility depends on the interaction between male sterile genes ($R + r$) and factors found in the cytoplasm of the female sex cell. The genes are derived from each parent in the normal Mendelian fashion, but the cytoplasm (and its factors) is provided by the egg only; therefore, the inheritance of cytosterility is determined by the female parent. All plants with fertile cytoplasm produce viable pollen, as do plants with sterile cytoplasm but at least one *R* gene; plants with sterile cytoplasm and two *r* genes are male sterile (produce defective pollen).

The production of F_1 hybrid seed between two strains is accomplished by interplanting a sterile version of one strain (say A) in an isolated field with a fertile version of another strain (B). Since strain A produces no viable pollen, it will be pollinated by strain B, and all seeds produced on strain A plants must therefore be F_1 hybrids between the strains. The F_1 hybrid seeds are then planted to produce the commercial crop. Much of the breeder's work in this process is in developing the pure-breeding sterile and fertile strains to begin the hybrid seed production.

Breeding Cross-pollinated Species

The most important methods of breeding cross-pollinated species are (1) mass selection; (2) development of hybrid varieties; and (3) development of synthetic varieties. Since cross-pollinated species are naturally hybrid (heterozygous) for many traits and lose vigour as they become purebred (homozygous), a goal of each of these breeding methods is to preserve or restore heterozygosity.

Mass Selection

Mass selection in cross-pollinated species takes the same form as in self-pollinated species; *i.e.*, a large number of superior appearing plants are selected and harvested in bulk and the seed used to produce the next generation. Mass selection has proved to be very effective in improving qualitative characters, and, applied over many generations, it is also capable of improving quantitative characters, including yield, despite the low heritability of such characters. Mass selection has long been a major method of breeding cross-pollinated species, especially in the economically less important species.

Hybrid Varieties

The outstanding example of the exploitation of hybrid vigour through the use of F_1 hybrid varieties has been with corn (maize). The production of a hybrid corn variety involves three steps: (1) the selection of superior plants; (2) selfing for several generations to produce a series of inbred lines, which although different from each other are each pure-breeding and highly uniform; and (3) crossing selected inbred lines. During the inbreeding process the vigour of the lines decreases drastically, usually to less than half that of field-pollinated varieties. Vigour is restored, however, when any two unrelated inbred lines are crossed, and in some cases the F_1 hybrids between inbred lines are much superior to open-pollinated varieties. An important consequence of the homozygosity of the inbred lines is that the hybrid between any two inbreds will always be the same. Once the inbreds that give the best hybrids have been identified, any desired amount of hybrid seed can be produced.

Pollination in corn (maize) is by wind, which blows pollen from the tassels to the styles (silks) that protrude from the tops of the ears. Thus, controlled cross-pollination on a field scale can be accomplished economically by interplanting two or three rows of the seed parent inbred with one row of the pollinator inbred and detasselling the former before it sheds pollen. In practice most hybrid corn is produced from "double crosses," in which four inbred lines are first crossed in pairs (A × B and C × D) and then the two F_1 hybrids are crossed again (A × B) × (C × D). The double-cross procedure has the advantage that the commercial F_1 seed is produced on the highly productive single cross A × B rather than on a poor-yielding inbred, thus reducing seed costs. In recent years cytoplasmic male sterility, described earlier, has been used to eliminate detasselling of the seed parent, thus providing further economies in producing hybrid seed.

Much of the hybrid vigour exhibited by F_1 hybrid varieties is lost in the next generation. Consequently, seed from hybrid varieties is not used for planting stock but the farmer purchases new seed each year from seed companies.

Perhaps no other development in the biological sciences has had greater impact on increasing the quantity of food supplies available to the world's population than has the development of hybrid corn (maize). Hybrid varieties in other crops, made possible through the use of male sterility, have also been dramatically successful and it seems likely that use of hybrid varieties will continue to expand in the future.

Synthetic Varieties

A synthetic variety is developed by intercrossing a number of genotypes of known superior combining ability—*i.e.*, genotypes that are known to give superior hybrid performance when crossed in all combinations. (By contrast, a variety developed by mass selection is made up of genotypes bulked together without having undergone preliminary testing to determine their performance in hybrid combination.) Synthetic varieties are known for their hybrid vigour and for their ability to produce usable seed for succeeding seasons. Because of these advantages, synthetic varieties have become increasingly favoured in the growing of many species, such as the forage crops, in which expense prohibits the development or use of hybrid varieties.

Distribution and Maintenance of New Varieties

The benefits of superior new varieties obviously cannot be realized until sufficient seed has been produced to permit commercial production. Although the primary function of the plant breeder is to develop new varieties, he usually also carries out an initial small-scale seed increase. Seed thus produced is called breeders seed. The next stage is the multiplication of breeders seed to produce foundation seed. Production of foundation seed is usually carried out by seed associations or institutes, whose work is regulated by government agencies. The third step is the production of certified seed, the progeny of foundation seed, produced on a large scale by specialized seed growers for general sale to farmers and gardeners. Certified seed must be produced and handled in such a way as to meet the standards set by the certifying agency (usually a seed association). Seed associations are also usually responsible for maintaining the purity of new varieties once they have been released for commercial production.

The distribution of new varieties developed by commercial plant-breeding companies is often through seed associations, but many reputable companies market their products without following the official certification process. In some countries, particularly in Europe, new varieties can be patented for periods up to 15 years or more, during which time the breeder has an exclusive right to reproduce and sell the variety.

Doubled Haploidy

The life cycle of angiosperms (higher plants) is characterized by alternating generations of sporophytes and gametophytes. The gametophytic phase arises when the diploid cells undergo meiosis (reduction division) to form male and female gametes. This phase is short lived as fertilization of the egg re-establishes the diploid sporophytic phase. The sporophytic phase characterized by chromosome number (2n) is the product of fertilization of male and female gametes, containing the haploid (n) set of chromosomes from each parent. Therefore, haploid is a generalized term for plants that contain the gametic chromosome number (n). This is in contrast to diploid plants, which contain two sets (2n) of chromosomes. Haploids are sexually sterile and, therefore, doubling of the chromosomes is required to produce fertile plants which are called doubled haploids or homozygous diploids.

Haploids have attracted great interest of plant physiologists, embryologists, geneticists and breeders. Spontaneous production of haploids usually occurs through the process of parthenogenesis (embryo development from an unfertilized egg). However, occasionally, they bear the characters of male parent only, suggesting their origin through 'ovule androgenesis' (embryo development inside the ovule by the activity of the male nucleus alone where elimination or inactivation of egg nucleus occurs before fertilization). Although *in vivo* occurrence of haploids has been reported in several species but at low and variable frequencies and was regarded as a special biological phenomenon. The low frequency of spontaneously arising these haploid plants severely limited the utilization of haploids for crop improvement and genetic studies.

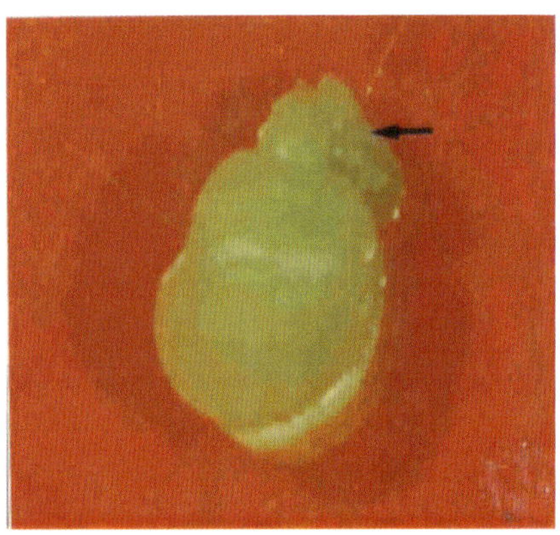

Figure: Twin embryos excised from a Neem seed. Smaller embryos (arrow marked) is a synergid embryo formed without fertilization and is haploid. The bigger embryo is zygotic embryo which is formed during fertilization of egg cell with the male gamete

In vitro Haploid plants can be obtained by triggering the male or female gametic cells to undergo sporophytic development. *In vitro* androgenesis (anther-microspore culture) is one of the most preferred techniques for obtaining haploids but, *in vitro* gynogenesis (unfertilized ovary-ovule culture) can prove to be a complementary technique in species where anther culture is inaccessible or less productive. It means that not only the microspore but, also the megaspore of angiosperms can be triggered *in vitro* to undergo sporophytic development. However, production of haploids via gynogenesis is more tedious, less efficient in comparison to androgenesis. In addition to the above techniques, *in situ* parthenogenesis (pollen irradiation and chemical treatment) can be employed for generation of haploid plants.

Haploid and diploid lines play a vital role in genomics and have been used for the purpose of physical mapping, genetic mapping and also for integration of genetic and physical maps. Additionally, haploid and doubled haploid plants are adapted for mutagenesis and genetic transformation experiments, presenting the advantage of immediate production of homozygous lines. It is also expected that, in the near future, haploid and doubled haploid plants will play an increasingly important role in whole genome sequencing projects, where homozygosity is of particular advantage. The technique has special significance for genetic improvement of the tree species where breeding is made difficult due to long generation intervals and highly heterozygous nature of these plants as a result of cross pollination.

Androgenesis

In androgenesis, the male gametophyte (microspore or immature pollen) produces haploid plants. The basic principle is to stop the development of pollen into a gamete (sex cell) and force it to develop into a haploid plant or sporophyte. The remarkable discovery that haploid embryos and plants can be produced by *in vitro* culture of anthers of brought renewed interest to haploidy. This method of androgenic haploid production was quickly attempted in many species to hasten the breeding programme in several economically important plants. Haploid production through anther/microspore culture scores higher over other methods due to the fact that anthers harbour large numbers of haploid microspores per anther and is a potentially efficient means to generate homozygous true-breeding progeny lines in plant breeding programs.

1. Methodologies

In androgenesis, immature pollen grains are induced to follow the sporophytic mode of development by the application of various physical and chemical stimuli. There are two methods for *in vitro* production of androgenic haploids –

(A) Anther culture, and (B) Isolated pollen (microspore) culture.

A. Anther culture

(a)	Select the flower buds from an elite plant and determine the stage of microspores by acetocarmine squashes of anthers or by staining them with fluorescent dye DAPI (4,6-diamidino-2-phenylindole).
(b)	Surface sterilize the selected size of buds (ca 2mm size flower bud) to initiate in vitro anther cultures.
(c)	Dissect the buds under a stereo-microscope, using pre-sterilized Petriplates, forceps and fine needles. Discard the damaged anthers, if any, and remove the filament gently.

(d)	Inoculate the anthers, bearing early-to-late uninucleate stage of microspores, in a nutrient medium and maintain the cultures in defined conditions.
(e)	As the anthers proliferate, they produce embryos/callus.
(f)	The callus/embryos formed can be transferred to a suitable medium to finally produce haploid plants and then diploidize them by colchicine to produce homozygous diploids.

B. Microspore (Pollen) culture

Haploid plants can also be produced from isolated immature pollens or microspores (male gametophytic cells):

(a) Select the flower buds from an elite plant and determine the stage of microspores by acetocarmine squashes of anthers or by staining them with fluorescent dye DAPI (4,6-diamidino-2-phenylindole).

(b) Surface sterilize the selected size of buds to initiate *in vitro* anther cultures.

(c) Extract the microspores by pressing and squeezing the buds with a glass rod against the sides of a beaker.

(d) Filter the pollen suspension to remove anther tissue debris.

(e) Wash and collect the viable and large pollen (smaller pollen do not regenerate) by filtration.

(f) Culture these microspores on a solid or liquid medium.

(g) As the microspores undergo multiple divisions, they produce multicellular and multinuclear structure.

(h) The callus/ embryos formed can be transferred to a suitable medium to finally produce a haploid plants and then diploid plants by colchicine treatment.

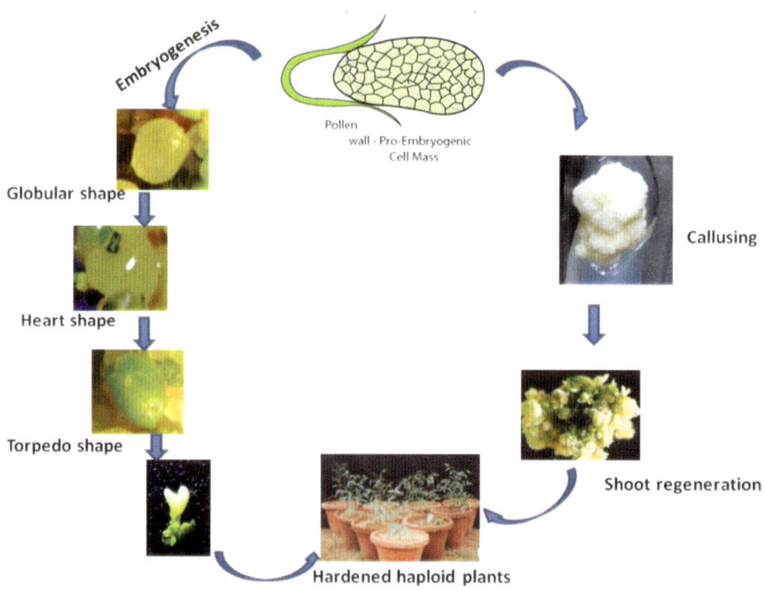

Figure: Androgenic haploid production pathways

2. Comparison Between Anther and Pollen Culture

Anther culture is an easy, quick and practicable approach. Anther walls act as conditioning factors and promote culture growth. Thus, anther cultures are reasonably efficient for haploid production. The major limitation is that the plants not only originate from pollen but also from other somatic parts of the anther. This results in the development of plants at different ploidy levels viz., haploids, diploids, aneuploids, as a result of which the final tissue, derived, may not be of purely gametophytic origin. Moreover, the plants arising from an anther would constitute a heterogenous population. It has been observed in some species that anther cultures show asynchronous pollen development, the older grains may suppress the androgenic potential of younger grains by releasing toxic substances . The disadvantages associated with anther culture can be overcome by pollen culture as it offers the following advantages:

- Undesirable effects of anther wall and associated tissues can be avoided.
- Androgenesis, starting from a single cell, can be better regulated.
- Isolated microspores (pollen) are ideal for various genetic manipulations like transformation, mutagenesis etc.
- The yield of haploid plants is relatively higher.

3. Pathways of Development

The early divisions in responding pollen grains may occur in any one of the following four pathways:

i) Pathway I - The uninucleate pollen grain may divide symmetrically to yield two equal daughter cells both of which undergo further divisions e.g. *Datura innoxia*.

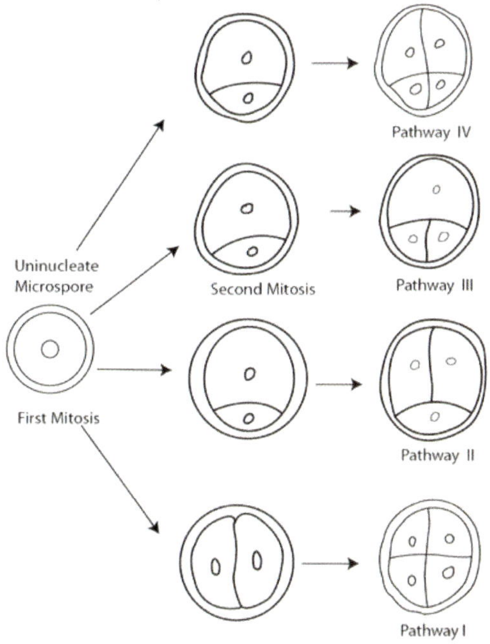

Figure: Pathways of development of microspores

ii) Pathway II - In some other cases e.g. *N.tabacum*, barley, wheat etc., the uninucleate pollen divides unequally. The generative cell degenerates, callus/embryo originates due to successive divisions of the vegetative cell.

iii) Pathway III - But in few species, the pollen embryos originate from the generative cell alone; the vegetative cell either does not divide or divides only to a limited extent forming a suspensor like structure.

iv) Pathway IV - Finally in few other species e.g. *Datura innoxia*, the uninucleate pollen grains divide unequally, producing generative and vegetative cells, but both these cells divide repeatedly to contribute to the developing embryo/callus.

Gynogenesis

Gynogenic development of plants from unfertilized cells of female gametophyte (embryo-sac) in ovary/ovule/young flower cultures is one of the available alternatives for haploid production. It was first reported in barley San Noeum. This method of haploid production is more tedious than androgenesis. The reasons for this being the indefinite numbers of microspores (male gametes) within the anther wall for androgenesis as against single egg cell (female gamete) per flower for gynogenic haploid production, which too, is deep seated within the embryo-sac (female gametophyte), thus making the entire process very cumbersome. The technique is very useful where anther culture has been unsuccessful, plants are male sterile or androgenesis is confronted with the problem of albino or non-haploid formation. The following techniques are generally used for production of haploids via *in vitro* gynogenesis either through direct embryogenesis or via callusing.

1. In situ parthenogenesis induced by irradiated pollen followed by in vitro embryo culture

Parthenogenesis induced *in vivo* by irradiated pollen, followed by *in vitro* culture of embryos can be an alternative method of obtaining haploids in fruit crops. Gynogenesis by *in situ* pollination with irradiated pollen has been successfully used for *Malus domestica* (L.) Borkh, *Pyrus communis* L., *Actinidia deliciosa* (A. Chev). This method is based on *in vitro* culture of immature seeds or embryos, obtained as a result of pollination by irradiated pollen with gamma rays from cobalt 60. The method is useful in those species in which *in vitro* anther culture has not been successful. Irradiation does not hinder pollen germination but prevents pollen fertilization and, thereby, stimulating the development of haploid embryos from ovules. The success of this technique is dependent on the choice of radiation dose, the developmental stage of the embryos at the time of culture, culture conditions and media requirements.

2. Ovary slice culture

Ovary slice culture technique involves culture of transverse sections of unpollinated ovaries on culture media. The following protocol was used to induce *in vitro* gynogenesis in Tea:

a. For Ovary slice culture in Tea, unopened and unpollinated mature flower-buds (6-10 mm) size were collected early in the morning. Some of the buds were fixed in FAA (5:5:90 v/v/v Formaldehyde: Acetic acid: 70% Ethanol), for 48 h, and then stored in 70% alcohol. Later on, the appropriate developmental stage of the embryo sac was determined by histological analysis.

b. The flower buds were surface sterilized with 0.1% $HgCl_2$ for 7 minutes, followed by rinsing with sterile distilled water at least thrice.

c. Carefully dissected transverse sections of ovaries were cultured on Murashige and Skoog's media supplemented with varying concentrations of Auxins and Cytokinins.

d. Six ovary slices containing unpollinated ovules were cultured in 60 mm X 15 mm pre sterilized disposable Petriplates containing 10 ml MS medium.

e. The sealed Petriplates were subjected to various regimes of temperature and light treatments.

3. Ovule culture

The unfertilized ovary is surface sterilized and the ovules were taken and placed into culture. Excision of ovule, followed by culture on specific media may be either extremely easy to accomplish, as in case of large-seeded species in which only a single ovule is present, or time-consuming and intricate, in small-seeded polyovulate species. Two types of ovule support systems have been developed. The filter paper support system involves culturing of the ovules on top of filter paper placed over liquid medium, whereas the vermiculite support technique demands placing the ovules on a sterile vermiculite/liquid media mixture (vermiculite support) with the micropylar side down. Unpollinated ovule culture has been used for haploid production in sugar beets and onions.

Factors affecting Haploid Androgenesis

There are numerous endogeneous and exogeneous factors that affect *in vitro* haploid production. These factors can be genetic, physiological, physical and chemical may also interact amongst each other to divert the microspores/egg cell to enter into a new developmental pathway. Some of the crucial factors affecting haploid production in plants have been discussed below:

1. Genotype of the donor plant

The genotype of the donor plants, i.e. genetic factor, has a great influence on the anther, ovary and ovule culture response. In earlier studies, significant difference in callus formation using varieties or crosses were observed. In some species only a few genotypes have responded of many tested. In fact genetic factors contribute in a major way to the differences in the number of haploid plants produced.

2. Physiological status of the donor plant

The physiological conditions of the donor plant, *i.e* the environmental conditions and age of the donor plant, directly affects both *in vitro* androgenesis and *in vitro* gynogenesis in almost all plant species. A correlation between plant age and anther response has also been demonstrated. Similar is the case with ovary culture. The frequency of androgenesis is usually higher in anthers harvested at the beginning of the flowering period and showed a gradual decline in relation to plant age. Varying temperature and light conditions during the growth of donor plants also affect anther response. In anther culture of grape, the induction frequency of embryoids derived from spring flowers was higher than that derived from summer flowers. The microscopical observations showed that some varieties of rubber tree often have a lot of degenerated and sterile microspores in their anthers in early spring or hot summer due to the influence of unfavourable climatic conditions. As a result no pollen embryoids were obtained from such anthers but only the somatic calli.

Stage of explants material at the time of inoculation:

1. Stage of microspores

The stage of microspores at the time of inoculation is one of the most critical factors in the induction of androgenesis. Detailed cytological studies conducted on poplars, rubber and apple have shown that androgenic callus and embryos were mainly induced through a deviation of the first pollen mitosis to produce two undifferentiated nuclei. Besides affecting the overall response, the microspore stage at culture also has a direct bearing on the nature of plants produced in anther culture. About 80% of the embryos obtained from binucleate microspores of *Datura innoxia*, a highly androgenic species, were non-haploids. In a vast majority of species where success has been achieved, anthers were cultured when microspores were at the uninucleate stage of microsporogenesis.

2. Stage of the embryo sac

It has been reported that the effect of ovule development on gynogenesis is profound as it harbours the embryo sac comprising of the egg cell. The stage of embryo sac is an important determining factor for *in vitro* gynogenesis in various plant species. However, it is difficult to know the stage of embryo sac at the time of inoculation. Several authors prefer to describe the inoculation stage according to the developmental stage of the flower bud or stage of pollen development. However, this could not be possible in several species, where male and female gametophytes do not mature simultaneously, a phenomenon known as protandry, the maturation of anthers before carpels (e.g., onion, leek, sunflower, sugar beet and carrot) and the opposite protogyny (e.g., pearl millet). In such cases, the stage of embryo-sac at culture can be determined by histological preparations of ovary/ovules that are at identical stage with that of cultured ovary/ovules.

Although a wide range of embryo sac stages are responsive to gynogenic development, but, in most cases nearly mature embryo-sac stage gave better results. This is quite contrary to anther culture in which mature pollen is non responsive to androgenesis. In Barley and rice, unfertilized ovary cultures with late staged mature embryo sacs gave good results while others reported success with ovary cultures containing uninucleate to mature embryo sacs.

3. Anther wall factor

One of the important research subject in anther culture of woody plants is to avoid the over proliferation of callus from anther wall tissues and to achieve a high yield of pollen embryoids and pollen calli. In anther culture of most woody plants, both pollen calli (or embryoids) and somatic calli from anther wall tissues grew simultaneously. The development of callus from somatic tissues of anther can be avoided by culture of isolated microspores. However, microspore culture is not very successful in woody plants. Pelletier and Ilami had introduced the concept of "Wall Factor", according to which the somatic tissues of anther play an important role in the induction of sporophytic divisions in pollen. In anther culture of rubber, 47% of the microspores in close contact with the surrounding somatic cells could develop into multicellular masses as compared to only 5% of microspores away from the wall. Anther wall callusing was regarded as a pre-requisite for the formation of androgenic haploids.

4. Chemical factors

The constituents of the basal medium and combination/s of growth regulators serve as important factors in eliciting successful androgenesis or gynogenesis. However, it is difficult to suggest one culture medium or one growth regulator for all the systems. The requirement of growth regulators and culture medium in terms of kind and concentration may vary with the system. Generally, there is an agreement that the source and amount of total nitrogen as well as a combination of a cytokinin and an auxin is necessary for pollen embryogenesis and pollen callusing in several woody plants. Sucrose has generally been used as the major carbohydrate source in the culture medium. The effect of sucrose on anther culture has been investigated in a number of species. Generally, sucrose is supplied at 2-3% concentration. However, increase in its concentration can lead to beneficial morphogenic potential by suppressing the proliferation of somatic tissues.

5. Effect of female flower position

Position of female flowers on the plant stem affected induction of embryos from ovule cultures of *Cucurbita pepo* L. One of the possible explanations for enhancing responses of tissue culture could be attributed to indigenous hormonal level.

Diploidization of Haploids

The haploids may grow normally up to flowering stage but in the absence of homologous chromosomes the meiosis is abnormal and consequently, viable gametes are not formed. To obtain fertile homozygous diploids, the chromosome complement of the haploids must be duplicated. For long colchicine has been used for the purpose and is more effective.

Another method to diploidise the haploids is by utilizing the tendency of *in vitro* growing callus cells to undergo endomitosis. Segments from vegetative parts of haploid plants are grown on a suitable culture medium and made to proliferate into calli. After some time many of the callus cells become diploid due to endomitosis. By transferring such calli to an appropriate medium fertile diploids shoot can be obtained.

Applications of haploid production

1. Development of pure homozygous lines

Homozygous, true breeding cultivars are highly important for screening of high yielding lines and to produce hybrid vigour as a method of crop improvement. Obtaining homozygous diploid plants by conventional methods is difficult in perennials. From several decades to over a hundred years are required to obtain a pure line by means of successive inbreeding throughout many generations. The seed set by inbreeding in many trees is very low, usually only a few of ten thousandth or sometimes no seed can be obtained at all; therefore, it is impractical to obtain pure lines by inbreeding. Moreover, conventional method of haploid production by inbreeding is impossible if the plant is strictly cross-pollinating in nature. On the other hand, homozygous diploid plants can be achieved in a single generation by diploidization of *in vitro* raised haploids by colchicine treatment.

2. Genetic studies

Because of the lack of accurate materials in research work, the progress in the study of genetics in trees is much slower than that in annual herbaceous plants. The genetics of a lot of important traits in economically important plant species has not been clearly demonstrated as yet. As a result, it is still unknown whether the desirable characters of the parents will appear in their progenies. Only when crossing between different homozygous diploid plants is carried out, we can gain a clear idea of dominance of genes controlling various characters and that these characters are either monogenic or polygenic. Furthermore, if we can use the haploid plants as samples of gametes, then we can obtain directly the recombination value between genes. Moreover, we can also use the haploid plants to study chromosome homology within genome or between genomes.

3. Gametoclonal variation

The "gametoclonal variation" arises among plants regenerated from cultured gametic cells consisting of differences in morphological and biochemical characteristics as well as in chromosome number and structures that are observed. Besides yielding haploids, *in vitro* androgenesis helps in the screening of gametophytic variation at plant level. Pollen grains within an anther form a highly heterogeneous population because they are the product of meiosis which involves recombination and segregation. Therefore, each pollen plant is genetically different from the other. The gametoclonal variants being hemizygous in nature expresses also the recessive characters in the R_0 plants. Different sources of variation can explain gametoclonal variation such as new genetic variation induced by cell culture procedures, from segregation and independant assortment, chromosome doubling procedures etc.

4. Induction of mutations

In general, majority of induced mutations are recessive and, therefore, are not expressed in diploid cells due to the presence of dominant allele. Since, haploid plants have only one set of chromosomes, their dominant and recessive characters can be seen simultaneously on separate plants. It is extremely advantageous to provide a convenient system for the induction of mutations and selection of mutants with desirable traits in the absence of their dominant counterparts.

5. Obtaining new genotypes with alien chromosomes

The technique of interspecific and intergeneric hybridization can be combined with anther culture techniques for obtaining new genotypes with alien chromosomes. Thus, new genotypes with various reconstructed chromosome complements can be obtained after their successful chromosome doubling.

6. Genetic manipulation

As microspore culture is a single cell system, it makes selection at the single cell level possible and, furthermore, offers new prospects for genetic manipulation like mutagenesis and transformation. Direct gene transfer by microinjection offers the possibility of transgenic plant formation by using isolated pollen culture having high regeneration efficiency. Moreover, if transgenes can be incorporated into the haploid microspore genome, prior to DNA synthesis and chromosome doubling, the doubled haploids may also be homozygous for the transgenes. Thus, isolated microspores not only provide a good target for bombardment but, also are readily amenable to transgene *in vitro* selection. Jahne et al. were the first to achieve plants homozygous for the transgenes using biolistic bombardment of barley microspores.

7. Genomics

Doubled haploids play a vital role in genomics, especially, in the integration of genetic and physical maps, thereby, providing precision in targeting candidate genes. Doubled haploids combined with marker assisted selection provides a short cut in backcross conversion, a plant breeding method for improving an elite line defective in a particular trait. Expressed sequence tags may help in identification of genes that determine any trait of interest.

Plant Transformation Vector

Vectors are the main tool to transfer transgenes into the plant genome. Vectors for plant transformation is created using the same concept as any other plasmid-derived cloning vector. Here most important features of a standard vector will be described, as well as insight into its development and application.

1. Common Features of Plasmid-derived Cloning Vector

– *Small size:* which makes the plasmids more difficult to damage, as well as enhances the yield. Small size also means there is less chance of having restriction sites outside the multiple cloning site (MCS), which makes design and integration much easier.

– Containing a *selectable marker* gene which offers the host a selection advantage. It will be useful for the next step where successfully transformed cells need to be isolated from others. Most popular ones are gene encoding for antibiotic resistance, very often ampicilin.

– Containing a *MCS*, which has many cleavage sites for a large range of restriction enzymes.

– Contaning a *reporter gene* which enable the easy detection of colonies containing recombinant plasmids. This is usually archieved by putting the small MCS inside the reporter gene's encoding sequence, much too often *lacZ*, gene for ß-galactosidase protein. The successful insertion of the gene of interest into the plasmid results in the insertion mutagenesis of *lacZ*, effectively disabling it. Cells without working lacZ is unable to degrade X-gal (a galactosid analog) and remain white, while cells with working lacZ can degrade X-gal and release a blue substance, which turns the colony in blue. Reporter gene is not 100% reliable, but it does dramatically reduce the number of colonies which need to be screened.

– *An origin of replication (oriC)* for replication of plasmids inside *E.coli* cells

– *The gene(s) of interest*

Plasmids used for plant transformation via *Agrobacterium* need additionally:

– The ability to replicate not only in E.Coli but also in *Agrobacterium*.

– The border sequences needed for integration into host's genome.

– Additional selectable markers for further screening of successfully transformed plant cells.

– Appropriate promoters and terminators allowing (particularly prokariotic) genes to be expressed by host's eukaryotic transcription machinery.

2. Promoters and Terminators

Many plant transgenes, reporter genes, marker genes are derived from bacteria, which need their own promoters in order to be expressed. Furthermore, suitable terminators are also required to ensure that transcription is stopped at the correct positon. Promoters and terminators are often also incorporated into the plasmid.

The genes from Ti-plasmid of *Agrobacterium* for opines synthesis, especially the nopaline synthase (*nos)* gene are used as suppyling source for both promoters and terminators used in vectors. Since they are in the T-DNA, they work without problems in plants.

The most widely used promoter is however the promoter of cauliflower mosaic virus 35S RNA gene (the *35S promoter*), a very strong constitutive one. This promoter can be used to drive expression in all tissues of transgenic plants, particularly in dicotylendons. The activity of 35S promoter can even more enhanced by the inclusion of one or more copies of enhancer region. Alternatively,- maize ubiquitin I or rice actin promoters are used in monocotyledons for high-level expression

Tissue-specific promoters are used to drive the gene expression only in certains tissues. Much effort has been put into isolation of promoters with this desired characteristic, with some success.

Inducible promoters are another actively researched field in plant biotechnology. These promoters only drive gene expression under certain conditions, allowing the timing of transgene expression to be controlled closely. There are three main categories: non-plant-derived systems, plant-derived systems that respond to environmental signals, and plant-derived systems based on developmental control of gene expression.

a) Non-plant-derived systems response to the application of an exogenous chemical. The common features of inducers are:

– Specificity: no gene expression takes place without inducer or with inducers from other types.

– Gene expression should be turned on and off rapidly with application/withdrawl of the inducer.

– Inducer should be non-toxic and cause no change in gene expression.

Some of typical inducers of this class are:

– *Tetracyline*: works in tandem with bacterial *tet* operator (from *tet*-Operon coding for tetracyline resistance). Without tetracyline, the repressor protein TetR binds closely to the operator and prevents the gene expression. Upon tetracyline binding, TetR is modified and releashed from the operator, allowing the transcription to begin. The regulatory machinery used in plants is a chimeric promoter consists of a 35S promoter and several copies of *tet* operator, while TetR is overexpressed from (another) 35S promoter.

Tetracyline can also be used as an inhibitor. This time, it's not TetR but another protein, tTA acting as an activator.

The drawbacks of this system are the issues with high level of TetR inside the plant cells, and the short half-life of tetracyline which means it must be applied constantly.

– *Alcohol-inducible promoters:* consists of a core 35S promoter and binding sites for the constitutively expressed AlcR transcription factor (from the fungus *Aspergillus nidulans*). Ethanol is applied as drench or vapour. This system is not very widespread due to the activation by other metabolites, as well as the side-effects of ethanol on plants.

– *Steroid-inducible promoter:* are again chimeric from 35S promoter and a modified transcription factor capable of binding a steroid hormone or its analogue.

– *Copper-inducible promoters:* are based on the yeast metallothionein regulatory system.

Non-plant-derived systems require the application of specific substances, some of them are quite expensive, and thus are unsuitable for any agricultural scale.

b) Plant-derived systems share the same concept as non-plant-derived systems, using chimeric promoter consists of a core promoter (usually 35S promoter) and binding sites for transactivating factors. These factors can either respond to environmental signals or be only expressed in specific stages of plant development.

– Systems based on response to environmental signals: include wound-inducible, pathogen-inducible and heat-shock-inducible.

– Systems base on developmental control of gene expression: involve senescence (aging) – specific gene expression, abscisic acid-inducible gene expression, auxin-inducible gene expression.

Plant-derived systems depends heavily on normal plant processes, but they are much simpler to use in agriculture. It's worth noting that no system is suitable for all situations.

3. Selectable markers are used to allow successfully transformed tissues to be effectively isolated, usually by conferring resistance to a toxic substance. Since transformation is a very-low-frequency event, selectable markers help reduce the number or samples for screening considerably. The most common selectable markers are antibiotics resistance genes, especially the *nptII* gene (conferring resistance to kanamycin). Though plants are eukaryotic, antibiotics still inhibit the protein synthesis in organelles, particularly in chloroplasts. Antibiotics other than kanamycin are used for plant with high degree of resistance to kanamycin (cereals), or on the other end, too sensitive to it (some soft fruits). Some prominent examples are: bleomycin, spectinomycin, hygromycin. Other than antibiotics, herbicides can also be used for selection. Genes conferring resistance to herbicides such as chlorsulphuron, bialaphos are already in use.

The use of any selectable markers needs to be controlled carefully. Too high concentration of toxic substance can lead to inability to regenerate, too low concentration can lead to high number of false-positive results. Furthermore, the use of antibiotics- and hercibide-resistance genes as selectable markers raises the concern about the spreading of antibiotics-resistance gene as well as the creation of the so-called "super weed". Much effort has been put to discover new types of selectable markers. Some of them involve the abitity to use mannose and xylose as carbon-source other than sucrose ability to express green fluroescent protein (GFP), the ability to grow in absense of cytokinin to name but a few.

4. Reporter genes are used to asses the expression by promoter analysis, and as indicators of transformation. The reporter genes confer an easily detectable phenotype on transformed plant tissues, allow them to be sorted out by simple non-destructive assays. Only a small numbers of reporter genes are widely used in plant transformation vectors. Each will be described below.

a) β-glucuronidase (GUS) is perhaps the most widely used reporter gene, because of its many advantages: GUS can be assayed sensitively, quickly and easily, both quantitative and qualitative data can be obtained, and little or no endogenous activity in most plant tissues has been reported.

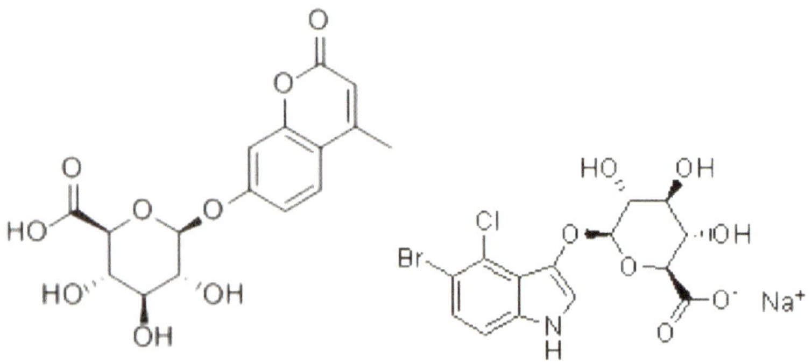

4-MUG X-gluc

GUS-assay with rice embryo

The substrate for GUS enzyme is 4-Methylumbelliferyl-b-D-glucuronide or 4-MUG. 4-MUG is hydrolysed to the flourescent 4-Methylumbelliferone (4-MU). The fluorimetrical measurement of 4-MU allows quantitative data to be obtained. Qualitative data is obtained from assay using 5-bromo-4-chloro-3-indolyl glucuronide (X-gluc). Hydrolysis with GUS results in an insoluble blue precipitation, effectively revealing the location of expression.

The *gus* gene in transformation vector is designed with an intron, allowing it only to be expressed in plants, in order to avoiding problem associated with GUS-expression in *Agrobacterium*.

b) Green fluorescent protein (GFP) has some advantages over GUS. It is easier to assayed, the assay is non-destructive, allowing some types of screening which are impossible with GUS, such as time-course experiments.

The gene is from the jellyfish Aequorea victoria, with some significantly modifications such as alternative splicing (results in shorter protein), more "plant-like" codon and prevention of accumulation in nucleoplasm.

c) Luciferases are firefly's emzymes catalyzing the ATP-dependent light emmiting oxidation of D-luciferin. Assays are highly sensitve allowing detection of low-level or highly localized gene expression, but the expression of luciferases in plants is difficult.

Other than firefly's luciferases, bacterial luciferases (LuxA and LuxB from *Vibrio harveyi*) can also be used. They catalyze the oxidation of long-chain fatty aldehydes, resulting in light emission.

d) Chloramphenicol acetyltransferase (CAT) was the first bacterial gene to be expressed in plants. It is also a widely used reporter gene in mamalian, but in plants newer reporter systems such as GUS or GFP are proved to be better.

5. Origin of replication: since both *E.coli* and *Agrobacterium* are Gram-negative bacteria, a single origin of replication with broad host range (e.g. the RK2 origin) can be used for replication in both. Unfortunately, this kind of origin often result in low plasmid copy number in E.coli, which reduces the yield and makes the DNA manipulation more difficult and time-consuming. Alternatively, two orginins are used: one for replication in *Agrobacterium* and one is from a high-copy-number cloning vector (e.g. pUC *ori*). The downside is the increase in plasmid size and with it associated problems.

6. Vector Initially, deletion derivatives of Ti plasmids, termed *co-integrative vectors*, were used for transformation. The DNA was first cloned into an intermediate vector, which was small and easy to manipulate *in vitro*, and then introduced into the *Agrobacterium* vector by means of recombination within the bacterium. Normally, the *vir* genes and T-DNA were on the same Ti plasmid.

Today, *binary vectors* are used instead. *vir* genes and T-DNA are located on separate plasmids, which allows to reduce the size of plasmids, making them easier to manipulate *in vitro*.

Gateway cloning: even modern binary vectors (e.g. pGreen, 4.6kb) are still too large, making cloning quite difficult and time-consuming. A new approach, named Gateway conversion technology, is developed by Invitrogen, an US company, using DNA recombination rather than classic cut-and-paste methods using restriction enzymes and ligases. DNA fragment that needs to be cloned is suitably flanked by recombination sites, which, in the presence of recombinase, recognize the recombination sites in the vector and insert the DNA fragment into it. The recombinase is provided by bacteriophage lambda.

7. Transgenes

Arrangement of genes in the vector: analysis of the efficiency of transgene expression allows some rules to be drawn up: 1) When there are more than one genes in the vector, each should have its own promoter and terminator, since using the same promoter / terminator for many genes can lead to gene silencing. 2) There should be enough space between genes, and genes should be located in the same orientation, to avoid adjacent inverted repeats which lead to plasmid instability and gene silencing.

Integration sites into the plant genome: The integration is basically a random process, since higher plants don't have an endogenous system for homologous recombination. Therefore, large differences in the level of transgene expression are expected. It has been tried using bacteriophage or yeast site-specific recombinases, with some success. Another approach is flanking the transgene with matrix-attachment regions (MAR), which keep the chromatin in open structure allowing gene expression.

Transgene modification differences between non-plant derived transgenes and plant genes can lead to poor level of transgene expression. Some transgene modifications are therefore needed:

– Changing the transgene's GC content.

– Removal/ replacement of the so-called cryptic introns (regions recognized by plants as introns and therefore spliced out, leading to deleterious mutation).

– Changing the upstream sequence of star codon to more plant-like.

– Adding export signal to avoid the accumulation of transgene's products in nucleoplasm.

8. Clean-gene Technology

Selectable markers are very useful in plant biotechnology, but together with them come several concerns about the health- and environmental issues such as the spreading of antibiotic- and herbicide resistance genes. It is therefore desirable to develop crops that contain no selectable markers. Some approaches are available as follows:

– Using other screening methods using no selectable markers, such as PCR. Proved to be quite laborious and expensive.

– Use selectable markers on separate T-DNA molecules.

– Excise the selectable markers out of plant genome by means of recombination.

References

- Plant-genetics-3143: maximumyield.com, Retrieved 17 May 2018
- Dna-in-plants: lifeofplant.blogspot.com, Retrieved 31 March 2018
- Plant-breeding, science: britannica.com, Retrieved 11 July 2018
- Vectors-for-plant-transformation: garung88.wordpress.com, Retrieved 19 June 2018

Chapter 5

Plant Evolution

The study of plant evolution is concerned with the investigation of the processes that resulted in the current diversity of plant life on Earth. The diverse topics elaborated in this chapter include the evolution of plants, their corresponding change in seed size and the process of polyploidy, which are vital to the understanding of plant evolution.

Modern classification systems, based largely on molecular evidence, divide living organisms into three domains: Bacteria (also called Eubacteria), Archaea, and Eukarya. Plants are classified as a kingdom (Plantae) within the Eukarya; organisms that possess a nucleus , mitochondria , an internal cytoskeleton , and, in photosynthetic species, chloroplasts. Most scientists recognize three other eukaryotic kingdoms: Protista (most of which are single-celled organisms), Fungi, and Animalia (animals). The fungi, plants, and animals are thought to have evolved from different groups of protists.

Plants are multicellular organisms that have evolved the ability to live on land. The vast majority can carry out photosynthesis, but they are not the only organisms with this ability: many protists can photosynthesize too, as can several important groups of bacteria.

Algae in Plant Evolution

Photosynthetic protists (commonly called algae) are a diverse group of organisms and are divided into several phyla. Many are unicellular, including most euglenoids (phylum Euglenophyta) and dinoflagellates (Dinophyta), and some diatoms (Bacillariophyta) and green algae (Chlorophyta). These, along with the cyanobacteria (often misleadingly called blue-green algae), form the phytoplankton of aquatic ecosystems . Others, including all brown algae (Phaeophyta), most red algae (Rhodophyta), and many green algae are multicellular. The large marine forms of these phyla are usually called seaweeds.

Plants are thought to have evolved from a class of freshwater green algae called the charophytes. Two particular groups of charophyte, the Coleochaetales and the Charales, resemble the earliest land plants (bryophytes) in a variety of ways, including the structure of their chloroplasts and sperm cells, and the way their cells divide during mitosis .

Importance of Vascular Tissue

Plants are classified into two main groups: the bryophytes (nonvascular plants) and the tracheophytes (vascular plants). Both groups have multicellular embryos, which indicates that they are closely related to each other and distinguishes them from the green algae. Indeed, true plants are often referred to as embryophytes because of this feature. The bryophytes consist of the liverworts, hornworts, and mosses, and as their name implies none of these plants possess vascular tissues.

All other plants, including the ferns, gymnosperms, and angiosperms, are classified as tracheophytes. These possess specialized vascular tissues—phloem and xylem —to transport sugars, water, and minerals throughout their bodies. The oldest known vascular plants appeared in the middle Silurian period (439–409 million years ago); the oldest known bryophytes appeared later, in the Devonian (409–354 million years ago). Despite this, most scientists believe that bryophytes evolved before vascular plants, and that the earliest bryophytes have not been found because they fossilize poorly. This belief is supported by a variety of evidence, including morphological traits, ultrastructural features visible under the electron microscope, and molecular information obtained from gene sequencing.

Bryophytes

Since bryophytes are land plants, they need to support themselves in air. However, because they lack lignified vascular tissues, this support must be provided largely by the turgor pressure of their cells. Consequently, they cannot grow to be very tall, and most bryophytes are small and rather inconspicuous. An additional important feature of their lifestyle is their reproductive system. The male gametes , produced by reproductive structures called antheridia, are free-swimming sperm cells that need water to transport them to the female gametes, which are enclosed within structures called archegonia. Because of the need for water, bryophytes are especially common in wet habitats such as bogs, streambanks, and in moist forests. However, they are not restricted to these habitats, and some mosses thrive in deserts, above the treeline, and in the Arctic tundra.

Among the living bryophytes, liverworts are probably most closely related to the earliest land plants, since unlike hornworts, mosses, and all vascular plants they do not possess stomata . Indeed, the fact that stomata first appeared in hornworts and mosses is evidence that vascular plants evolved from one of these two groups. Vascular plants appear to be more closely related to mosses than to hornworts, because some mosses possess food-conducting cells (leptoids) and water-conducting cells (hydroids) that resemble the phloem and xylem of vascular plants.

Early Vascular Plants

The first detailed vascular plant fossils appear in rocks from middle Silurian, about 425 million years ago. The oldest of these, including a plant called *Aglaophyton,* appear to have possessed conducting cells similar to the hydroids of mosses. These ancient plants, which are sometimes called prototracheophytes, may have been an evolutionary link between the bryophytes and the true tracheophytes. Early vascular plants possessed two features that made them especially well adapted to life on land. First, their vascular tissues transported sugars, nutrients, and water far more efficiently than the conducting cells of mosses. Second, they evolved the ability to synthesize lignin , which made the cell walls of their vascular tissues rigid and supportive. Taken together, these features allowed them to grow much larger than their bryophyte ancestors and considerably reduced their dependence on moist habitats.

There are three major groups of tracheophytes: seedless vascular plants, gymnosperms, and angiosperms. Since the first appearance of tracheophytes in the Silurian, the fossil record shows three major evolutionary transitions, in each of which a group of plants that were predominant before the transition is largely replaced by a different group that becomes predominant afterward. The first such transition occurred in the late Devonian, approximately 375 million years ago. Prior to this time the most common plants were simple, seedless vascular plants in various phyla, several of which are now extinct. However, one phylum from this time, the Psilophyta, still has two living genera, including a greenhouse weed called *Psilotum*.

From the late Devonian until the end of the Carboniferous period (290 million years ago) larger, more complex seedless plants were predominant. The main phyla were the Lycophyta, the Sphenophyta, and the Pterophyta. All three groups contain living relatives, including club mosses (Lycopodiaceae) in the Lycophyta, *Equisetum* (the only living genus of sphenophytes), and ferns, which are pterophytes. Only the ferns, which have about 11,000 living species, are common today, but in the Carboniferous these three phyla comprised a large fraction of the vegetation on the planet. Many grew to the size of trees and dominated the tropical and subtropical swamps that covered much of the globe at this time.

The second major transition was the decline of the lycophytes, sphenophytes, and pterophytes at the end of the Carboniferous and their replacement by gymnosperms in the early Permian. Gymnosperms dominated the vegetation of the land for the next 200 million years until they themselves began to decline and were replaced by angiosperms in the middle of the Cretaceous. Although one group of gymnosperms (the conifers) is still abundant, the angiosperms have been the most diverse and widespread group of plants on Earth for the last 100 million years.

Gymnosperms

The gymnosperms probably evolved from an extinct phylum of seedless vascular plants, the progymnosperms, that appeared about 380 million years ago. The fossils of these plants, some of which were large trees, appear to form a link between the trimerophytes (another extinct phylum of seedless vascular plants) and true gymnosperms. Progymnosperms reproduced by means of spores like the former, but their vascular tissues were very similar to those of living conifers. The oldest true gymnosperms, which produce seeds rather than spores, first appeared about 365 million years ago. The evolution of seeds, with their hard, resilient coats, was almost certainly a key factor in the success of the group. A second factor was the evolution of pollen grains to protect and transport the male gametes. As a consequence of this, gymnosperms, unlike seedless vascular plants, were no longer dependent on water for successful fertilization and could broadcast their male gametes on the wind.

Several early gymnosperm groups are now extinct, but there are four phyla with living representatives: the cycads, the gnetophytes, the conifers, and one phylum (Ginkgophyta) that has only a single living species, the ginkgo tree (*Ginkgo biloba*). Of these, the conifers are by far the most abundant and diverse, and many species are of considerable ecological and economic importance. Most conifers are well adapted to dry environments, particularly in their leaf morphology , and some can withstand severe cold. These features may have enabled them to thrive in the Permian, when Earth became much drier and colder than it had been in the Carboniferous.

Angiosperms

The angiosperms, or flowering plants, are all members of the phylum Anthophyta. There are at least 250,000 species, making the group easily the most diverse of all plant phyla. They share a number of features that distinguish them from other plant groups. The most obvious of these is the possession of flowers, highly modified shoots that carry the male and female reproductive structures. They also carry out a process called double fertilization, in which two male gametes (sperm nuclei) are released from the pollen tube into the ovule . One of these sperm nuclei fuses with an egg cell in a similar way to gymnosperms. The second nucleus (which degenerates in most gymnosperms) fertilizes other cells in the ovule called polar nuclei. Most commonly, two polar nuclei fuse with the sperm nucleus to form a triploid endosperm nucleus. The tissue that forms from this fusion is called endosperm, which in most angiosperms provides nutrients for the developing embryo.

A third feature that separates angiosperms from gymnosperms is that angiosperm embryos are protected by an ovary wall, which develops into a fruit after fertilization has taken place. In contrast, gymnosperm embryos are held relatively unprotected on the surfaces of ovule-bearing scales in the female cones.

Angiosperm Evolution

Angiosperms first appear in the fossil record about 130 million years ago, and by 90 million years ago they had become the predominant group of plants on the planet. English naturalist Charles Darwin considered the sudden appearance of angiosperms to be an "abominable mystery," and scientists have debated about the origin of the group for many years. Comparative studies of living species suggest that angiosperms evolved from the gnetophytes, a group of gymnosperms with three living genera of rather strange plants: *Ephedra, Gnetum,* and *Welwitschia.* Double fertilization has been shown to occur in both *Ephedra* and *Gnetum,* and the reproductive structures (strobili) of all three genera are similar to the flowering stalks of some angiosperms. Some gene sequencing studies also indicate that gnetophytes and angiosperms are closely related to each other and to an extinct group of gymnosperms called the Bennettitales. However, more recent molecular studies suggest that gnetophytes are more closely related to conifers than they are to angiosperms.

In 1998, the discovery of an angiosperm-like fossil called *Archaefructus,* which apparently existed 145 million years ago, also cast some doubt on the idea that angiosperms descended from gnetophytes or Bennettitales. Although a great deal of information has been obtained since the time of Darwin, the origin of angiosperms is still something of a mystery.

Early Angiosperms, Monocots, and Eudicots

The oldest known angiosperms were a diverse group of plants called magnoliids. Some of these were herbs with simple flowers; others were woody plants with more complex flowers that were very similar to living magnolias. Magnoliids, probably those with small, inconspicuous flowers, gave rise to the two main groups of angiosperms, monocots and eudicots , although a few angiosperm families, including the water lilies, may have evolved earlier.

These plants possessed a number of adaptations that were probably crucial to their eventual success. Their vascular tissues were particularly efficient, their embryos were enclosed in a protective seed coat, their leaves were resistant to desiccation , and they were pollinated by insects, rather than by the wind. This last feature made pollen transfer much more efficient and was almost certainly a key innovation in the diversification of the group, as coevolution of plants and their pollinators, particularly bees, gave rise to increasing specialization of both flowers and insects.

The orchid family contains some of the most specialized insect-pollinated flowers of all and has more species (at least 24,000) than any other plant family. Other groups of angiosperms re-evolved the ability to be pollinated by wind. One of these groups—the grasses—appeared about 50 million years ago, diversified rapidly, and became the dominant plants over many regions of the planet. They still thrive and are crucial to human well-being. Approximately 54 percent of the food eaten by people is provided by grain (seed) from cultivated varieties of just three grasses: rice, wheat, and corn.

Evolution of Life Cycles

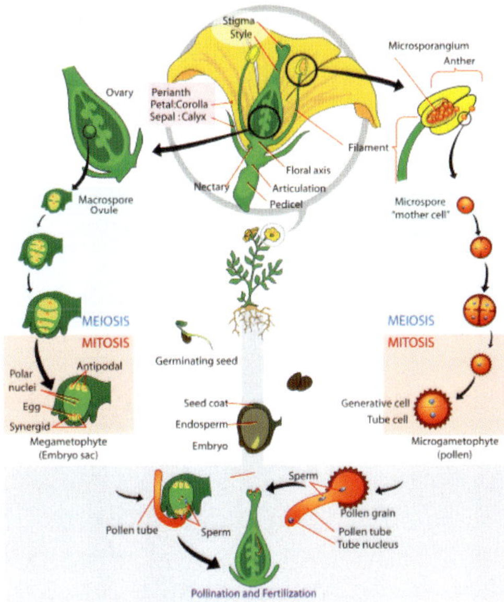

Angiosperm life cycle

All multicellular plants have a life cycle comprising two generations or phases. The gametophyte phase has a single set of chromosomes (denoted 1*n*), and produces gametes (sperm and eggs). The sporophyte phase has paired chromosomes (denoted 2*n*), and produces spores. The gametophyte and sporophyte phases may be homomorphic, appearing identical in some algae, such as *Ulva lactuca*, but are very different in all modern land plants, a condition known as heteromorphy.

The pattern in plant evolution has been a shift from homomorphy to heteromorphy. The algal ancestors of land plants were almost certainly haplobiontic, being haploid for all their life cycles, with a unicellular zygote providing the 2N stage. All land plants (i.e. embryophytes) are diplobiontic – that is, both the haploid and diploid stages are multicellular. Two trends are apparent: bryophytes (liverworts, mosses and hornworts) have developed the gametophyte as the dominant phase of the life cycle, with the sporophyte becoming almost entirely dependent on it; vascular plants have developed the sporophyte as the dominant phase, with the gametophytes being particularly reduced in the seed plants.

It has been proposed that the basis for the emergence of the diploid phase of the life cycle as the dominant phase, is that diploidy allows masking of the expression of deleterious mutations through genetic complementation. Thus if one of the parental genomes in the diploid cells contains mutations leading to defects in one or more gene products, these deficiencies could be compensated for by the other parental genome (which nevertheless may have its own defects in other genes). As the diploid phase was becoming predominant, the masking effect likely allowed genome size, and hence information content, to increase without the constraint of having to improve accuracy of replication. The opportunity to increase information content at low cost is advantageous because it permits new adaptations to be encoded. This view has been challenged, with evidence showing that selection is no more effective in the haploid than in the diploid phases of the lifecycle of mosses and angiosperms.

There are two competing theories to explain the appearance of a diplobiontic lifecycle.

The interpolation theory (also known as the antithetic or intercalary theory) holds that the interpolation of a multicellular sporophyte phase between two successive gametophyte generations was an innovation caused by preceding meiosis in a freshly germinated zygote with one or more rounds of mitotic division, thereby producing some diploid multicellular tissue before finally meiosis produced spores. This theory implies that the first sporophytes bore a very different and simpler morphology to the gametophyte they depended on. This seems to fit well with what is known of the bryophytes, in which a vegetative thalloid gametophyte nurtures a simple sporophyte, which consists of little more than an unbranched sporangium on a stalk. Increasing complexity of the ancestrally simple sporophyte, including the eventual acquisition of photosynthetic cells, would free it from its dependence on a gametophyte, as seen in some hornworts (*Anthoceros*), and eventually result in the sporophyte developing organs and vascular tissue, and becoming the dominant phase, as in the tracheophytes (vascular plants). This theory may be supported by observations that smaller *Cooksonia* individuals must have been supported by a gametophyte generation. The observed appearance of larger axial sizes, with room for photosynthetic tissue and thus self-sustainability, provides a possible route for the development of a self-sufficient sporophyte phase.

The alternative hypothesis, called the transformation theory (or homologous theory), posits that the sporophyte might have appeared suddenly by delaying the occurrence of meiosis until a fully developed multicellular sporophyte had formed. Since the same genetic material would be employed by both the haploid and diploid phases they would look the same. This explains the behaviour of some algae, such as *Ulva lactuca*, which produce alternating phases of identical sporophytes and gametophytes. Subsequent adaption to the desiccating land environment, which makes sexual reproduction difficult, might have resulted in the simplification of the sexually active gametophyte, and elaboration of the sporophyte phase to better disperse the waterproof spores. The tissue of sporophytes and gametophytes of vascular plants such as *Rhynia* preserved in the Rhynie chert is of similar complexity, which is taken to support this hypothesis. By contrast, with the exception of *Psilotum* modern vascular plants have heteromorphic sporophytes and gametophytes in which the gametophytes rarely have any vascular tissue.

Evolution of Plant Anatomy

Arbuscular Mycorrhizal Symbiosis

There is no evidence that early land plants of the Silurian and early Devonian had roots, although fossil evidence of rhizoids occurs for several species, such as *Horneophyton*. The earliest land plants did not have vascular systems for transport of water and nutrients either. *Aglaophyton*, a rootless vascular plant known from Devonian fossils in the Rhynie chertwas the first land plant discovered to have had a mycorrhizal relationship with fungi which formed arbuscules, literally "tree-like fungal roots", in a well-defined cylinder of cells (ring in cross section) in the cortex of its stems. The fungi fed on the plant's sugars, in exchange for nutrients generated or extracted from the soil (especially phosphate), to which the plant would otherwise have had no access. Like other rootless land plants of the Silurian and early Devonian *Aglaophyton* may have relied on arbuscular mycorrhizal fungi for acquisition of water and nutrients from the soil.

The fungi were of the phylum Glomeromycota, a group that probably first appeared 1 billion years ago and still forms arbuscular mycorrhizal associations today with all major land plant groups from bryophytes to pteridophytes, gymnosperms and angiosperms and with more than 80% of vascular plants.

Evidence from DNA sequence analysis indicates that the arbuscular mycorrhizal mutualism arose in the common ancestor of these land plant groups during their transition to landand it may even have been the critical step that enabled them to colonise the land. Appearing as they did before these plants had evolved roots, mycorrhizal fungi would have assisted plants in the acquisition of water and mineral nutrients such as phosphorus, in exchange for organic compounds which they could not synthesize themselves. Such fungi increase the productivity even of simple plants such as liverworts.

Xylem

To photosynthesise, plants must absorb CO_2 from the atmosphere. However, this comes at a price, since making the tissues available for CO_2 to enter allows water to evaporate. Water is lost much faster than CO_2 is absorbed, so plants need to replace it. Early land plants transported water apoplastically, within the porous walls of their cells. Later, they evolved the ability to control water loss (and CO_2 acquisition) through the use of a waterproof outer covering or cuticle perforated by stomata, variable apertures that could open and close to regulate evapotranspiration. Specialised water transport vascular tissues subsequently evolved, first in the form of hydroids, then tracheids and secondary xylem, followed by vessels in flowering plants. As water transport mechanisms and waterproof cuticles evolved, plants could survive without being continually covered by a film of water. This transition from poikilohydry to homoiohydry opened up new potential for colonisation.

The high CO_2 concentrations of the Silurian and early Devonian, when plants were first colonising land, meant that the need for water was relatively low. As CO_2 was withdrawn from the atmosphere by plants, more water was lost in its capture, and more elegant water acquisition and transport mechanisms evolved. Plants then needed a robust internal structure that contained long narrow channels for transporting water from the soil to all the different parts of the above-soil plant, especially to the parts where photosynthesis occurred. By the end of the Carboniferous, when CO_2 concentrations had been reduced to something approaching today's, around 17 times more water was lost per unit of CO_2 uptake. However, even in these "easy" early days, water was at a premium, and had to be transported to parts of the plant from the wet soil to avoid desiccation. Even today, water transport takes advantage of the cohesion-tension property of water. Water can be wicked along a fabric with small spaces, and in narrow columns of water, such as those within the plant cell walls or in tracheids, when molecules evaporate from one end, they pull the molecules behind them along the channels. Therefore, transpiration alone provides the driving force for water transport in plants. However, without dedicated transport vessels, the cohesion-tension mechanism can cause negative pressures sufficient to collapse the water conducting cells, limiting the transport water to no more than a few cm, and therefore limiting the size of the earliest plants.

A banded tube from the Late Silurian/Early Devonian. The bands are difficult to see on this specimen, as an opaque carbonaceous coating conceals much of the tube. Bands are just visible in places on the left half of the image. Scale bar: 20 μm

To be free from the constraints of small size and constant moisture that the parenchymatic transport system inflicted, plants needed a more efficient water transport system. During the early Silurian, they developed specialized xylem cells, with walls that were strengthened by bands of lignification (or similar chemical compounds) This process was followed by cell death, allowing the cell contents to be emptied and water to be passed through them. These wider, dead, empty cells, the xylem tracheids were much more conductive than the inter-cell pathway, and more resistant to collapse under the tension caused by water stress, giving the potential for transport over longer distances.

The early Devonian pretracheophytes *Aglaophyton* and *Horneophyton* have unreinforced water transport tubes with wall structures very similar to the hydroids of modern moss sporophytes, but they grew alongside several species of tracheophytes, such as *Rhynia gwynne-vaughanii* that had well-reinforced xylem tracheids. The earliest macrofossils known to have xylem tracheids are small, mid-Silurian plants of the genus *Cooksonia*. Plants continued to innovate ways of reducing the resistance to flow within their cells, thereby increasing the efficiency of their water transport. Thickened bands on the walls of tubes are apparent from the early Silurian onwards are adaptations to increase the resistance to collapse under tension. and, when they form single celled conduits, are referred to as tracheids. These, the "next generation" of transport cell design, have a more rigid structure than hydroids, preventing their collapse at higher levels of water tension. Tracheids may have a single evolutionary origin, possibly within the hornworts, uniting all tracheophytes (but they may have evolved more than once).

Water transport requires regulation, and dynamic control is provided by stomata. By adjusting the rate of gas exchange, they can restrict the amount of water lost through transpiration. This is an important role where water supply is not constant, and indeed stomata appear to have evolved before tracheids, since they are present in the sporophytes of mosses and the non-vascular hornworts.

An endodermis may have evolved in the earliest plant roots during the Devonian, but the first fossil evidence for such a structure is Carboniferous. The endodermis in the roots surrounds the water transport tissue and regulates ion exchange (and prevents unwanted pathogens etc. from entering the water transport system). The endodermis can also provide an upwards pressure, forcing water out of the roots when transpiration is not enough of a driver.

Once plants had evolved this level of controlled water transport, they were truly homoiohydric, able to extract water from their environment through root-like organs rather than relying on a film of surface moisture, enabling them to grow to much greater size. As a result of their independence from their surroundings, they lost their ability to survive desiccation – a costly trait to retain.

During the Devonian, maximum xylem diameter increased with time, with the minimum diameter remaining pretty constant. By the Middle Devonian, the tracheid diameter of some plant lineages such as the Zosterophyllophytes had plateaued. Wider tracheids allow water to be transported faster, but the overall transport rate depends also on the overall cross-sectional area of the xylem bundle itself.

While wider tracheids with robust walls make it possible to achieve higher water transport flow rates, this increases the problem of cavitation that occurs when the cohesive tension of the water column is broken, resulting in the formation of a bubble. Pits in tracheid walls have very small diameters, preventing air bubbles from passing through to adjacent tracheids., but at the cost of restricted flow rates. By the Carboniferous, Gymnosperms had developed bordered pits, valve-like structures that seal the pits when one side of a tracheid is depressurized.

Growing to height also employed another trait of tracheids – the support offered by their lignified walls. Defunct tracheids were retained to form a strong, woody stem, produced in most instances by a secondary xylem. However, in early plants, tracheids were too mechanically vulnerable, and retained a central position, with a layer of tough sclerenchyma on the outer rim of the stems. Even when tracheids do take a structural role, they are supported by sclerenchymatic tissue.

Tracheids end with walls, which impose a great deal of resistance on flow; vessel members have perforated end walls, and are arranged in series to operate as if they were one continuous vessel. The function of end walls, which were the default state in the Devonian, was probably to avoid embolisms. An embolism is where an air bubble is created in a tracheid. This may happen as a result of freezing, or by gases dissolving out of solution. Once an embolism is formed, it usually cannot be removed; the affected cell cannot pull water up, and is rendered useless.

End walls excluded, the tracheids of prevascular plants were able to operate under the same hydraulic conductivity as those of the first vascular plant, *Cooksonia*.

The size of tracheids is limited as they comprise a single cell; this limits their length, which in turn limits their maximum useful diameter to 80 μm. Conductivity grows with the fourth power of diameter, so increased diameter has huge rewards; vessel elements, consisting of a number of cells, joined at their ends, overcame this limit and allowed larger tubes to form, reaching diameters of up to 500 μm, and lengths of up to 10 m.

Vessels first evolved during the dry, low CO_2 periods of the Late Permian, in the horsetails, ferns and Selaginellales independently, and later appeared in the mid Cretaceous in angiosperms and gnetophytes. Vessels allow the same cross-sectional area of wood to transport around a hundred times more water than tracheids! This allowed plants to fill more of their stems with structural fibres, and also opened a new niche to vines, which could transport water without being as thick as the tree they grew on. Despite these advantages, tracheid-based wood is a lot lighter, thus cheaper to make, as vessels need to be much more reinforced to avoid cavitation.

Evolution of Plant Morphology

Leaves

The lycopod *Isoetes* bears microphylls (leaves with a single vascular trace).

The branching pattern of megaphyll veins may indicate their origin as webbed, dichotomising branches.

Leaf lamina. The megaphyllous leaf architecture arose multiple times in different plant lineages

Leaves are the primary photosynthetic organs of a modern plant. The origin of leaves was almost certainly triggered by falling concentrations of atmospheric CO_2 during the Devonian period, increasing the efficiency with which carbon dioxide could be captured for photosynthesis.

Leaves certainly evolved more than once. Based on their structure, they are classified into two types: microphylls, which lack complex venation and may have originated as spiny outgrowths known as enations, and megaphylls, which are large and have complex venation that may have arisen from the modification of groups of branches. It has been proposed that these structures arose independently. Megaphylls, according to Walter Zimmerman's telome theory, have evolved from plants that showed a three-dimensional branching architecture, through three transformations—overtopping, which led to the lateral position typical of leaves, planation, which involved formation of a planar architecture, webbing or fusion, which united the planar branches, thus leading to the formation of a proper leaf lamina. All three steps happened multiple times in the evolution of today's leaves.

It is widely believed that the telome theory is well supported by fossil evidence. However, Wolfgang Hagemann questioned it for morphological and ecological reasons and proposed an alternative theory. Whereas according to the telome theory the most primitive land plants have a three-dimensional branching system of radially symmetrical axes (telomes), according to Hagemann's alternative the opposite is proposed: the most primitive land plants that gave rise to vascular plants were flat, thalloid, leaf-like, without axes, somewhat like a liverwort or fern prothallus. Axes such as stems and roots evolved later as new organs. Rolf Sattler proposed an overarching process-oriented view that leaves some limited room for both the telome theory and Hagemann's alternative and in addition takes into consideration the whole continuum between dorsiventral (flat) and radial (cylindrical) structures that can be found in fossil and living land plants. This view is supported by research in molecular genetics. Thus, James concluded that "it is now widely accepted that radiality [characteristic of axes such as stems] and dorsiventrality [characteristic of leaves] are but extremes of a continuous spectrum. In fact, it is simply the timing of the KNOX gene expression."

From the point of view of the telome theory, it has been proposed that before the evolution of leaves, plants had the photosynthetic apparatus on the stems. Today's megaphyll leaves probably became commonplace some 360mya, about 40my after the simple leafless plants had colonized the land in the Early Devonian. This spread has been linked to the fall in the atmospheric carbon dioxide concentrations in the Late Paleozoic era associated with a rise in density of stomata on leaf surface. This would have resulted in greater transpiration rates and gas exchange, but especially at high CO_2 concentrations, large leaves with fewer stomata would have heated to lethal temperatures in full sunlight. Increasing the stomatal density allowed for a better-cooled leaf, thus making its spread feasible, but increased CO_2 uptake at the expense of decreased water use efficiency.

The rhyniophytes of the Rhynie chert consisted of nothing more than slender, unornamented axes. The early to middle Devonian trimerophytes may be considered leafy. This group of vascular plants are recognisable by their masses of terminal sporangia, which adorn the ends of axes which may bifurcate or trifurcate. Some organisms, such as *Psilophyton*, bore enations. These are small, spiny outgrowths of the stem, lacking their own vascular supply.

Around the same time, the zosterophyllophytes were becoming important. This group is recognisable by their kidney-shaped sporangia, which grew on short lateral branches close to the main axes. They sometimes branched in a distinctive H-shape. The majority of this group bore pronounced spines on their axes. However, none of these had a vascular trace, and the first evidence of vascularised enations occurs in the Rhynie genus *Asteroxylon*. The spines of *Asteroxylon* had a primitive vascular supply – at the very least, leaf traces could be seen departing from the central protostele towards each individual "leaf". A fossil clubmoss known as *Baragwanathia* had already appeared in the fossil record about 20 million years earlier, in the Late Silurian. In this organism, these leaf traces continue into the leaf to form their mid-vein. One theory, the "enation theory", holds that the leaves developed by outgrowths of the protostele connecting with existing enations, but it is also possible that microphylls evolved by a branching axis forming "webbing".

Asteroxylon and *Baragwanathia* are widely regarded as primitive lycopods, a group still extant today, represented by the quillworts *Isoetes*, the spikemosses and the club mosses. Lycopods bear distinctive microphylls, defined as leaves with a single vascular trace. Microphylls could grow to some size, those of Lepidodendrales reaching over a meter in length, but almost all just bear the one vascular bundle. An exception is the rare branching in some *Selaginella* species.

The more familiar leaves, megaphylls, are thought to have originated four times independently, in the ferns, horsetails, progymnosperms and seed plants. They appear to have originated by modifying dichotomising branches, which first overlapped (or "overtopped") one another, became flattened or planated and eventually developed "webbing" and evolved gradually into more leaf-like structures. Megaphylls, by Zimmerman's telome theory, are composed of a group of webbed branches and hence the "leaf gap" left where the leaf's vascular bundle leaves that of the main branch resembles two axes splitting. In each of the four groups to evolve megaphylls, their leaves first evolved during the Late Devonian to Early Carboniferous, diversifying rapidly until the designs settled down in the mid Carboniferous.

The cessation of further diversification can be attributed to developmental constraints, but why did it take so long for leaves to evolve in the first place? Plants had been on the land for at least 50 million years before megaphylls became significant. However, small, rare mesophylls are known from the early Devonian genus *Eophyllophyton* – so development could not have been a barrier to their appearance. The best explanation so far incorporates observations that atmospheric CO_2 was declining rapidly during this time – falling by around 90% during the Devonian. This required an increase in stomatal density by 100 times to maintain rates of photosynthesis. When stomata open to allow water to evaporate from leaves it has a cooling effect, resulting from the loss of latent heat of evaporation. It appears that the low stomatal density in the early Devonian meant that evaporation and evaporative cooling were limited, and that leaves would have overheated if they grew to any size. The stomatal density could not increase, as the primitive steles and limited root systems would not be able to supply water quickly enough to match the rate of transpiration. Clearly, leaves are not always beneficial, as illustrated by the frequent occurrence of secondary loss of leaves, famously exemplified by cacti and the "whisk fern" *Psilotum*.

Secondary evolution can also disguise the true evolutionary origin of some leaves. Some genera of ferns display complex leaves which are attached to the pseudostele by an outgrowth of the vascular bundle, leaving no leaf gap. Further, horsetail (*Equisetum*) leaves bear only a single vein, and appear to be microphyllous; however, both the fossil record and molecular evidence indicate that their forebears bore leaves with complex venation, and the current state is a result of secondary simplification.

Deciduous trees deal with another disadvantage to having leaves. The popular belief that plants shed their leaves when the days get too short is misguided; evergreens prospered in the Arctic circle during the most recent greenhouse earth. The generally accepted reason for shedding leaves during winter is to cope with the weather – the force of wind and weight of snow are much more comfortably weathered without leaves to increase surface area. Seasonal leaf loss has evolved independently several times and is exhibited in the ginkgoales, some pinophyta and certain angiosperms. Leaf loss may also have arisen as a response to pressure from insects; it may have been less costly to lose leaves entirely during the winter or dry season than to continue investing resources in their repair.

Factors Influencing Leaf Architectures

Various physical and physiological factors such as light intensity, humidity, temperature, wind speeds etc. have influenced evolution of leaf shape and size. High trees rarely have large leaves, because they are damaged by high winds. Similarly, trees that grow in temperate or taiga regions have pointed leaves, presumably to prevent nucleation of ice onto the leaf surface and reduce water loss due to transpiration. Herbivory, by mammals and insects, has been a driving force in leaf evolution. An example is that plants of the New Zealand genus *Aciphylla* have spines on their laminas, which probably functioned to discourage the extinct Moas from feeding on them. Other members of *Aciphylla*, which did not co-exist with the moas, do not have these spines.

At the genetic level, developmental studies have shown that repression of KNOX genes is required for initiation of the leaf primordium. This is brought about by *ARP* genes, which encode transcription factors. Repression of KNOX genes in leaf primordia seems to be quite conserved, while expression of KNOX genes in leaves produces complex leaves. The *ARP* function appears to have arisen early in vascular plant evolution, because members of the primitive group Lycophytes also have a functionally similar gene. Other players that have a conserved role in defining leaf primordia are the phytohormones auxin, gibberelin and cytokinin.

The diversity of leaves

The arrangement of leaves or phyllotaxy on the plant body can maximally harvest light and might be expected to be genetically robust. However, in maize, a mutation in only one gene called *ABPHYL (ABnormal PHYLlotaxy)* is enough to change the phyllotaxy of the leaves, implying that mutational adjustment of a single locus on the genome is enough to generate diversity.

Once the leaf primordial cells are established from the SAM cells, the new axes for leaf growth are defined, among them being the abaxial-adaxial (lower-upper surface) axes. The genes involved in defining this, and the other axes seem to be more or less conserved among higher plants. Proteins of the *HD-ZIPIII* family have been implicated in defining the adaxial identity. These proteins deviate some cells in the leaf primordium from the default abaxial state, and make them adaxial. In early plants with leaves, the leaves probably just had one type of surface — the abaxial one, the underside of today's leaves. The definition of the adaxial identity occurred some 200 million years after the abaxial identity was established.

How the wide variety of observed plant leaf morphology is generated is a subject of intense research. Some common themes have emerged. One of the most significant is the involvement of KNOX genes in generating compound leaves, as in the tomato. But, this is not universal. For example, the pea uses a different mechanism for doing the same thing. Mutations in genes affecting leaf curvature can also change leaf form, by changing the leaf from flat, to a crinkly shape, like the shape of cabbage leaves. There also exist different morphogen gradients in a developing leaf which define the leaf's axis and may also affect the leaf form. Another class of regulators of leaf development are the microRNAs.

Roots

The roots (bottom image) of Lepidodendrales (Stigmaria) are thought to be developmentally equivalent to the stems (top), as the similar appearance of "leaf scars" and "root scars" on these specimens from different species demonstrates.

Roots are important to plants for two main reasons: Firstly, they provide anchorage to the substrate; more importantly, they provide a source of water and nutrients from the soil. Roots allowed plants to grow taller and faster.

The evolution of roots had consequences on a global scale. By disturbing the soil and promoting its acidification (by taking up nutrients such as nitrate and phosphate), they enabled it to weather more deeply, injecting carbon compounds deeper into soils with huge implications for climate. These effects may have been so profound they led to a mass extinction.

While there are traces of root-like impressions in fossil soils in the Late Silurian, body fossils show the earliest plants to be devoid of roots. Many had prostrate branches that sprawled along the ground, with upright axes or thalli dotted here and there, and some even had non-photosynthetic subterranean branches which lacked stomata. The distinction between root and specialised branch is developmental. differing in their branching pattern and in possession of a root cap. So while Siluro-Devonian plants such as *Rhynia* and *Horneophyton* possessed the physiological equivalent of roots, roots – defined as organs differentiated from stems – did not arrive until later. Unfortunately, roots are rarely preserved in the fossil record, and our understanding of their evolutionary origin is sparse.

Rhizoids – Small structures performing the same role as roots, usually a cell in diameter – probably evolved very early, perhaps even before plants colonised the land; they are recognised in the Characeae, an algal sister group to land plants. That said, rhizoids probably evolved more than once; the rhizines of lichens, for example, perform a similar role. Even some animals (*Lamellibrachia*) have root-like structures. Rhizoids are clearly visible in the Rhynie chert fossils, and were present in most of the earliest vascular plants, and on this basis seem to have presaged true plant roots.

More advanced structures are common in the Rhynie chert, and many other fossils of comparable early Devonian age bear structures that look like, and acted like, roots. The rhyniophytes bore fine rhizoids, and the trimerophytes and herbaceous lycopods of the chert bore root-like structure penetrating a few centimetres into the soil. However, none of these fossils display all the features borne by modern roots. Roots and root-like structures became increasingly common and deeper penetrating during the Devonian, with lycopod trees forming roots around 20 cm long during the

Eifelian and Givetian. These were joined by progymnosperms, which rooted up to about a metre deep, during the ensuing Frasnian stage. True gymnosperms and zygopterid ferns also formed shallow rooting systems during the Famennian.

The rhizophores of the lycopods provide a slightly different approach to rooting. They were equivalent to stems, with organs equivalent to leaves performing the role of rootlets. A similar construction is observed in the extant lycopod *Isoetes*, and this appears to be evidence that roots evolved independently at least twice, in the lycophytes and other plants, a proposition supported by studies showing that roots are initiated and their growth promoted by different mechanisms in lycophytes and euphyllophytes.

A vascular system is indispensable to rooted plants, as non-photosynthesising roots need a supply of sugars, and a vascular system is required to transport water and nutrients from the roots to the rest of the plant. Rooted plants are little more advanced than their Silurian forebears, without a dedicated root system; however, the flat-lying axes can be clearly seen to have growths similar to the rhizoids of bryophytes today.

By the Middle to Late Devonian, most groups of plants had independently developed a rooting system of some nature. As roots became larger, they could support larger trees, and the soil was weathered to a greater depth. This deeper weathering had effects not only on the aforementioned drawdown of CO_2, but also opened up new habitats for colonisation by fungi and animals.

Roots today have developed to the physical limits. They penetrate as much as 60 metres of soil to tap the water table. The narrowest roots are a mere 40 µm in diameter, and could not physically transport water if they were any narrower. The earliest fossil roots recovered, by contrast, narrowed from 3 mm to under 700 µm in diameter; of course, taphonomy is the ultimate control of what thickness can be seen.

Tree Form

The trunk of early tree fern *Psaronius*, showing internal structure. The top of the plant would have been to the left of the image

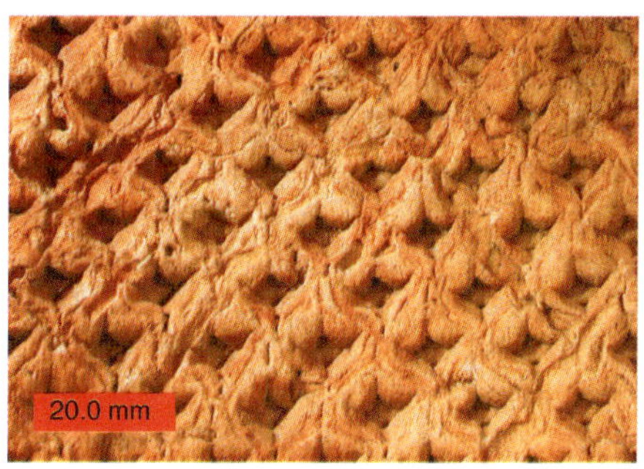

External mold of *Lepidodendron* trunk showing leaf scars from the Upper Carboniferous of Ohio

The early Devonian landscape was devoid of vegetation taller than waist height. Greater height provided a competitive advantage in the harvesting of sunlight for photosynthesis , overshadowing of competitors and in spore distribution, as spores (and, later, seeds) could be blown for greater distances if they started higher. An effective vascular system was required in order to achieve greater heights. To attain arborescence, plants had to develop woody tissue that provided both support and water transport, and thus needed to evolve the capacity for secondary growth. The stele of plants undergoing secondary growth is surrounded by a vascular cambium, a ring of meristematic cells which produces more xylem on the inside and phloem on the outside. Since xylem cells comprise dead, lignified tissue, subsequent rings of xylem are added to those already present, forming wood.

The first plants to develop secondary growth and a woody habit, were apparently the ferns, and as early as the Middle Devonian one species, *Wattieza*, had already reached heights of 8 m and a tree-like habit.

Other clades did not take long to develop a tree-like stature. The Late Devonian *Archaeopteris*, a precursor to gymnosperms which evolved from the trimerophytes, reached 30 m in height. The progymnosperms were the first plants to develop true wood, grown from a bifacial cambium. The first appearance of one of them, *Rellimia*, was in the Middle Devonian. True wood is only thought to have evolved once, giving rise to the concept of a "lignophyte" clade.

Archaeopteris forests were soon supplemented by arborescent lycopods, in the form of Lepidodendrales, which exceeded 50m in height and 2m across at the base. These arborescent lycopods rose to dominate Late Devonian and Carboniferous forests that gave rise to coal deposits. Lepidodendrales differ from modern trees in exhibiting determinate growth: after building up a reserve of nutrients at a lower height, the plants would "bolt" as a single trunk to a genetically determined height, branch at that level, spread their spores and die. They consisted of "cheap" wood to allow their rapid growth, with at least half of their stems comprising a pith-filled cavity. Their wood was also generated by a unifacial vascular cambium – it did not produce new phloem, meaning that the trunks could not grow wider over time.

The horsetail *Calamites* appeared in the Carboniferous. Unlike the modern horsetail *Equisetum*, *Calamites* had a unifacial vascular cambium, allowing them to develop wood and grow to heights in excess of 10 m and to branch repeatedly.

While the form of early trees was similar to that of today's, the Spermatophytes or seed plants, the group that contain all modern trees, had yet to evolve. The dominant tree groups today are all seed plants, the gymnosperms, which include the coniferous trees, and the angiosperms, which contain all fruiting and flowering trees. No free-sporing trees like *Archaeopteris* exist in the extant flora. It was long thought that the angiosperms arose from within the gymnosperms, but recent molecular evidence suggests that their living representatives form two distinct groups. The molecular data has yet to be fully reconciled with morphological data, but it is becoming accepted that the morphological support for paraphyly is not especially strong. This would lead to the conclusion that both groups arose from within the pteridosperms, probably as early as the Permian.

The angiosperms and their ancestors played a very small role until they diversified during the Cretaceous. They started out as small, damp-loving organisms in the understorey, and have been diversifying ever since the mid-Cretaceous, to become the dominant member of non-boreal forests today.

Seeds

Fossil seed *Trigonocarpus*

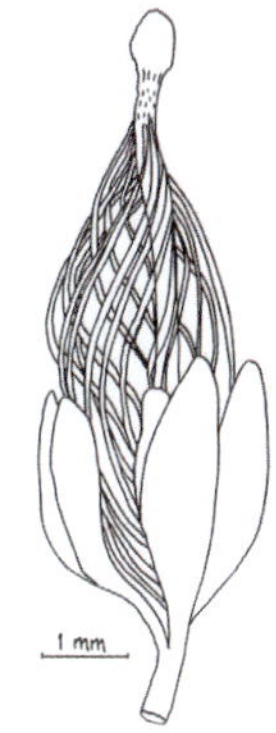

Transitional fossil *Runcaria*

Early land plants reproduced in the fashion of ferns: spores germinated into small gametophytes, which produced eggs and/or sperm. These sperm would swim across moist soils to find the female organs (archegonia) on the same or another gametophyte, where they would fuse with an egg to produce an embryo, which would germinate into a sporophyte.

Heterosporic plants, as their name suggests, bear spores of two sizes – microspores and megaspores. These would germinate to form microgametophytes and megagametophytes, respectively. This system paved the way for ovules and seeds: taken to the extreme, the megasporangia could bear only a single megaspore tetrad, and to complete the transition to true ovules, three of the megaspores in the original tetrad could be aborted, leaving one megaspore per megasporangium.

The transition to ovules continued with this megaspore being "boxed in" to its sporangium while it germinates. Then, the megagametophyte is contained within a waterproof integument, which forms the bulk of the seed. The microgametophyte – a pollen grain which has germinated from a microspore – is employed for dispersal, only releasing its desiccation-prone sperm when it reaches a receptive megagametophyte.

Lycopods and sphenopsids got a fair way down the path to the seed habit without ever crossing the threshold. Fossil lycopod megaspores reaching 1 cm in diameter, and surrounded by vegetative tissue, are known (Lepidocarpon, Achlamydocarpon);– these even germinate into a megagametophyte *in situ*. However, they fall short of being ovules, since the nucellus, an inner spore-covering layer, does not completely enclose the spore. A very small slit (micropyle) remains, meaning that the megasporangium is still exposed to the atmosphere. This has two consequences – firstly, it means it is not fully resistant to desiccation, and secondly, sperm do not have to "burrow" to access the archegonia of the megaspore.

A Middle Devonian precursor to seed plants from Belgium has been identified predating the earliest seed plants by about 20 million years. *Runcaria*, small and radially symmetrical, is an integumented megasporangium surrounded by a cupule. The megasporangium bears an unopened distal extension protruding above the multilobed integument. It is suspected that the extension was involved in anemophilous pollination. *Runcaria* sheds new light on the sequence of character acquisition leading to the seed. *Runcaria* has all of the qualities of seed plants except for a solid seed coat and a system to guide the pollen to the ovule.

The first spermatophytes (literally: "seed plants") – that is, the first plants to bear true seeds – are called pteridosperms: literally, "seed ferns", so called because their foliage consisted of fern-like fronds, although they were not closely related to ferns. The oldest fossil evidence of seed plants is of Late Devonian age, and they appear to have evolved out of an earlier group known as the progymnosperms. These early seed plants ranged from trees to small, rambling shrubs; like most early progymnosperms, they were woody plants with fern-like foliage. They all bore ovules, but no cones, fruit or similar. While it is difficult to track the early evolution of seeds, the lineage of the seed ferns may be traced from the simple trimerophytes through homosporous Aneurophytes.

This seed model is shared by basically all gymnosperms (literally: "naked seeds"), most of which encase their seeds in a woody cone or fleshy aril (the yew, for example), but none of which fully enclose their seeds. The angiosperms ("vessel seeds") are the only group to fully enclose the seed, in a carpel.

Fully enclosed seeds opened up a new pathway for plants to follow: that of seed dormancy. The embryo, completely isolated from the external atmosphere and hence protected from desiccation, could survive some years of drought before germinating. Gymnosperm seeds from the Late Carboniferous have been found to contain embryos, suggesting a lengthy gap between fertilisation and germination. This period is associated with the entry into a greenhouse earth period, with an associated increase in aridity. This suggests that dormancy arose as a response to drier climatic conditions, where it became advantageous to wait for a moist period before germinating. This evolutionary breakthrough appears to have opened a floodgate: previously inhospitable areas, such as dry mountain slopes, could now be tolerated, and were soon covered by trees.

Seeds offered further advantages to their bearers: they increased the success rate of fertilised gametophytes, and because a nutrient store could be "packaged" in with the embryo, the seeds could germinate rapidly in inhospitable environments, reaching a size where it could fend for itself more quickly. For example, without an endosperm, seedlings growing in arid environments would not have the reserves to grow roots deep enough to reach the water table before they expired from dehydration. Likewise, seeds germinating in a gloomy understory require an additional reserve of energy to quickly grow high enough to capture sufficient light for self-sustenance. A combination of these advantages gave seed plants the ecological edge over the previously dominant genus *Archaeopteris*, thus increasing the biodiversity of early forests.

Despite these advantages, it is common for fertilized ovules to fail to mature as seeds. Also during seed dormancy (often associated with unpredictable and stressful conditions) DNA damage accumulates. Thus DNA damage appears to be a basic problem for survival of seed plants, just as DNA damage is a major problem for life in general.

Flowers

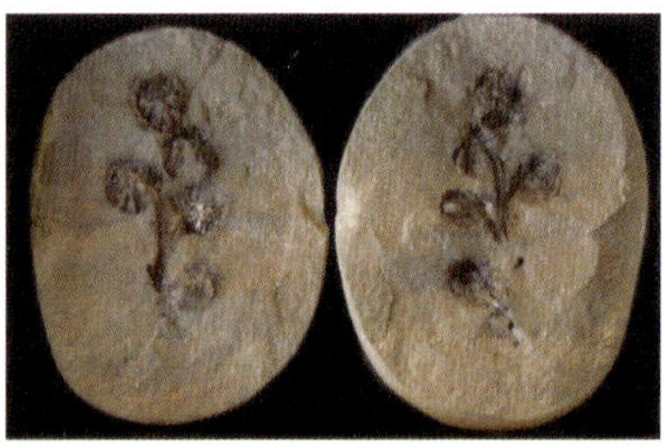

Pollen bearing organs of the early "flower" *Crossotheca*

Flowers are modified leaves possessed only by the angiosperms, which are relatively late to appear in the fossil record. The group originated and diversified during the Early Cretaceous and became ecologically significant thereafter. Flower-like structures first appear in the fossil records some ~130 mya, in the Cretaceous.

Colorful and/or pungent structures surround the cones of plants such as cycads and Gnetales, making a strict definition of the term "flower" elusive.

The main function of a flower is reproduction, which, before the evolution of the flower and angiosperms, was the job of microsporophylls and megasporophylls. A flower can be considered a powerful evolutionary innovation, because its presence allowed the plant world to access new means and mechanisms for reproduction.

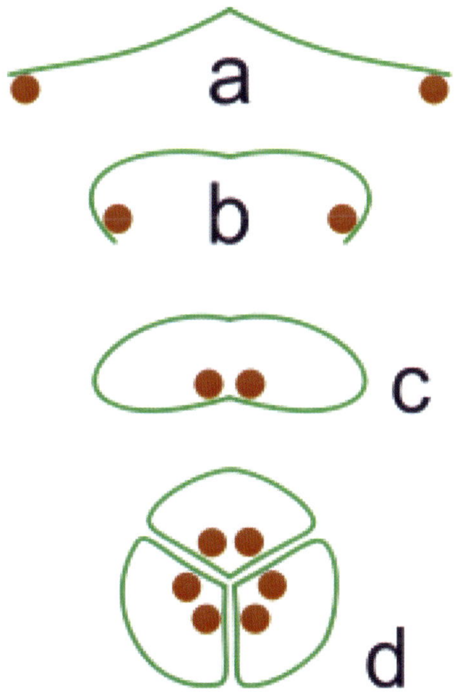

Evolution of syncarps.
a: sporangia borne at tips of leaf
b: Leaf curls up to protect sporangia
c: leaf curls to form enclosed roll
d: grouping of three rolls into a syncarp

The flowering plants have long been assumed to have evolved from within the gymnosperms; according to the traditional morphological view, they are closely allied to the Gnetales. However, as noted above, recent molecular evidence is at odds with this hypothesis, and further suggests that Gnetales are more closely related to some gymnosperm groups than angiosperms, and that extant gymnosperms form a distinct clade to the angiosperms, the two clades diverging some 300 million years ago.

The relationship of stem groups to the angiosperms is important in determining the evolution of flowers. Stem groups provide an insight into the state of earlier "forks" on the path to the current state. Convergence increases the risk of misidentifying stem groups. Since the protection of the megagametophyte is evolutionarily desirable, probably many separate groups evolved protective encasements independently. In flowers, this protection takes the form of a carpel, evolved from a leaf and recruited into a protective role, shielding the ovules. These ovules are further protected by a double-walled integument.

Penetration of these protective layers needs something more than a free-floating microgametophyte. Angiosperms have pollen grains comprising just three cells. One cell is responsible for drilling down through the integuments, and creating a conduit for the two sperm cells to flow down. The megagametophyte has just seven cells; of these, one fuses with a sperm cell, forming the nucleus of the egg itself, and another joins with the other sperm, and dedicates itself to forming a nutrient-rich endosperm. The other cells take auxiliary roles. This process of "double fertilisation" is unique and common to all angiosperms.

Inflorescences of the Bennettitales are strikingly similar to flowers

In the fossil record, there are three intriguing groups which bore flower-like structures. The first is the Permian pteridosperm *Glossopteris*, which already bore recurved leaves resembling carpels. The Mesozoic *Caytonia* is more flower-like still, with enclosed ovules – but only a single integument. Further, details of their pollen and stamens set them apart from true flowering plants.

The Bennettitales bore remarkably flower-like organs, protected by whorls of bracts which may have played a similar role to the petals and sepals of true flowers; however, these flower-like structures evolved independently, as the Bennettitales are more closely related to cycads and ginkgos than to the angiosperms.

However, no true flowers are found in any groups save those extant today. Most morphological and molecular analyses place *Amborella*, the nymphaeales and Austrobaileyaceae in a basal clade called "ANA". This clade appear to have diverged in the early Cretaceous, around 130 million years ago – around the same time as the earliest fossil angiosperm, and just after the first angiosperm-like pollen, 136 million years ago. The magnoliids diverged soon after, and a rapid radiation had produced eudicots and monocots by 125 million years ago. By the end of the Cretaceous 66 million years ago, over 50% of today's angiosperm orders had evolved, and the clade accounted for 70% of global species. It was around this time that flowering trees became dominant over conifers.

The features of the basal "ANA" groups suggest that angiosperms originated in dark, damp, frequently disturbed areas. It appears that the angiosperms remained constrained to such habitats throughout the Cretaceous – occupying the niche of small herbs early in the successional series.

This may have restricted their initial significance, but given them the flexibility that accounted for the rapidity of their later diversifications in other habitats.

Phylogeny of anthophytes and gymnosperms,

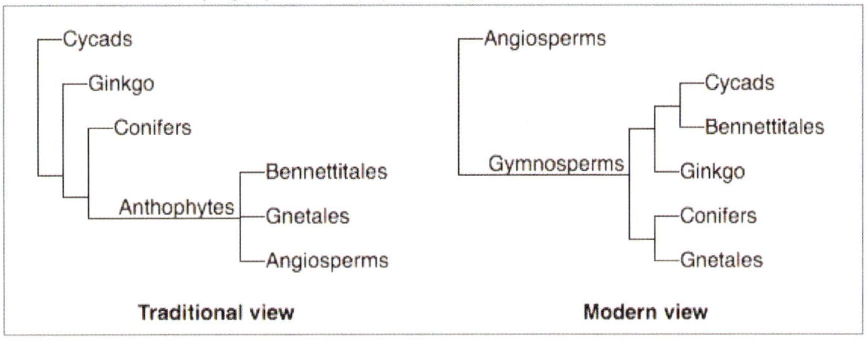

Some propose that the Angiosperms arose from an unknown Seed Fern, Pteridophyte, and view Cycads as living Seed Ferns with both Seed-Bearing and sterile leaves (Cycas revoluta).

In August 2017, scientists presented a detailed description and 3D image of a reconstruction of possibly the first flower that lived about 140 million years ago.

Origins of the Flower

A male flower of *Amborella trichopoda*. Amborellaceae is considered the sister family of all other flowering plants.

The family Amborellaceae is regarded as being the sister clade to all other living flowering plants. The complete genome of *Amborella trichopoda* is still being sequenced as of March 2012. By comparing its genome with those of all other living flowering plants, it will be possible to work out the most likely characteristics of the ancestor of *A. trichopoda* and all other flowering plants, i.e. the ancestral flowering plant.

It seems that on the level of the organ, the leaf may be the ancestor of the flower, or at least some floral organs. When some crucial genes involved in flower development are mutated, clusters of leaf-like structures arise in place of flowers. Thus, sometime in history, the developmental program leading to formation of a leaf must have been altered to generate a flower. There probably also exists an overall robust framework within which the floral diversity has been generated. An example of that is a gene called *LEAFY (LFY)*, which is involved in flower development in *Arabidopsis thaliana*. The homologs of this gene are found in angiosperms as diverse as tomato, snapdragon, pea, maize and even gymnosperms. Expression of *Arabidopsis thaliana* LFY in distant plants like poplar and citrus also results in flower-production in these plants. The *LFY* gene regulates the expression of some genes belonging to the MADS-box family. These genes, in turn, act as direct controllers of flower development.

Evolution of the MADS-box Family

The members of the MADS-box family of transcription factors play a very important and evolutionarily conserved role in flower development. According to the ABC Model of flower development, three zones — A, B and C — are generated within the developing flower primordium, by the action of some transcription factors, that are members of the MADS-box family. Among these, the functions of the B and C domain genes have been evolutionarily more conserved than the A domain gene. Many of these genes have arisen through gene duplications of ancestral members of this family. Quite a few of them show redundant functions.

The evolution of the MADS-box family has been extensively studied. These genes are present even in pteridophytes, but the spread and diversity is many times higher in angiosperms. There appears to be quite a bit of pattern into how this family has evolved. Consider the evolution of the C-region gene *AGAMOUS (AG)*. It is expressed in today's flowers in the stamens, and the carpel, which are reproductive organs. Its ancestor in gymnosperms also has the same expression pattern. Here, it is expressed in the strobili, an organ that produces pollen or ovules. Similarly, the B-genes' *(AP3 and PI)* ancestors are expressed only in the male organs in gymnosperms. Their descendants in the modern angiosperms also are expressed only in the stamens, the male reproductive organ. Thus, the same, then-existing components were used by the plants in a novel manner to generate the first flower. This is a recurring pattern in evolution.

Factors Influencing Floral Diversity

There is enormous variation in floral structure in plants, typically due to changes in the MADS-box genes and their expression pattern. For example, grasses possess unique floral structures. The carpels and stamens are surrounded by scale-like lodicules and two bracts, the lemma and the palea, but genetic evidence and morphology suggest that lodicules are homologous to eudicot petals. The palea and lemma may be homologous to sepals in other groups, or may be unique grass structures.

Another example is that of *Linaria vulgaris*, which has two kinds of flower symmetries-radial and bilateral. These symmetries are due to epigenetic changes in just one gene called *CYCLOIDEA*.

Large number of petals in roses is the result of human selection

Arabidopsis thaliana has a gene called *AGAMOUS* that plays an important role in defining how many petals and sepals and other organs are generated. Mutations in this gene give rise to the floral meristem obtaining an indeterminate fate, and proliferation of floral organs in double-flowered forms of roses, carnations and morning glory. These phenotypes have been selected by horticulturists for their increased number of petals. Several studies on diverse plants like petunia, tomato, Impatiens, maize etc. have suggested that the enormous diversity of flowers is a result of small changes in genes controlling their development.

The Floral Genome Project confirmed that the ABC Model of flower development is not conserved across all angiosperms. Sometimes expression domains change, as in the case of many monocots, and also in some basal angiosperms like *Amborella*. Different models of flower development like the *Fading boundaries model*, or the *Overlapping-boundaries model* which propose non-rigid domains of expression, may explain these architectures. There is a possibility that from the basal to the modern angiosperms, the domains of floral architecture have become more and more fixed through evolution.

Flowering Time

Another floral feature that has been a subject of natural selection is flowering time. Some plants flower early in their life cycle, others require a period of vernalization before flowering. This outcome is based on factors like temperature, light intensity, presence of pollinators and other environmental signals: genes like *CONSTANS (CO)*, *Flowering Locus C* (*FLC*) and *FRIGIDA* regulate integration of environmental signals into the pathway for flower development. Variations in these loci have been associated with flowering time variations between plants. For example, *Arabidopsis thaliana* ecotypes that grow in the cold, temperate regions require prolonged vernalization before they flower, while the tropical varieties, and the most common lab strains, don't. This variation is due to mutations in the *FLC* and *FRIGIDA* genes, rendering them non-functional.

Many of the genes involved in this process are conserved across all the plants studied. Sometimes though, despite genetic conservation, the mechanism of action turns out to be different. For example, rice is a short-day plant, while *Arabidopsis thaliana* is a long-day plant. Both plants have the proteins *CO* and *FLOWERING LOCUS T (FT)*, but, in *Arabidopsis thaliana*, *CO* enhances *FT* production, while in rice, the *CO* homolog represses *FT* production, resulting in completely opposite downstream effects.

Theories of Flower Evolution

The Anthophyte theory was based on the observation that a gymnospermic group Gnetales has a flower-like ovule. It has partially developed vessels as found in the angiosperms, and the megasporangium is covered by three envelopes, like the ovary structure of angiosperm flowers. However, many other lines of evidence show that Gnetales is not related to angiosperms.

The *Mostly Male theory* has a more genetic basis. Proponents of this theory point out that the gymnosperms have two very similar copies of the gene *LFY*, while angiosperms just have one. Molecular clock analysis has shown that the other *LFY* paralog was lost in angiosperms around the same time as flower fossils become abundant, suggesting that this event might have led to floral evolution. According to this theory, loss of one of the *LFY* paralog led to flowers that were more male, with the ovules being expressed ectopically. These ovules initially performed the function of attracting pollinators, but sometime later, may have been integrated into the core flower.

Mechanisms and Players in Evolution of Plant Morphology

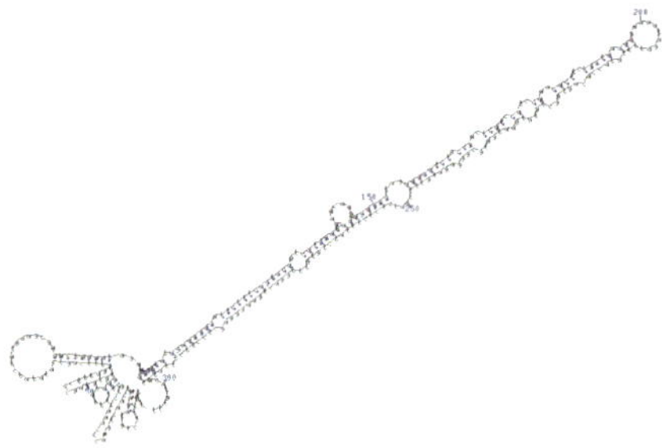

Stem-loop secondary structure of a pre-microRNA from *Brassica oleracea*

While environmental factors are significantly responsible for evolutionary change, they act merely as agents for natural selection. Change is inherently brought about via phenomena at the genetic level: mutations, chromosomal rearrangements, and epigenetic changes. While the general types of mutations hold true across the living world, in plants, some other mechanisms have been implicated as highly significant.

Genome doubling is a relatively common occurrence in plant evolution and results in polyploidy, which is consequently a common feature in plants. It is estimated that at least half (and probably all) plants have seen genome doubling in their history. Genome doubling entails gene duplication, thus generating functional redundancy in most genes. The duplicated genes may attain new function, either by changes in expression pattern or changes in activity. Polyploidy and gene duplication are believed to be among the most powerful forces in evolution of plant form; though it is not known why genome doubling is such a frequent process in plants. One probable reason is the production of large amounts of secondary metabolites in plant cells. Some of them might interfere in the normal process of chromosomal segregation, causing genome duplication.

Top: teosinte, bottom: maize, middle: maize-teosinte hybrid

In recent times, plants have been shown to possess significant microRNA families, which are conserved across many plant lineages. In comparison to animals, while the number of plant miRNA families are lesser than animals, the size of each family is much larger. The miRNA genes are also much more spread out in the genome than those in animals, where they are more clustered. It has been proposed that these miRNA families have expanded by duplications of chromosomal regions. Many miRNA genes involved in regulation of plant development have been found to be quite conserved between plants studied.

Domestication of plants like maize, rice, barley, wheat etc. has also been a significant driving force in their evolution. Research concerning the origin of maize has found that it is a domesticated derivative of a wild plant from Mexico called teosinte. Teosinte belongs to the genus *Zea*, just as maize, but bears very small inflorescence, 5–10 hard cobs and a highly branched and spread out stem.

Cauliflower – *Brassica oleracea* var. *botrytis*

Crosses between a particular teosinte variety and maize yields fertile offspring that are intermediate in phenotype between maize and teosinte. QTL analysis has also revealed some loci that, when mutated in maize, yield a teosinte-like stem or teosinte-like cobs. Molecular clock analysis of these genes estimates their origins to some 9,000 years ago, well in accordance with other records of maize domestication. It is believed that a small group of farmers must have selected some maize-like natural mutant of teosinte some 9,000 years ago in Mexico, and subjected it to continuous selection to yield the familiar maize plant of today.

The edible cauliflower is a domesticated version of the wild plant *Brassica oleracea*, which does not possess the dense undifferentiated inflorescence, called the curd, that cauliflower possesses.

Cauliflower possesses a single mutation in a gene called *CAL*, controlling meristem differentiation into inflorescence. This causes the cells at the floral meristem to gain an undifferentiated identity and, instead of growing into a flower, they grow into a dense mass of inflorescence meristem cells in arrested development. This mutation has been selected through domestication since at least the time of the Greek empire.

Evolution of Photosynthetic Pathways

The C_4 metabolic pathway is a valuable recent evolutionary innovation in plants, involving a complex set of adaptive changes to physiology and gene expression patterns.

Photosynthesis is a complex chemical pathway facilitated by a range of enzymes and co-enzymes. The enzyme RuBisCO is responsible for “fixing” CO_2 – that is, it attaches it to a carbon-based molecule to form a sugar, which can be used by the plant, releasing an oxygen molecule. However, the enzyme is notoriously inefficient, and, as ambient temperature rises, will increasingly fix oxygen instead of CO_2 in a process called photorespiration. This is energetically costly as the plant has to use energy to turn the products of photorespiration back into a form that can react with CO_2.

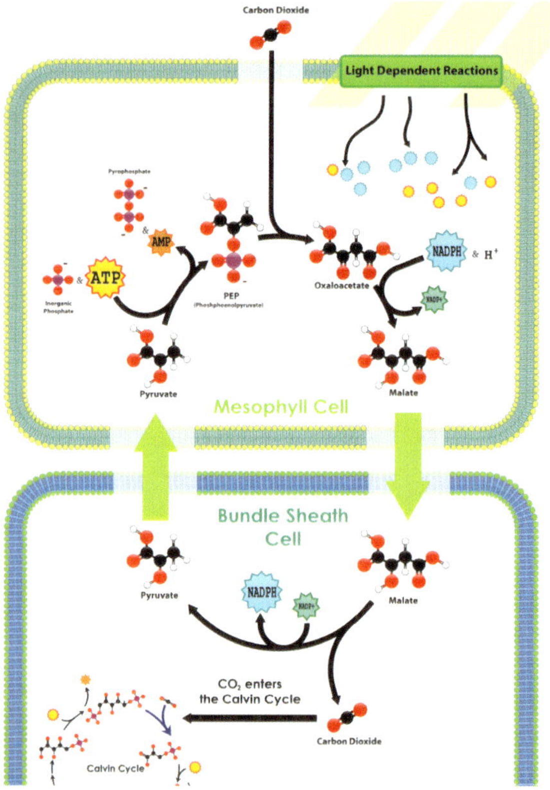

The C_4 carbon concentrating mechanism

Concentrating Carbon

C_4 plants evolved carbon concentrating mechanisms that work by increasing the concentration of CO_2 around RuBisCO, and excluding oxygen, thereby increasing the efficiency of photosynthesis by decreasing photorespiration. The process of concentrating CO_2 around RuBisCO requires more energy than allowing gases to diffuse, but under certain conditions – i.e. warm temperatures (>25 °C), low CO_2 concentrations, or high oxygen concentrations – pays off in terms of the decreased loss of sugars through photorespiration.

One type of C_4 metabolism employs a so-called Kranz anatomy. This transports CO_2 through an outer mesophyll layer, via a range of organic molecules, to the central bundle sheath cells, where the CO_2 is released. In this way, CO_2 is concentrated near the site of RuBisCO operation. Because RuBisCO is operating in an environment with much more CO_2 than it otherwise would be, it performs more efficiently.

A second mechanism, CAM photosynthesis, temporally separates photosynthesis from the action of RuBisCO. RuBisCO only operates during the day, when stomata are sealed and CO_2 is provided by the breakdown of the chemical malate. More CO_2 is then harvested from the atmosphere when stomata open, during the cool, moist nights, reducing water loss.

Evolutionary Record

These two pathways, with the same effect on RuBisCO, evolved a number of times independently – indeed, C_4 alone arose 62 times in 18 different plant families. A number of 'pre-adaptations' seem to have paved the way for C4, leading to its clustering in certain clades: it has most frequently been innovated in plants that already had features such as extensive vascular bundle sheath tissue. Many potential evolutionary pathways resulting in the C_4 phenotype are possible and have been characterised using Bayesian inference, confirming that non-photosynthetic adaptations often provide evolutionary stepping stones for the further evolution of C_4.

The C_4 construction is used by a subset of grasses, while CAM is employed by many succulents and cacti. The C_4 trait appears to have emerged during the Oligocene, around 25 to 32 million years ago; however, they did not become ecologically significant until the Miocene, 6 to 7 million years ago. Remarkably, some charcoalified fossils preserve tissue organised into the Kranz anatomy, with intact bundle sheath cells, allowing the presence C_4 metabolism to be identified. Isotopic markers are used to deduce their distribution and significance. C_3 plants preferentially use the lighter of two isotopes of carbon in the atmosphere, ^{12}C, which is more readily involved in the chemical pathways involved in its fixation. Because C_4 metabolism involves a further chemical step, this effect is accentuated. Plant material can be analysed to deduce the ratio of the heavier ^{13}C to ^{12}C. This ratio is denoted $\delta^{13}C$. C_3 plants are on average around 14‰ (parts per thousand) lighter than the atmospheric ratio, while C_4 plants are about 28‰ lighter. The $\delta^{13}C$ of CAM plants depends on the percentage of carbon fixed at night relative to what is fixed in the day, being closer to C_3 plants if they fix most carbon in the day and closer to C_4 plants if they fix all their carbon at night.

Original fossil material in sufficient quantity to analyse the grass itself is scarce, but horses provide a good proxy. They were globally widespread in the period of interest, and browsed almost exclusively on grasses. There's an old phrase in isotope paleontology, "you are what you eat (plus a little bit)" – this refers to the fact that organisms reflect the isotopic composition of whatever they eat, plus a small adjustment factor. There is a good record of horse teeth throughout the globe, and their $\delta^{13}C$ record shows a sharp negative inflection around 6 to 7 million years ago, during the Messinian that is interpreted as resulting from the rise of C_4 plants on a global scale.

Advantages of C_4 Plant

While C_4 enhances the efficiency of RuBisCO, the concentration of carbon is highly energy intensive. This means that C_4 plants only have an advantage over C_3 organisms in certain conditions: namely, high temperatures and low rainfall. C_4 plants also need high levels of sunlight to thrive. Models suggest that, without wildfires removing shade-casting trees and shrubs, there would be no space for C_4 plants. But, wildfires have occurred for 400 million years – why did C_4 take so long to arise, and then appear independently so many times? The Carboniferous (~300 million years ago) had notoriously high oxygen levels – almost enough to allow spontaneous combustion – and very low CO_2, but there is no C_4 isotopic signature to be found. And there doesn't seem to be a sudden trigger for the Miocene rise.

During the Miocene, the atmosphere and climate were relatively stable. If anything, CO_2 increased gradually from 14 to 9 million years ago before settling down to concentrations similar to the Holocene. This suggests that it did not have a key role in invoking C_4 evolution. Grasses themselves (the group which would give rise to the most occurrences of C_4) had probably been around for 60 million years or more, so had had plenty of time to evolve C_4, which, in any case, is present in a diverse range of groups and thus evolved independently. There is a strong signal of climate change in South Asia; increasing aridity – hence increasing fire frequency and intensity – may have led to an increase in the importance of grasslands. However, this is difficult to reconcile with the North American record. It is possible that the signal is entirely biological, forced by the fire- (and elephant?)- driven acceleration of grass evolution – which, both by increasing weathering and incorporating more carbon into sediments, reduced atmospheric CO_2 levels. Finally, there is evidence that the onset of C_4 from 9 to 7 million years ago is a biased signal, which only holds true for North America, from where most samples originate; emerging evidence suggests that grasslands evolved to a dominant state at least 15Ma earlier in South America.

Evolution of Transcriptional Regulation

Transcription factors and transcriptional regulatory networks play key roles in plant development and stress responses, as well as their evolution. During plant landing, many novel transcription factor families emerged and are preferentially wired into the networks of multicellular development, reproduction, and organ development, contributing to more complex morphogenesis of land plants.

Evolution of Secondary Metabolism

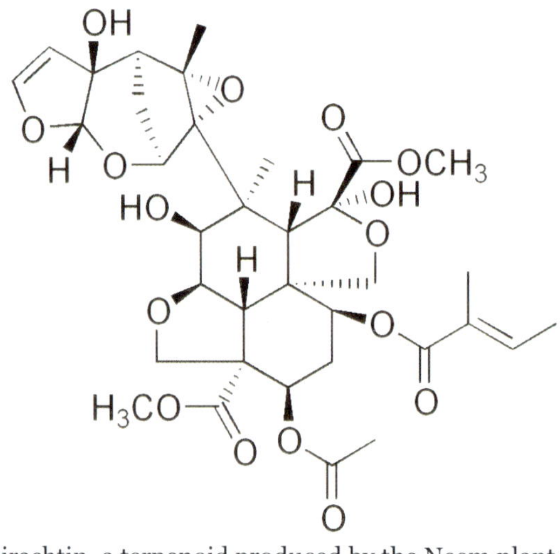

Structure of Azadirachtin, a terpenoid produced by the Neem plant, which helps ward off microbes and insects. Many secondary metabolites have complex structures

Secondary metabolites are essentially low molecular weight compounds, sometimes having complex structures, that are not essential for the normal processes of growth, development, or reproduction. They function in processes as diverse as immunity, anti-herbivory, pollinator attraction, communication between plants, maintaining symbiotic associations with soil flora, or enhancing the rate of fertilization, and hence are significant from the evo-devo perspective. Secondary metabolites are structurally and functionally diverse, and it is estimated that hundreds of thousands of enzymes might be involved in the process of producing them, with about 15–25% of the genome coding for these enzymes, and every species having its unique arsenal of secondary metabolites. Many of these metabolites, such as salicylic acid are of medical significance to humans.

The purpose of producing so many secondary metabolites, with a significant proportion of the metabolome devoted to this activity is unclear. It is postulated that most of these chemicals help in generating immunity and, in consequence, the diversity of these metabolites is a result of a constant arms race between plants and their parasites. Some evidence supports this case. A central question involves the reproductive cost to maintaining such a large inventory of genes devoted to producing secondary metabolites. Various models have been suggested that probe into this aspect of the question, but a consensus on the extent of the cost has yet to be established; as it is still difficult to predict whether a plant with more secondary metabolites increases its survival or reproductive success compared to other plants in its vicinity.

Secondary metabolite production seems to have arisen quite early during evolution. In plants, they seem to have spread out using mechanisms including gene duplications or the evolution of novel genes. Furthermore, research has shown that diversity in some of these compounds may be positively selected for. Although the role of novel gene evolution in the evolution of secondary metabolism is clear, there are several examples where new metabolites have been formed by small changes in the reaction. For example, cyanogen glycosides have been proposed to have evolved multiple times in different plant lineages. There are several such instances of convergent evolution. For example, enzymes for synthesis of limonene – a terpene – are more similar between angiosperms and gymnosperms than to their own terpene synthesis enzymes. This suggests independent evolution of the limonene biosynthetic pathway in these two lineages.

Coevolution of Plants and Fungal Parasites

An additional contributing factor in some plants leading to evolutionary change is the force due to coevolution with fungal parasites. In an environment with a fungal parasite, which is common in nature, the plants must make adaptation in an attempt to evade the harmful effects of the parasite.

Whenever a parasitic fungus is siphoning limited resources away from a plant, there is selective pressure for a phenotype that is better able to prevent parasitic attack from fungi. At the same time, fungi that are better equipped to evade the defenses of the plant will have greater fitness level. The combination of these two factors leads to an endless cycle of evolutionary change in the host-pathogen system.

Because each species in the relationship is influenced by a constantly changing symbiont, evolutionary change usually occurs at a faster pace than if the other species was not present. This is true of most instances of coevolution. This makes the ability of a population to quickly evolve vital to its survival. Also, if the pathogenic species is too successful and threatens the survival and reproductive success of the host plants, the pathogenic fungi risk losing their nutrient source for future generations. These factors create a dynamic that shapes the evolutionary changes in both species generation after generation.

Genes that code for defense mechanisms in plants must keep changing to keep up with the parasite that constantly works to evade the defenses. Genes that code for attachment mechanisms are the most dynamic and are directly related to the evading ability of the fungi. The greater the changes in these genes, the more change in the attachment mechanism. After selective forces on the resulting phenotypes, evolutionary change that promotes evasion of host defenses occurs.

Fungi not only evolve to avoid the defenses of the plants, but they also attempt to prevent the plant from enacting the mechanisms to improve its defenses. Anything the fungi can do to slow the evolution process of the host plants will improve the fitness of future generations because the plant will not be able to keep up with the evolutionary changes of the parasite. One of the main processes by which plants quickly evolve in response to the environment is sexual reproduction. Without sexual reproduction, advantageous traits could not be spread through the plant population as quickly allowing the fungi to gain a competitive advantage. For this reason, the sexual reproductive organs of plants are targets for attacks by fungi. Studies have shown that many different current types of obligate parasitic plant fungi have developed mechanisms to disable or otherwise affect the sexual reproduction of the plants. If successful, the sexual reproduction process slows for the plant, thus slowing down evolutionary change or in extreme cases, the fungi can render the plant sterile creating an advantage for the pathogens. It is unknown exactly how this adaptive trait developed in fungi, but it is clear that the relationship to the plant forced the development of the process.

Some researchers are also studying how a range of factors affect the rate of evolutionary change and the outcomes of change in different environments. For example, as with most evolution, increases in heritability in a population allow for a greater evolutionary response in the presence of selective pressure. For traits specific to the plant-fungi coevolution, researchers have studied how the virulence of the invading pathogen affects the coevolution. Studies involving *Mycosphaerella graminicola* have consistently showed that virulence of a pathogen does not have a significant impact on the evolutionary track of the host plant.

There can be other factors in that can affect the process of coevolution. For example, in small populations, selection is a relatively weaker force on the population due to genetic drift. Genetic drift increases the likelihood of having fixed alleles which decreases the genetic variance in the population. Therefore, if there is only a small population of plants in an area with the ability to reproduce together, genetic drift may counteract the effects of selection putting the plant in a disadvantageous position to fungi which can evolve at a normal rate. The variance in both the host and pathogen population is a major determinant of evolutionary success compared to the other species. The greater the genetic variance, the faster the species can evolve to counteract the other organism's avoidance or defensive mechanisms.

Due to the process of pollination for plants, the effective population size is normally larger than for fungi because pollinators can link isolated populations in a way that the fungus is not able. This means positive traits that evolve in non-adjacent but close areas can be passed to nearby areas. Fungi must individually evolve to evade host defenses in each area. This is obviously a clear competitive advantage for the host plants. Sexual reproduction with a broad, high variance population leads to fast evolutionary change and higher reproductive success of offspring.

Environment and climate patterns also play a role in evolutionary outcomes. Studies with oak trees and an obligate fungal parasite at different altitudes clearly show this distinction. For the same species, different altitudinal positions had drastically different rates of evolution and changes in the response to the pathogens due to the organism also in a selective environment due to their surroundings.

Coevolution is a process that is related to the red queen hypothesis. Both the host plant and parasitic fungi have to continue to survive to stay in their ecological niche. If one of the two species in the relationship evolves at a significantly faster rate than the other, the slower species will be at a competitive disadvantage and risk the loss of nutrients. Because the two species in the system are so closely linked, they respond to external environment factors together and each species affects the evolutionary outcome of the other. In other words, each species exerts selective pressure on the other. Population size is also a major factor in the outcome because differences in gene flow and genetic drift could cause evolutionary changes that do not match the direction of selection expected by forces due to the other organism. Coevolution is an important phenomenon necessary for understanding the vital relationship between plants and their fungal parasites.

Evolution of Seed Size

Seed mass affects many aspects of plant ecology. Small-seeded species are able to produce more seeds for a given amount of energy than are large-seeded species, whereas large-seeded species have seedlings that are better able to tolerate many of the stresses encountered during seedling establishment. Seed mass is also correlated with the environmental conditions under which species establish and with traits such as plant size, dispersal syndrome, plant life-span, and the ability to form a persistent seed bank. Present-day species have seed masses ranging over 11.5 orders of magnitude, from the dust-like seeds of orchids (some of which weigh just 0.0001 mg) to the 20-kg seeds of the double coconut. Improving our knowledge of the changes in seed mass that occurred as the angiosperms radiated out of the tropics, colonized a wide range of habitats, developed a range of growth forms and dispersal strategies, and became the most abundant and diverse group of plants on earth will greatly enhance our understanding of the ecological history of plants.

Mechanism

Modern seed sizes range from 0.0001 mg in orchid seeds to 42 kg in double coconuts. Larger seeds have larger quantities of metabolic reserves in their embryo and endosperm available for the seedling than smaller seeds, and often aid establishment under low resource availability. However, smaller seeds can be produced in larger quantities which has the potential to produce more offspring and have better chances of some of the seeds dispersing into suitable habitat. This seed size-number trade off has led to the evolution of a wide range in size and number of seeds in response to environmental selection pressures.

Selective Pressures

No single event, such as a large divergence in the phylogeny of seeded plants, is seen as the cause of major divergences in seed size. Rather, small events are thought to occur fairly consistently through time with minor evolutionary influence.

Shade

Species growing in shaded environments tend to produce larger seeds and larger seeded species have higher seedling survivorship in low-light conditions. The increased metabolic reserves of larger seeds allows the first shoots to grow taller and leaves to grow broader more quickly in order to compete for what little sunlight is available. A few large seeded trees that occur in closed canopy wooded areas such as old-growth forests are the many oak species, hickory, pecan, and butternut trees.

Drought

Small seeds are seen to be predominant in arid, desert environments. In some desert systems the vast majority of annual seeds weigh between zero and two milligrams. small seed size may be a favorable adaptation in dessert plants for a couple reasons. Small seeds have been found to have the ability to store in dry environments for several years without desiccating. Also, in many cases, deserts have rainy seasons that provide opportunity for small seeds to germinate under conditions with ample external resources available. Due to the great importance that seeds germinate when water is available, seeds often sense the presence of water and use it as a cue to germinate. Also, many desert plants have evolved the ability to produce a fraction of their seeds to not germinate at the same time as the rest of the plant's seeds as a safe guard known as bet hedging in which if the majority of a plant's seeds germinate at one time and then die due to rain followed by drought, the potential for the plant to have successful offspring is not completely lost.

Predation

Granivors (those that feed on seeds and grains) can selectively eat either smaller or larger seeds, favoring seeds on the opposite side of the spectrum. Commonly, granivorous predation by rodents, which selectively feed on larger seeds, leads to higher fitness of smaller seeds (e.g. kangaroo rats in desert systems selectively forage on the larger seeds in the seed bank. Similarly, sometimes smaller seeds are selectively preyed upon such as with Australian granivorous ants which are only capable of carrying smaller seeds.

Seedling-Seedling Competition

Competition between seedlings for limited resources can result selective pressures on seed size. In dense mats of competing seedlings, those from larger seeds have higher survivorship due to their ability to more quickly grow taller shoots, broader leaves, and thus out-compete smaller seeded seedlings for resources. Germinated seedlings from larger seeds could also possibly outlive the smaller seeded seedlings which cannot live as long off their stored energy reserves.

Optimal Dormancy Depth

If there is a selective pressure favoring the survival of seeds buried deeper in the soil, larger seed size may evolve because of their larger reserves of energy required to emerge from further depths. One such pressure causing this type of selection is the recurrence of fires (e.g. in prairies the heat from a fire can damage or kill seeds near the surface of the soil but leave seeds buried deeper unharmed).

Dispersal

The smaller the seed, the further they can disperse, which can be beneficial for avoiding competition with siblings and the parent as well as having better chances of some of the seeds dispersing into suitable habitat. Dispersal may also lead to greater fitness in future generations if further dispersed individuals are more likely to cross pollinate with an unrelated individuals, leading to greater genetic variation. The type of seed dispersal evolved has been highly correlated to seed size in floras across the world. In general, seeds smaller than 0.1 mg are often unassisted (wind dispered), seeds larger than 100 mg are often dispersed by vertebrates or by water, and seeds between 0.1 and 100 mg are dispersed by a large variety of dispersal modes including dispersal by a great variety of animals.

Evolution of Seed Size

Since the evolution of the first seeded plants ~370 million years ago, the largest change in seed size was found to be at the divergence of gymnosperms and angiosperms ~325 million years ago, but overall, the divergence of seed size appears to take place relatively consistently through evolutionary time. Seed mass has been found to be phylogenetically conservativewith most differences in mean seed mass within types of seed dispersal (dispersal modes) being phylogenetic. This type of information gives us clues about how seed size evolved. Dating fossilized seeds of various sizes and comparing them with the presence of possible animal dispersers and the environmental conditions of the time is another technique used to study the evolution of seed size. Environmental conditions appear to have had a larger influence on the evolution of seed size compared to the presence of animal dispersers. One example of seed size evolving to environmental conditions is thought to have been abundant, closed forest vegetation selecting for larger seed sizes during the Eocene epoch. A general increase or decrease in seed size through time has not been found, but instead a fluctuation in seed size following the environmental conditions of the Maastrichtian, Paleocene, Eocene, Oligocene, Miocene, and Pliocene epochs. Today we also see a pattern with seed size distribution and global environmental conditions where the largest mean seed size is found in tropical forests and a steep decrease in seed size takes places globally as vegetation type changes to non-forest.

Polyploidy

An organism having more than two sets of homologous chromosomes is known as polyploid and the phenomenon polyploidy. It was discovered by Lutz. It is rarely found in animals but is of general occurrence in plants. A survey of the chromosome numbers of the species in a family shows that these species generally fall in to polyploid series e.g., 2n, 3n, 4n, 5n, 6n etc.

The species are grouped together under a taxonomic head due to certain morphological similarities and relationships. They may or may not be crossable at all, with one another. However, the chromosome numbers of the species show a general relationship i.e., they form a multiple of common basic number like 7 in Triticum, Hordeum, Pisum, Lathyrus.

Thus, the different species of Solanaceae form a polyploid series with a basic chromosome number of 12.

The basic chromosome number is usually expressed by the symbol x. Examples of Oryza are:

Oryza, x = 12

Oryza glaberrima = 24

Oryza sativa = 24

Oryza coarctata = 48

Oryza perennis = 24

Oryza officinalis = 24

Types of Polyploids

A typical diploid is rare in nature. Most of the so-called diploids are really polyploids but their behavior is like diploids. Many of our cultivated plants are considered diploids but actually these are polyploids of two types namely primary and secondary polyploids.

1. Primary Polyploids:

Primary polyploids are of two types:

(i) Autopolyploid

(ii) Allopolyploids which may be distinguished on the basis of the source of the chromosome.

(i) Allopolyploids:

Autopolyploid are those polyploids, which have the same basic set of chromosomes multiplied. For example, if a diploid species has two similar sets of chromosomes or genomes (AA) an auto-tri-polyploid will have three similar genomes (AAA) and an autotetraploid will have four such genomes (AAAA). In most cases tetraploids are normal.

One of the most important effects of auto-polyploidy is that it often reduces fertility. Morphologically the tetraploids are larger and vigorous than the normal diploid plants. Autotetraploid condition occurs when diploid gametes fuse. One of the very common examples of natural autoploidy relevant to Northern India pertains to 'doob' grass (Cynodon dactylon).

Auto-triploids are also known in water lemons, sugar beet, tomato, banana, grapes etc. Likewise, auto-tetraploids are known in rye (Secale cereale), berseem (Trifolium alexandrium), marigolds (Tagetes), corn (Zea mays), apples, Phlox and Oenothera lamarckiana (an American plant, a giant mutant), etc.

(a) Auto-triploids:

These are some times more vigorous than diploids. These are comparatively more leafy and tending to perenniality. In some cases, disturbances may be present in floral parts. The plants usually are highly sterile and rarely any seeds set.

In nature, seeds propagated triploids do not occur. On the other hand, vegetatively propagated plants can exist as triploids. This advantage is taken of in Horticulture specially for ornamental plants. It was first reported by Gates in Oenothera.

Cytological behavior of autopolyploid is briefly discussed as below:

(i) No irregularities are noticed in somatic divisions. At telophase 3 nucleoli are formed.

(ii) Due to abnormalities in meiotic cell division, functional gametes are rarely formed in triploids. They are mostly sterile and rarely set seed on selfing or out-crossing.

(b) Autotetraploids:

These usually show greater vigour, increased cell size, mainly in stomata and guard cells. Depending on the genie make up of diploid 'gigas' characters may be expressed. The autotetraploidy leads the plant to perenniality and may show reduced fertility.

Autotetraploids are slower in growth, have greater adaptability, variability and some times show disease resistance. Because of their greater economic importance and breeding possibility, auto-tetraploids are now induced artificially.

Auto-tetiaploids have been reported in Sorghum, wheat, rice, maize, chili, red gram, black gram, green gram, bengal gram, cotton, guava, coffee etc. It may arise by somatic doubling and somatic doubling generally happens by the failure of first meiotic division in the zygote.

(c) Auto-pentaploids:

These behave like tetraploids. The phenotype may differ. At meiosis pentavalents, quadrivalents, trivalents, bivalents and univalents may be formed. Gametes produced are fully sterile as meiosis is quite irregular.

(d) Auto-hexaploids:

These are more stable. Meiosis may be more regular and fertility high. Multivalents already occur.

(ii) Allopolyploids:

Polyploidy may also result from doubling of chromosome number of F_1 hybrid which is derived from two distinctly different species. This will bring two sets of chromosomes in F_t hybrid. Let A represent a set of chromosome (= genome = a complete set of genes present in a haploid set of chromosomes) in species X, and let B represent another genome in a species Y. The F_1 then will have one A genome and another B genome.

The doubling of chromosome in this F_1 hybrid (AB) will give rise to a tetraploid with two A and two B genomes. Such a polyploid is called an allopolyploid or amphidiploid (form derived from a hybrid between two diploids, so that the homologues come from different sources).

An allotetraploid has been produced by the Russian Geneticist G.D. Karpechenko by crossing Raphanus sativus (2n = 18) and Brassica oleracea (2n = 18). The hybrid formed by crossing these two species is itself a diploid (2n = 18). It contains only one set of radish chromosome (n = 9) and one set of cabbage (n = 9) chromosomes. The hybrid differs from both the parents and showed many characters of both.

It is almost sterile, because radish and cabbage chromosome are so different that they do not pair or fail to pair at meiosis I. But the hybrid forms an occasional gamete which contains one complete set of radish chromosomes and one complete set of cabbage chromosomes.

When such two gametes combine they produce a plant which contains two sets of radish chromosome and two sets of cabbage chromosomes (18+18 = 36). These F_2 progenies were fertile and tetraploids. At meiosis, pairing was regular and the plant was fertile. This plant showed foliage like radish and root like cabbage. The fruit was peculier.

It resembled the cabbage in its lower portion and the radish in its apical portion. The allotetraploid bred true, hence of no practical value. As it combines characters of both radish and cabbage, therefore, has been named Raphanobrassica.

	Raphanus sativa	×	*Brassica oleracea*
	(radish)	↓	(cabbage)
	2n = 18		2n = 18
Gametes	n = 9		n = 9
F_1		*Raphanobrassica*	
		2n = 18	
		(Diploid hybrid)	
		Doubling of chromosome number by colchicine	
		4n = 36 (*Raphanobrassica*)	

(Artificial synthesis of *Raphanobrassica* by intergeneric cross)

However, the experiment was significant and it demonstrated a method by which fertile interspecific hybrids can be obtained. Attempts to produce a fertile hybrid between potato and tomato, with underground parts like potato and aerial parts like tomato, have also been unsuccessful so far.

Some of the synthetic allotetraploids resemble closely with the existing species. Various species like wheat, cotton, tobacco etc. might have developed by this method. During the recent years a new genus Triticale has been synthesized by combining the chromosome of Triticum duram and Secale cereale (rye). This new genus Triticale is a very useful allopolyploid (2n = 56).

2. Aneuploidy or Secondary Polyploidy = (Heteroploidy):

Another type of difference in the chromosome number is known as aneuploidy (not true ploidy in which the chromosome number is some other or different than the exact multiple of the basic number). This difference may be due to addition or loss of a single chromosome or several chromosomes. Aneuploidy are, therefore, unbalanced individuals and show phenotypic differences.

Blakeslee discovered in Datura stramonium, the first case of aneuploidy. When a diploid individual having one chromosome represented three or four times instead of twice is called as polysomic.

When the chromosome complement is increased by one chromosome, it is called trisomic (2n + 1). These are found in Drosophila and more common in plants. Organisms containing 2n—1 chromosomes are called monosomic but they are neither fertile and nor vigorous.

If the increase is in two or more different chromosomes, each having one extra homologue, then it is called double trisomic (2n + 1 + 1). When both the chromosome of a given pair are missing, the individual is called a nullisomic (2n—2). These are inviable in some species but viable in others.

If there are two homologues added to a chromosome pair, then it is called tetrasomic (2n + 2) and when it is 2n + 2+ 2 and so on, it is called double tetrasomic.

Polysomics form trivalents, quadrivalents according to the duplications. Trisomy has led to pollen sterility.

A summary of the terms used to describe polyploidy/heteroploidy (variation in chromosome number).

	Term	Type of change	Symbol
A.	Euploidy	No. of genomes more than	2
	Haploid	Gametic chromosome complement	n
1-	Monoploid	One genome	x
2-	Autopolyploid	Genomes similar with each otherq	
	Autotriploid	Three genomes	3x
	Autotetraploid	Four genomes	4x
	Autopentaploid	Five genomes	5x
	Autohexaploid	Six genomes	6x
3-	Allopolyploid	Two or more distinct genomes (Generally each genome has two copies)	
	Allotetraploid	Two distinct genome from different species	
	Allohexaploid	Three distinct genomes	
B.	Aneuploids	One or a few chromosome extra or missing from 2n.	2n+ few
	Trisomic	One chromosome extra	2n+1
	Double trisomic	One chromosome from each of two different chromosome pair extra	2n+1+1
	Monosomic	One chromosome missing	2n-1
	Double mono-somic	One chromosome from each of two different chromosomes pairs missing	2n-1-1
	Nullisomic	One chromosome pair missing	2n-2
	Tetrasomic	One chromosome pair extra	2n+2
	2n=	Somatic chromosome number	
	n=chromosome number of gamete (sex cell) of any species, whether diploid or polyploid.		

Different kinds of numerical changes in chromosome (x= basic chromosome number; 2n = somatic chromosome number).

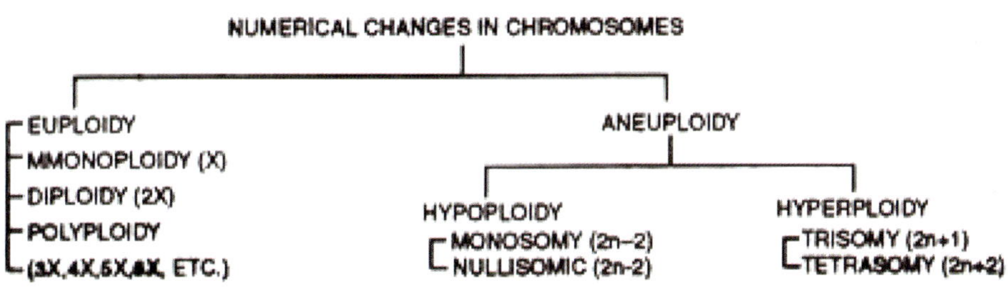

Hypoploidy. Loss of one or more chromosomes
Hyperploidy. Addition of one or more chromosomes.

Euploidy means that the organism possess one or more full sets of chromosomes where as aneuploidy means presence of chromosome number which is different than a multiple of basic chromosome number. Let us suppose that 7 is the basic chromosome number (x) in a particular class of individuals where diploid number (2n) is 14.

In this case, chromosome number 2n = 13 and 2n = 15 would be aneuploidy, while those having 2n = 7, 21, 28, 35, 42 would be euploids. A classification of different kinds of numerical changes in chromosomes is given above.

Causes or Origin of Polyploidy

Polyploids might originate by one of the several methods given below:

(i) Doubling of the chromosomes during early stage of development of the embryo.

(ii) Union or fusion of the gametes, one or both of which due to some reason may be unreduced in chromosome number.

(iii) Two male gametes occasionally or sometimes unite with a single egg cell thus forming a triploid.

Despite above, it has been already described that aberration in cell division may lead to polyploidy. Sometimes it is caused by a large number of mutations which occur in the nature. It may also arise by manipulation of the breeder. Irregularities in cell division which occur in mitosis and meiosis, are given as below-

Mitosis:

(i) Failure of formation of cell wall under cytokinesis after the division of nucleus (karyokinesis).

(ii) At anaphase, the spindle mechanism may fail to separate the sister chromatids or daughter chromosomes to the different poles, consequently the chromosome number is doubled.

Meiosis:

Polyploid gametes may result by aberrations during meiosis:

(i) If the first, second or both the divisions fail, leading to the formation of restitution nucleus (A nucleus) which contains all the chromosomes present in a cell following the failure of cell to undergo a complete division.

(ii) By the failure of mitotic division immediately preceding meiosis, pollen mother cell giving rises to the formation of two nuclei.

(iii) If the fusion of homotypic spindle in second mitotic or equational division takes place.

(iv) If double division of chromosomes occur during both the division of meiosis.

Techniques of Inducing Polyploidy:

1. Decapitation:

It has been found in various seedlings that if their tip is removed or cut off by a sharp knife the callus is produced which give rise to some polyploids.

2. Graft combinations:

It has been observed that callus formation occurs during the graft combinations i.e., 7% (fusion of stock and scion) which may lead to some extent polyploidy – Winkler, 1916.

3. Radiations:

Irradiation of vegetative and floral buds with X-rays, gamma rays or ultra-violet rays, polyploidy may be brought in some frequencies.

4. Temperature:

Application of heat and cold shocks to flowers at or near the time of first division of zygote brings about polyploidy.

5. Hybridization:

It also to some extent brings about polyploidy.

6. Chemicals:

There are various chemicals like chloral hydrate, acenaphthylene, coumarin, vertanine sulphate, cavadin, vernatrine, ethyl mercury chloride, vitamin sulphate, granosan, hydroxyquinoline and nitrous oxide, colchicine etc. Of these the most effective results have been obtained by colchicine and this is now being widely used on all plant species.

Nebel and Ruttle working on Tradescantia, Petunia, Marigolds & Blakeslee and Avery working on Portulaca, Datura and Cucurbita noticed that the alkaloid colchicine was most effective in doubling the chromosome number.

This alkaloid is extracted from the plant Colchicum autumnale (Family: Liliaceae) that is available in temperate regions. Colchicum luteum is an allied species found in the western Himalayans. Gloriosa superba (Liliaceae) is reported by Parthasarthy to contain the alkaloid colchicine.

In many cases partially functional spindle may be formed and incipient anaphase and even telophase may be evidenced but in no case cell wall is formed.

The final effect may be brought about in any one of the following three ways:

(i) Complete failure of spindle formation.

(ii) Fusion of nuclei after pseudo-anaphase.

(iii) Formation of sticky chromosome bridges.

Reports from Russian work show that the chemical 'GRANOSAN' is as effective as colchicine. Colchicine has been used with great success in a large number of crop species belonging to both dicot and monocot groups.

Colchicine is a poisonous chemical isolated from seeds (0.2-0.8%) and bulbs (0.1-0.5%). It is readily soluble in alcohol, cold water and chloroform, but is relatively less soluble in hot water.

Pure colchicine is $C_{22} H_{25} O_6 N$. It blocks spindle formation and thus prevents the movement of sister chromatids or daughter chromosomes to the opposite poles. The resulting restitution nucleus includes all the chromatids; as a result, the chromosome number is just doubled.

Colchicine affects only meristematic or dividing cells. The method of colchicine application varies considerably. It is usually applied in aqueous solution, dilute alcohol, appropriate emulsion, lanoline paste, agar solution, glycerine, cold water. It may be used up in the range of 0.0006% —1% concentration.

It is used in three ways:

(i) Seed treatment:

The seeds are immersed in the solution of 0.2 -1.6% concentration and are sown before germination has initiated.

(ii) Seedling treatment:

Here the radicle is immersed in the solution of 0.02%-0.1 concentration for 3-24 hours. The after-effect is visible subsequently and then sown in the field.

(iii) Treatment of Young shoots and buds:

The buds or tips of shoots, when are treated with colchicine (.5-1% concentration) the satisfactory results come out. Injection of colchicine solution even is effective in buds. It has already exhibited most significant result.

Evolution of Polyploidy

Polyploidy is believed to be a method for the origin of new species. As a general rule, species with a higher chromosome number are regarded more advanced than those with lower number from evolutionary point.

Plants with a higher chromosome number are considered to have released from either by direct increase of the lower number or by crossing with the other species, hence the three species of wheat, Triticum monococcum ($n = 7$), Triticum duram ($n = 14$) and Triticum vulgare ($n = 21$), have been evolved by polyploidy. Several species, which already remain in nature, have been synthesized in the laboratory by inducing polyploidy.

According to Danish geneticist O.Winje 1/2 to 1/3 species of angiosperms are polyploids. The maximum number (about 70%) is found in Gramineae family. In Cruciferae 42% and in Leguminosae 23% are available as polyploids. Despite it, other families where polyploids are found in plenty are notable e.g., Malvaceae, Cyperaceae etc. All these species are produced by hybridization.

Any organism where the chromosome number is highly increased, the genetic differences also are increased. This is due to increase in number of genes on the chromosomes. Hence, polyploidy is of great importance in the development of taxonomic series. The varieties, which are produced in this way, are very important for production of crops.

An example for clear understanding of polyploidy role in evolution is given below:

♂ *Triticum aegilopoides* × *Aegilops speltoides* ♀

AA, 2X = 14 BB, 2X = 14

♀ *Aegilops speltoides* × *Triticum aegilopoides* ♂

AA, 2X = 14 BB, 2X = 14

Hybridization ↓

Hybrid AB

Tetraploid wheat ↓

AABB

Cultivated *Aegilops squarrosa*

T. dicoccum AABB × DD

T. dicoccoides

T. durum 4X = 28 ↓ 2X = 14

Hybridization

ABD

↓ Chromosome doubling

Hexaploid wheat AABBDD

(Bread wheat) 6X = 42

Triticum aestivum Cultivated common wheat

T. spelta cultivated spelta wheat

Evolution of wheat

It has been found that completive ability is more in polyploids. It has got more adaptability and their geographical distribution is found maximum.

Practical Value of Polyploidy

Induced polyploidy has important practical application. It is possible to produce disease resistant plants having desirable qualities by induced polyploidy. Polyploids are bigger in size, produce large seeds and fruits, exhibit gigas characters. These are also more vigorous. Blakeslee points out that tetraploids are of horticultural value because of the unusually large size of flowers and of the plants.

The most important use of polyploidy is to overcome or remove the sterility of hybrids which are borne by hybridization or distant cross, interspecific or inter-generic because no pairing occurs in F_1 and the gametes which are formed are abnormal. By doubling the number of chromosomes in F_1 allotetraploids (= amphidiploids) are produced which are now fertile.

Due to hybridization species of new characters may be produced. Raphanobrassica, Triticale are inter-generic cross, Inter-varietal and interspecific crosses have become most successful. The plants which are propagated asexually, gigas characters may be developed in them by application of chemicals like colchicine etc.

Under gigas character plants show larger cells, leaves, stomata, flowers, pollen grains and seeds. These exhibit slower growth, late flowering and maturity etc. These also show the size larger than diploids. The best example is sugarcane and many ornamental plants.

Polyploids are better adjusted to tolerate unfavorable climates. Some of the most important polyploid (allopolyploid) crops are wheat, cotton, oats, tobacco, Brassica species etc. Allopolyploidy has played an important role in the evolution of plant species e.g., in allotetraploid cabbage and tomato ascorbic acid content is high.

Role of Polyploidy in Plant Breeding

The first new polyploid variety which came in the light was Galeopsis tetraliita, 2n = 32, produced from the genomes of Galeopsis speciosa (2n = 16) and Galeopsis pubescens (2n = 16).

Similarly, there are several autopolyploid like orchard grass (Dactylis glomerata), Alfa-alfa (Medicago sativa), potato (Solanum tuberosum), Barley (Hordeum bulbosum), Agropyron cristatum, Phleum pratense, Biscutella laevigata, Tradescantia species etc. These are all allopolyploids and perennials asexually propagated plants. Due to irregularities in the meiosis, auto-polyploids survive least in the nature.

Polyploidy is believed to be a means for the origin of new species, especially in the plants in which this phenomenon is more common. As a general rule, species with a higher chromosome number are regarded more advanced from an evolutionary point than those with lower number.

Plants with a higher chromosome number are thought to have evolved either by the direct increase of the lower number, or by crossing with the other species. Thus, as described earlier the three species of wheat, Triticum monococcum (n = 7), T. durum (n = 14), and T. vulgare (n = 21) have been evolved by polyploidy. Several species, which already exist in nature, have been synthesized in the laboratory by inducing polyploidy.

Induced polyploidy has important practical application. By artificially induced polyploidy, it is possible to produce disease resistant plants having desirable qualities. Polyploids are larger in size and produce large seeds and fruits. They are more vigorous too.

Some Artificially Produced Polyploids:

1. Wheat:

↓ Chromosome doubling

Emmer wheats
4n = 28
(AABB)

(*ii*) Parents : *Emmer wheat* × *Aegilops squarrosa*
4n = 28 2n = 14
(AABB) (CC)

↓

Infertile hybrid
3n = 21
(ABC)

↓ Chromosome **doubling**

Triticum aestivum
6n = 42 (Hexaploid wheat)
(AABBCC)

2. Cotton:

Parents : *Gossypium arboreum or herbaceum* × *Gossypium raimondii*
2n = 26 2n = 26
AA BB

↓

F_1 AB
2n = 26

↓ doubling

AABB
G. barbadense or hirsutum
4n = 52

3. Tobacco:

(*i*) Parents : *Nicotiana sylvestris* × *Nicotiana tomentosa*
2n = 24 ↓ 2n = 24
F_1 Sterile
2n = 24
↓ Doubling
4n = 48
Nicotiana tabacum

(*ii*) Parents : *N. tabacum* × *N.glutinosa*
4n = 48 ↓ 2n = 24
Sterile
F_1 3n = 36
↓ Doubling
N.digluta
6n = 72

4. Primula:

Parents : *Primula floribunda* × *P. verticillata*
2n = 18 ↓ 2n = 18
F_1 Sterile
2n = 18
↓ Doubling
P.kewensis
2n = 36

5. Pentaploid:

The variety B.C. 201 of Gossypium was evolved by crossing American and Asian cotton at Surat Agriculture Research Centre which chromosome number is 65 (2n).

Parents : *Gossypium barbadense* × *G.herbaceum*
4n = 52 ↓ 2n = 26
AABB AA
F_1 Sterile (AAB)
3n = 39
Doubling
Allohexaploid × *G.barbadense*
AAAABB ↓ AABB
6n = 78 4n = 52
AAABB
B.C.201
5n = 65

6. Hexaploid:

In 1944, a variety of Gossypium which was triploid was made hexaploid after doubling chromosome number. It was found more stable and fertile and exhibited many irregularities in meiosis. The artificial induction of chromosome doubling by the colchicine technique has opened new fields for the production of polyploid plants.

More successful results are obtained in cross fertilized plants and those in which the chromosome number is high or that have fewer chromosomes than self fertilized plants. Superior auto-polyploids have been obtained in sugar beet, mustard, rye, turnip, red clover etc.

On account of the larger number of genes, polyploidy possess a greater potential for variability arising from mutation. Structural changes in the chromosomes are also likely to be more frequent in polyploids than in diploids and a greater variability is likely to result from crossing with different species.

Tetraploids grapes, produced in somatic tissues are now cultivated in U.S.A., England and Japan. Although the number of fruits in a cluster was low in the tetraploids but these far exceeded in weight and size than diploid and contained fewer seeds.

The first example of synthetic crop made by man with commercially valuable characters in Triticale (rye wheat) produced by Muntzing by crossing Secale cereale (rye) and Triticum aestivum (wheat) and subsequent chromosome doubling in the hybrid. Further breeding of Triticale and selection by Borlaugh in Mexico and at the IARI, Delhi have resulted in forms that are promising from an agronomic points of view.

The Russian scientist Pissorev developed day-neutral and semi-dwarf varieties of this new species. A new variety of this rye-wheat hybrid has been evolved at the IARI, New Delhi for cultivation in the desert area of Rajasthan. This new cereal is more drought resistant than other varieties of barley and wheat and resembles in production and resistance to disease and grain character. It also grows well in hill and rain-fed areas.

It is also rich in protein, but its glutein content is low that is why the chapattis are not prepared good. The research work is going on at Pantnagar to evolve commercial useful variety of this synthetic crop which may be acceptable as a food grain. Borlaugh is making the possibilities of hybridising wheat and rice cells at the cellular level to produce 'whrice', a combination of wheat & rice.

References

- Stewart, W.N.; Rothwell, G.W. (1993). Paleobotany and the evolution of plants (2 ed.). Cambridge University Press. ISBN 978-0-521-23315-6
- Rothwell, G. W.; Scheckler, S. E.; Gillespie, W. H. (1989). "Elkinsia gen. nov., a Late Devonian gymnosperm with cupulate ovules". Botanical Gazette. 150 (2): 170–189. doi:10.1086/337763
- evolution-plants, botany-general, plants-and-animals: encyclopedia.com, Retrieved 14 April 2018
- "First Land Plants and Fungi Changed Earth's Climate, Paving the Way for Explosive Evolution of Land Animals, New Gene Study Suggest". science.psu.edu. Retrieved 9 April2018
- Sattler, R (1992). "Process morphology: structural dynamics in development and evolution". Canadian Journal of Botany. 70 (4): 708–714. doi:10.1139/b92-091
- A-Brief-History-of-Seed-Size-8051515: researchgate.net, Retrieved 29 March 2018
- Raven, J.A.; Edwards, D. (2001). "Roots: evolutionary origins and biogeochemical significance". Journal of Experimental Botany. 52 (90001): 381–401. doi:10.1093/jexbot/52.suppl_1.381. PMID 11326045
- Polyploidy-meaning-causes-and-evolution-chromosomal-aberration-35573: biologydiscussion.com, Retrieved 20 June 2018
- Donoghue, M.J. (2005). "Key innovations, convergence, and success: macroevolutionary lessons from plant phylogeny" (abstract). Paleobiology. 31 (2): 77–93. doi:10.1666/0094-8373(2005)031[0077:KICASM]2.0.CO;2. ISSN 0094-8373. Retrieved 2008-04-07

Chapter 6

Plant Physiology, Anatomy and Morphology

Plant physiology is a branch of botany that is concerned with the study of the physiology and functioning of plants. Some of the important parts of a plant are leaf, root, stem, flower, etc. This chapter discusses in detail about such important parts of the plant for an in-depth understanding of plant physiology, anatomy and morphology.

Root

Primary and secondary roots in a cotton plant

In vascular plants, the root is that organ of a plant body that typically lies below the surface of the soil (though not always) and whose major functions are to absorb water and inorganic nutrients and to anchor the plant body to the substrate. A root can also be aerial (that is, growing above the ground) or aerating (that is, growing up above the ground or especially above water).

Similarly, a stem called a rhizome grows horizontally below the surface of the soil, but it differs from a root in that it contains nodes from which roots sprout, and it does not absorb nutrients or water.

An analogy is sometimes drawn between roots, which function in absorbing nutrients for the plant, with the human digestive system, which functions in absorbing food for the human body. Similarly, plant leaves have been compared to the human lungs, and the xylem and phloem of the vascular system to the veins and arteries of the human circulatory system. These analogies are more than chance comparisons, since some religions, particularly Western theologies, commonly view human beings as "microcosms of creation". Such theologies hold that human beings resemble the Supreme Being, while all life resembles the image and likeness of the human character and form.

Root Structure

Roots of a hydroponically grown plant

At the tip of every growing root is a conical covering of tissue called the root cap. It usually is not visible to the naked eye. It consists of undifferentiated soft tissue (parenchyma) with unthickened walls covering the apical meristem, tissue at the root tip responsible for root growth. The root cap provides mechanical protection to the meristem cells as the root advances through the soil, its cells worn away but quickly replaced by new cells generated by cell division within the meristem. The root cap is also involved in the production of mucigel, a sticky mucilage that coats the new formed cells. These cells contain statoliths, starch grains that move in response to gravity and thus control root orientation.

The outside surface of a root is the epidermis, an outer single-layered group of cells covering the young tissues and leaves of a plant. Recently produced epidermal cells absorb water from the surrounding environment and produce outgrowths called root hairs that greatly increase the cell's absorptive surface. Root hairs are very delicate and generally short-lived, remaining functional for only a few days. However, as the root grows, new epidermal cells emerge and these form new root hairs, replacing those that die. The process by which water is absorbed into the epidermal cells from the soil is known as osmosis. Because of the nature of osmosis, saline water is more difficult for most plant species to absorb.

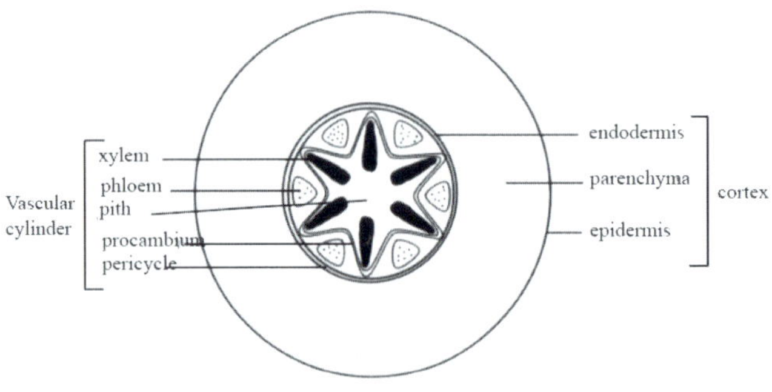

Cross section of the root of a dicotyledon

Beneath the epidermis is the cortex, which comprises the bulk of the root. Its main function is storage of starch. Intercellular spaces in the cortex aerate cells for respiration. An endodermis is a thin layer of small cells forming the innermost part of the cortex and surrounding the vascular tissues deeper in the root. The tightly packed cells of the endodermis contain a substance known as suberin and create an impermeable barrier of sorts. Water can only flow in one direction through the endodermis: In towards the center of the root, rather than outward from the stele (the central part of the root) to the cortex.

The stele consists of the cells inside the endodermis. The outer part, known as the pericycle, surrounds the actual vascular tissue. In monocotyledonous plants, the xylem and phloem cells are arranged in a circle around a pith or center, whereas in dicotyledons, the xylem cells form a central "hub" with lobes, and phloem cells fill in the spaces between the lobes.

Root Growth

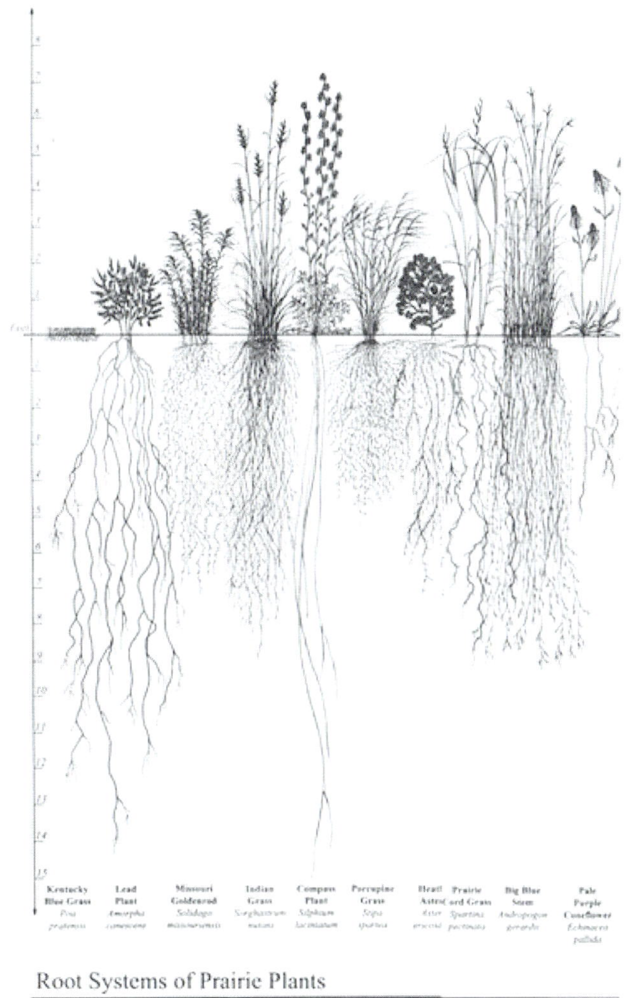

Root systems of prairie plants

Early root growth is a function of the apical meristem located near the tip of the root. The meristem cells more or less continuously divide, producing more meristem, root cap cells (these sacrificed to protect the meristem), and undifferentiated root cells. The latter will become the primary tissues of the root, first undergoing elongation, a process that pushes the root tip forward in the growing medium. Gradually these cells differentiate and mature into specialized cells of the root tissues.

Roots will generally grow in any direction where the correct environment of air, nutrients, and water exists to meet the plant's needs. Roots will not grow in dry soil. Over time, given the right conditions, roots can crack foundations, snap water lines, and lift sidewalks. At germination, roots grow downward due to gravitropism, the growth mechanism of plants that also causes the shoot to grow upward. In some plants (such as ivy), the "root" actually clings to walls and structures; this is known as thigmotropism, or response to touch.

Most plants experience growth only along the apical meristems; this is known as primary growth, which encompasses all vertical growth. On the other hand, secondary growth encompasses all lateral growth, a major component of woody plant tissues. Secondary growth occurs at the lateral meristems, namely the vascular cambium and cork cambium. The former forms secondary xylem and secondary phloem, while the latter forms the periderm, found only in woody plants.

In woody plants, the vascular cambium, originating between the xylem and the phloem, forms a cylinder of tissue along the stem and root. The cambium layer forms new cells on both the inside and outside of the cambium cylinder, with those on the inside forming secondary xylem cells, and those on the outside forming secondary phloem cells. As secondary xylem accumulates, the "girth" (lateral dimensions) of the stem and root increases. As a result, tissues beyond the secondary phloem (including the epidermis and cortex, in many cases) tend to be pushed outward and are eventually "sloughed off" (shed).

At this point, the cork cambium (this process only occurs in woody plants) begins to form the periderm, consisting of protective cork cells containing suberin. In roots, the cork cambium originates in the pericycle, a component of the vascular cylinder.

Stilt roots in the Amazon Rainforest support a tree in very soft, wet soil conditions.

The vascular cambium produces new layers of secondary xylem annually. This dead tissue is responsible for most water transport through the vascular tissue (systems and roots).

Types of Roots

A true root system consists of a primary root and secondary roots (or lateral roots).

The primary root originates in the radicle (first part of a seedling—a growing plant embryo—to emerge from the seed during germination) of the seedling. During its growth it rebranches to form the lateral roots. Generally, two categories are recognized:

Taproot

A taproot is typically a long and somewhat thick root that goes deep down into the soil. It is the first root to appear from the seed and remain the largest, central root of the plant. Lateral roots will branch off from the taproots and then more lateral roots will form from the initial lateral roots, but the central taproot will remain the largest and will burrow down into the soil the deepest. A good example is a common carrot. The part we eat is the taproot, but you will also notice smaller roots all along the central root.

Carrots are an example of conical taproots, but taproots do not have to be straight or even tapered. Radishes are also taproots, but theirs are wide in the middle and taper at the bottom and often at the top. They are a "fusiform" taproot. Then there are the "napiform" taproots, like beets, that are wide across the top, becoming very thin at the bottom. The shape may change, but the function remains the same, to keep the root deep enough in the soil to access water.

Benefits of Plants with Taproots

Plants with taproots tend to be very drought tolerant. Many desert plants can send roots down more than 75 ft., allowing them to find water, even in dry climates or conditions.

Taproots can also serve to store food reserves, making them even more self-sufficient and resilient.

Drawbacks of Tap Roots - Transplanting and Dividing

Because the taproot goes so deeply into the soil, it can be very hard to dig and lift a tap-rooted plant. Think of the dandelions in the yard.

Dividing tap-rooted plants is another challenge. You can't simply break off parts of the crown, with roots attached, as you would with something like daylilies or coreopsis because you need to get a piece of that tap root with each division. But again, think back to the dandelion and you'll understand that it's not impossible.

Taproots often form offshoots near the crown, called necks. If your plant has these, you can cut off each neck that has some smaller roots attached to it and replants with good success. Buttery bush *(Buddleia davidii)* can often be divided this way.

If no offshoots have formed, you can still try taking a small piece of the actual taproot, with at least one eye and some smaller roots attached, and replant that. Much like your beheaded dandelion, it will send up a new shoot.

Young seedlings of taproot plants are much easier to transplant. Plants such as butterfly weed (*Asclepias tuberosa*) will set many seedlings, so you shouldn't need to disturb the original plant. Don't wait too long, are the additional stress will make moving them harder.

Plants with Taproots

It's hard to give a definitive list of tap-rooted plants because of many plants, like most trees, start out with taproots, but will switch to sending out lateral roots closer to the soil surface, once they are established. Tomato plants grown from seed tend to send down a tap root, but those grown from cuttings will not.

Root vegetables, as mentioned above and also including jicama, parsnips, salsify and turnips are considered taproots.

Some common garden flowers and herbs with taproots include:

- Balloon flower (*Platycodon Grandi floras* and CVS)
- Bugbane (*Cimicifuga racemosa*)
- Butterfly weed (*Asclepias tuberosa* and CVS.)
- Comfrey
- Cushion spurge (*Euphorbia polychrome* and CVS.)
- Dill
- False blue indigo (*Baptisia* and CVS.)
- Horseradish
- Lupine (*Lupinus polyphyllous*)
- Oriental poppy (*Papaver orientale* and CVS.)
- Parsley
- Sea holly (*Eryngium*)

And several weeds survive with the help of tap roots, like plantain and kudzu.

Fibrous Root

A fibrous root is a root that consists of groups of roots of similar size and length. They do not penetrate as deeply into the soil as does a taproot.

The fibrous root is a feature of plants that are monocotyledons. Unlike the tap root, the primary root produced during development does not remain, and instead roots, known as adventitious roots, are produced from the stem of the plant.

All these roots forming the fibrous root collectively are of equal size and length.

A fibrous root system does not penetrate deeply into the soil but rather creates a thick network of roots that are good at holding the soil together.

Many types of grasses have fibrous roots, including plants related to grass such as corn.

Advantages of fibrous root systems include that they allow the plant to absorb water and minerals over a large surface area closer to the surface of the soil.

They are also useful in helping prevent or reduce soil erosion since these root systems help hold the soil particles together.

Some plants, have roots with modifications that adapt them for performing specific functions as well as the absorption of water and minerals in solution.

Food-Storage Roots

Most roots and stems store some food, but in certain plants, the roots are enlarged and store large quantities of starch and other carbohydrates, which may later be used for extensive growth. In sweet potatoes and yams, for example, extra cambial cells develop in parts of the xylem of branch roots and produce large numbers of parenchyma cells. As a result, the organs swell and provide storage areas for large amounts of starch and other carbohydrates. Similar food-storage roots are found in the deadly poisonous water hemlocks, in dandelions, and in salsify. In carrots, beets, turnips, and radishes, the food-storage tissues are actually a combination of root and stem. Although the external differences are not obvious, approximately 2 centimeters (0.8 inch) at the top of an average carrot is derived from stem tissue that merges with the root tissue.

Figure: A sweet potato (Ipomoea) plant. Note the food-storage roots.

Figure: Q A manroot (Marah) water-storage root that weighed over 25.3 kilograms (60 pounds)

Water-Storage Roots

Some members of the Pumpkin Family (Cucurbitaceae) produce huge water-storage roots. This is particularly characteristic of those that grow in arid regions or in those areas where there may be no precipitation for several months of the year. In certain manroots (Marah), for example, roots weighing 30 kilograms (66 pounds) or more are frequently produced, and a major root of one calabazilla plant (Cucurbita perennis) was found to weigh 72.12 kilograms (159 pounds). The water in the roots is apparently used by the plants when the supply in the soil is inadequate.

Propagative Roots

Many plants produce adventitious buds (buds appearing in places other than stems) along the roots that grow near the surface of the ground. The buds develop into aerial stems called suckers, which have additional rootlets at their bases. The rooted suckers can be separated from the original root and grown individually. Cherries, apples, pears, and other fruit trees often produce suckers. The adventitious roots of rice-paper plants (*Tetrapanax papyrifera*) and tree-of-heaven (*Ailanthus altissima*) can become a nuisance in gardens, often producing propagative roots 10 meters (33 feet) or more from the parent plant. Horseradish (*Rorippa armoracia*), Canada thistle (*Cirsium arvense*), and some other weeds have a remarkable facility to reproduce in this fashion as well as by means of seeds. In the past, this capacity has made it difficult to control them, but some biological controls being investigated may be an answer to the problem in the future.

Pneumatophores

Water, even after air has been bubbled through it, contains less than one-thirtieth the amount of free oxygen found in the air. Accordingly, plants growing with their roots in water may not have enough oxygen available for normal respiration in their root cells. Some swamp plants, such as the black mangrove (Avicennia nitida) and the yellow water weed.

Figure: Pneumatophores (foreground) of tropical mangroves rising above the sand at low tide. The pneumatophores are spongy outgrowths from the roots beneath the surface. Pneumatophores facilitate the exchange of oxygen and carbon dioxide for the roots, which grow in areas where little oxygen is otherwise available to them.

Ludwigia repens develop special spongy roots, called pneumatophores, which extend above the water's surface and enhance gas exchange between the atmosphere and the subsurface roots to which they are connected. The woody "knees" of the bald cypress (Taxodium distichum), which occurs in southern swamps, were in the past believed to be pneumatophores, but there is no conclusive evidence for this theory.

Aerial Roots

Velamen roots of orchids, prop roots of corn and banyan trees, adventitious roots of ivies, and photosyn-thetic roots of certain orchids are among various kinds of aerial roots produced by plants. It was formerly assumed that the epidermis of velamen roots, which is several cells thick, aided in the absorption of rain water. It appears, however, it may function more in preventing loss of moisture from the root. Corn prop roots, produced toward the base of the stems, support the plants in a high wind. Some tropical plants, including the screw pines and various mangroves, produce sizable prop roots extending for several feet above the surface of the ground or water. Debris collects between them and helps to create additional soil.

Figure: The aerial (velamen) roots of orchids have a thick epidermis that reduces water loss from internal tissues.

Many of the tropical figs or banyan trees produce roots that grow down from the branches until they contact the soil. Once they are established, they continue secondary growth and look just like additional trunks. Banyan trees may live for hundreds of years and can become very large. In India and Southeast Asia, there are several banyan trees that have almost 1,000 root-trunks and have circumferences approaching 450 meters (1,476 feet). The oldest is estimated to be about 2,000 years old.

The vanilla orchid, from which we obtain vanilla flavoring, produces chlorophyll in its aerial roots and, through photosynthesis, can manufacture food with them. The adventitious roots of English ivy, Boston ivy, and Virginia creeper appear along the stem and aid the plants in climbing.

Contractile Roots

Some herbaceous dicots and monocots have contractile roots that pull the plant deeper into the soil. Many lily bulbs are pulled a little deeper into the soil each year as new sets of contractile roots are developed. The bulbs continue to be pulled down until an area of relatively stable.

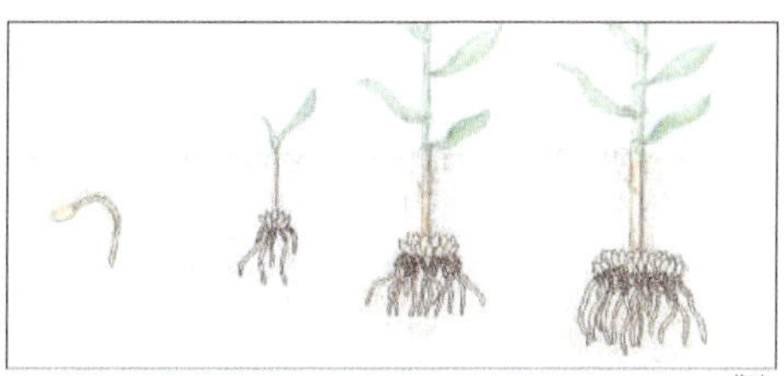

How a lily bulb over three seasons is pulled deeper into the soil by the action of contractile roots. A. A seed germinates. B. Contractile roots pull a newly formed bulb down several millimeters during the first season. C. The bulb is pulled down farther the second season. D. The bulb is pulled down even farther the third season. The bulb will continue to be pulled down in succeeding seasons until it reaches an area of relatively stable soil temperatures.

Figure: Buttress roots of a tropical fig tree

Temperatures is reached. Plants such as dandelions always seem to have the leaves coming out of the ground as the top of the stem is pulled down a small amount each year when the root contracts. The contractile part of the root may lose as much as two-thirds of its length within a few weeks as stored food is used and the cortex collapses.

Parasitic plants are capable of manufacturing at least some of their own food through photosynthesis. Other plants lacking chlorophyll (e.g., Indian pipes) are not parasitic at all. Instead, these plants are saprophytic, obtaining all the nutrients they require from organic materials in the soil.

Buttress Roots

Some tropical trees growing in shallow soils produce huge buttresslike roots toward the base of the trunk, giving them great stability. Except for their angular appearance, these roots look like a part of the trunk.

Parasitic Roots

Some plants, including dodders, broomrapes, and pine-drops, have no chlorophyll (necessary for photosynthesis) and have become dependent on chlorophyll-bearing plants for their nutrition. They parasitize their host plants via peglike projections called haustoria (singular: haustorium), which develop along the stem in contact with the host. The haustoria penetrate the outer tissues and establish connections with the water-conducting and food-conducting tissues. Some green plants, including Indian warrior and the mistletoes, also form haustoria. These haustoria, however, apparently aid primarily in obtaining water and dissolved minerals from the host plants, since the partially.

Primary or Main Functions of Roots:

1. Anchorage:

Roots take part in fixation of the plant and supporting the aerial shoot system.

2. Absorption of Water:

Roots absorb water from soil.

3. Absorption of Minerals:

Roots absorb mineral salts from soil.

4. Prevention of Soil Erosion:

Roots hold the soil particles firmly to prevent soil erosion.

5. Transport:

They take part in transport of absorbed water and minerals to shoot system. Similarly, root has channels for the flow of organic food from aerial parts.

Rootless Plants:

Many plants growing in aquatic habitats do not possess roots because there is little requirement for absorption of water and mineral salts, e.g., Wolffia, Utricularia, Myriophyllum, Ceratophyllum. In other aquatic plants roots develop only for balancing (e.g., Lemna, Pistia) and fixation (e.g., Hydrilla).

Secondary or Accessory Functions of Roots:

1. Storage:

It occurs in fleshy roots.

Examples of fleshy tap roots are:

(i) Conical, e.g., Carrot

(ii) Fusiform, e.g., Radish

(iii) Napiform, e.g., Beet, Turnip and

(iv) Tuberous, e.g., Mirabilis.

Depending upon their shape and grouping, adventitious fleshy roots are of six types:

(a) Tuberous (root tubers), e.g., Sweet Potato

(b) Fasciculated, e.g., Asparagus, Dahlia

(c) Palmate, e.g., Orchis

(d) Nodulose, e.g., Curcuma

(e) Moniliform, e.g., Momordica

(f) Annuiated, e.g., Ipecac.

2. Extra or Mechanical Support:

It is provided by several types of roots:

(i) Buttress roots (plank roots), e.g., Bombax

(ii) Prop roots, e.g., Banyan, Rhizophora

(iii) Stilt roots, e.g., Maize, Sugarcane, Pandanus.

3. Climbing:

Roots help some of the weak stemmed plants to cling and hence climb up a support, e.g., Ivy, Tecoma, Betel, and Money Plant.

4. Perennation:

In many biennial and perennial plants e.g., Trichosanthes.

5. Nitrogen Fixation:

Nodulated roots of Pea, Bean, Gram, Methi, etc.

6. Aeration:

Prop roots, knee roots and pneumatophores of Mangrove plants, e.g., Rhizophora, Heritiera, Sonneratia.

7. Reproduction:

By bearing adventitious buds on both tap roots (e.g., Dalbergia) and adventitious roots (e.g., Sweet Potato).

8. Hygroscopic Roots:

The roots absorb water from air, e.g., epiphytic roots, young prop roots of Banyan.

9. Floating:

By storing air (e.g., Jussiaea = Ludwigia) some of the roots function as floats.

10. Balancing:

Free floating plants usually possess a cluster of adventitious roots from their nodes and bases to help in balancing the plants over the water surface, e.g., Lemna, Pistia, Eichhornia.

11. Photosynthesis:

As in Trapa, Taeniophyllum, Tinospora, Podostemum.

12. Haustoria (Haustorial Roots):

They occur in some parasitic plants for absorption of food from the host, e.g., Cuscuta, Mistletoe.

13. Mycorrhiza:

The roots of several plants are associated with fungi to form mycorrhiza. Mycorrhiza helps the plant in absorption of phosphorus and nitrogen from organic sources and protects the plants from attack of pathogenic fungi.

14. Growth Regulators:

Roots synthesize certain growth regulations like cytokinins.

Stem

In Botany, stem is the plant axis that bears buds and shoots with leaves and, at its basal end, roots.

The stem conducts water, minerals, and food to other parts of the plant; it may also store food, and green stems themselves produce food. In most plants the stem is the major vertical shoot, in some it is inconspicuous, and in others it is modified and resembles other plant parts (e.g., underground stems may look like roots).

The primary functions of the stem are to support the leaves; to conduct water and minerals to the leaves, where they can be converted into usable products by photosynthesis; and to transport these products from the leaves to other parts of the plant, including the roots. The stem conducts water and nutrient minerals from their site of absorption in the roots to the leaves by means of certain vascular tissues in the xylem. The movement of synthesized foods from the leaves to other plant organs occurs chiefly through other vascular tissues in the stem called phloem. Food and water are also frequently stored in the stem. Examples of food-storing stems include such specialized forms as tubers, rhizomes, and corms and the woody stems of trees and shrubs. Water storage is developed to a high degree in the stems of cacti, and all green stems are capable of photosynthesis.

Ball cactus (*Parodia magnifica*).

Growth and Anatomy

The first rudiment of the young stem, or shoot, of an embryonic plant appears from the seed after the root has first protruded. The growing portion at the apex of the shoot is the terminal bud of the plant, and by the continued development of this bud and its adjacent tissues, the stem increases in height. Lateral buds and leaves grow out of the stem at intervals called nodes; the intervals on the stem between the nodes are called internodes. The number of leaves that appear at a node depends on the species of plant; one leaf per node is common, but two or or more leaves may grow at the nodes of some species. When a leaf drops off a stem at the end of a growing season, it leaves a scar on the stem because of the severing of the vascular (conducting) bundles that had connected stem and leaf. As the stem continues to grow, lateral buds are produced that develop into lateral shoots more or less resembling the parent stem, and these ultimately determine the branching of the plant. In trees the lateral shoots develop into branches, from which other lateral shoots, called branchlets, or twigs, arise. The point at which a leaf diverges in axis from a stem is called the axil. A bud formed in the axil of a previously formed leaf is called an axillary bud, and it, like the leaves, is produced from the tissues of the stem. During the development of such buds, vascular bundles are formed within them that are continuous with those of the stem.

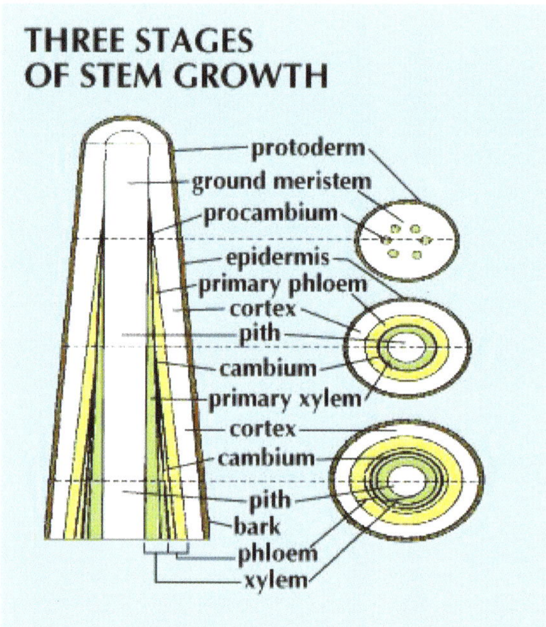

A longitudinal section, left, and cross section, right, of a growing stem show the organization of various tissues for younger, top, and older, bottom, parts of the stem

In the stems of young dicotyledons (angiosperms with two seed leaves) and gymnosperms, the vascular bundles (xylem and phloem) are arranged in a circle around a central core of spongy ground tissue called the pith. Surrounding the vascular bundles is a layer that varies in thickness in different species and is called the cortex. Surrounding this and comprising the exterior surface of the stem is a layer called the epidermis. In plants with woody stems, a variety of secondary tissues are added to these primary tissues. Among the most important of these is a ring of meristematic cells that in turn give rise to the vascular cambium. This tissue arises between the primary xylem and phloem and gives rise to secondary phloem on the outside and secondary xylem on the inside; the latter tissue is the wood of trees.

Stem Types and Modifications

Many plants are annuals and complete their life cycles in one growing season, after which the entire plant, including the stem, dies. In biennial plants the lower part of the stem, often modified for food storage, persists after the first growing season and bears buds from which an erect stem arises during the second growing season. In perennial plants the short stem may produce new shoots for many years. Plants producing woody stems are called trees and shrubs; the latter produce branches from or near the ground, while the former have conspicuous trunks.

Joshua trees (*Yucca brevifolia*) in Joshua Tree National Park, California, U.S.

In general, the habit of a stem is erect or ascending, but it may lie prostrate on the ground, as in the sweet potato and strawberry. A stem may climb on rocks or plants by means of rootlets, as in ivy; other vines have twining stems that twist around a supporting plant in a spiral manner, as in the honeysuckle and hop. In other cases, climbing plants are supported by tendrils that may be specialized stems, as in the grape and passion-flower. In tropical climates twining plants often form thick woody stems and are called lianas, while in temperate regions they are generally herbaceous vines. A stolon is a stem that curves toward the ground and, on reaching a moist spot, takes root and forms an upright stem and ultimately a separate plant. Among the subterranean stems are the rhizome, corm, and tuber. In some plants the stem does not elongate during its early development but instead forms a short conical structure from which a crown of leaves arises. These may form a bulb (as in the onion and lily), a head (cabbage, lettuce), or a rosette (dandelion, plantain).

Twining stem of a decorative vine

Sprouted potato tuber. The “eyes” of a potato are clusters of buds in the axils of the scalelike leaves, each of which can grow into a new plant

Functions of Stem

Primary or main functions:

1. It supports and holds leaves, flowers and fruits.

2. Leaves are borne on stem in such a fashion that they are able to carry on the important function efficiently just like to receive the light and to carry on the gaseous exchange.

3. The stem conducts the water and minerals from roots to leaves and fruits.

4. Stem bears flowers and fruits in position to facilitate the processes of pollination and fertilization.

Secondary or Accessory functions:

1. Perennation

2. Storage of food

3. Photosynthesis

4. Vegetative propagation

5. As climbers

There are three types of stem: Underground stem, Aerial stem and Sub- aerial stem.

1. Underground stem:

Stems of some plants remain in the ground and serve the function of perennation and storage of food. They produce aerial shoots annually. They resemble roots superficially but are distinguishable by the presence of scale leaves and buds at nodes. Such stem also act as a means of vegetative propagation. The modified underground stems are the following:

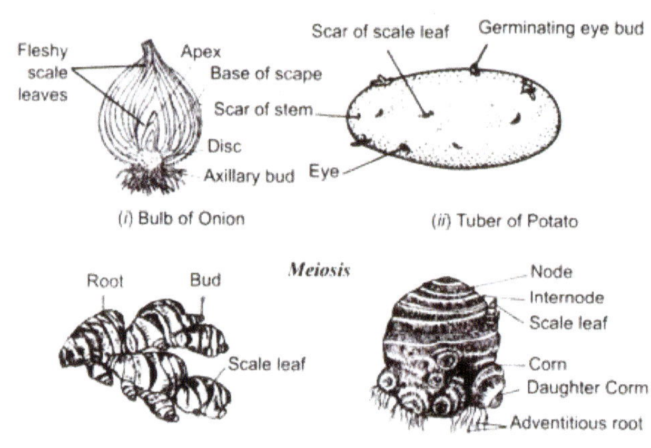

(i) Bulb of Onion (ii) Tuber of Potato

(iii) Rhizome of Ginger (iv) Corm of Colocasia

(i) Rhizome:

It is a thickened, prostrate, underground stem having distinct nodes and internodes, scaly leaves at the nodes, axillary and terminal buds present; may be branched or un-branched; sometimes adventitious roots also arise, e.g. Ginger.

(ii) Tuber:

The underground stem becomes enlarged at the growing tips by the accumulation of stored food, commonly starch, tubers are produced e.g. Potato. The eyes of potato are nodes at each of which 1-3 buds are produced in the axils of small scaly like leaves.

(iii) Bulb:

Bulb is a short underground stem with fleshy leaf base called scales. Stem is very much reduced and becomes disc like. The discoid stem in convex or conical in shape and bears highly compressed internodes. These nodes bear fleshy scales. On the upper side, disc bears terminal bud surrounded by number of leaves. The axillary buds are present between the axis of leaves. The adventitious roots are borne on the lower side of the disc e .g. Onion.

(iv) Corm:

Corm is short, thick and unbranched underground stem with stored food material. It grows vertically and covered by thin sheathing leaf bases of dead leaves called scales. The corm bears buds at their nodes. These buds are responsible for giving off adventitious roots. Corm serves the functions of food storage, vegetative propagation and perennation. Corm is more or less rounded in shape or often somewhat flattened from top to bottom, e.g. Colocasia.

2. Sub – aerial stems:

Lower buds of the stem in some plants grow out into short, lateral branches. These are named according to their origin, nature and mode of reproduction:

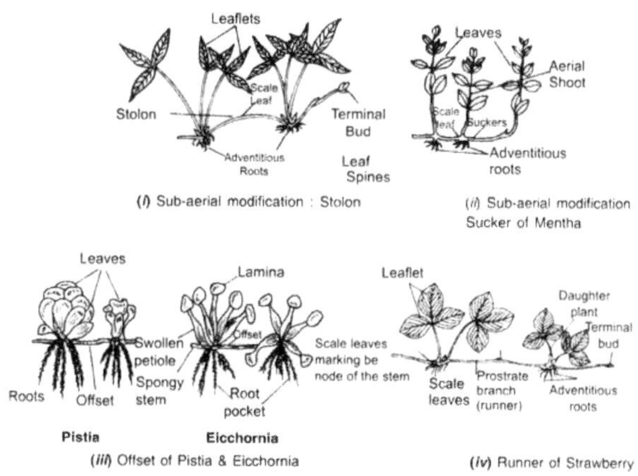

(i) Sub-aerial modification : Stolon

(ii) Sub-aerial modification Sucker of Mentha

(iii) Offset of Pistia & Eicchornia

(iv) Runner of Strawberry

(i) Runner:

It grows prostrate in all directions above the soil level. Nodes bear scale leaves. It has a creeping stem with long internodes. On the lower sides, nodes bear adventitious roots. Runner develops from the axils of lower leaves of aerial stem which sends slender horizontal branches in the form of runners. When older parts of plant die, the branches separate from parent plant and form independent plants e.g. Doob grass.

(ii) Stolon:

It is a slender lateral branch which appears from the lower part of main axis. This lateral branch grows aerially for some distance and becomes arched and finally touches the ground to give rise to new shoot with the help of its terminal bud. It also bears roots to get fixed with the soil e.g. Jasmine.

(iii) Offset:

Offset is shorter and thicker. It is usually found in aquatic plants like water hyacinth and Pistia. It bears a cluster of leaves near the water or ground level and gives adventitious roots inside water or ground from all nodes, e.g. Pistia.

(iv) Sucker:

Like the stolon the sucker is also a lateral branch but it grows obliquely upwards and gives rise to a new plants e.g. Mentha.

3. Aerial stems:

These modified aerial stems perform unusual functions. Different forms of these stems are the following:

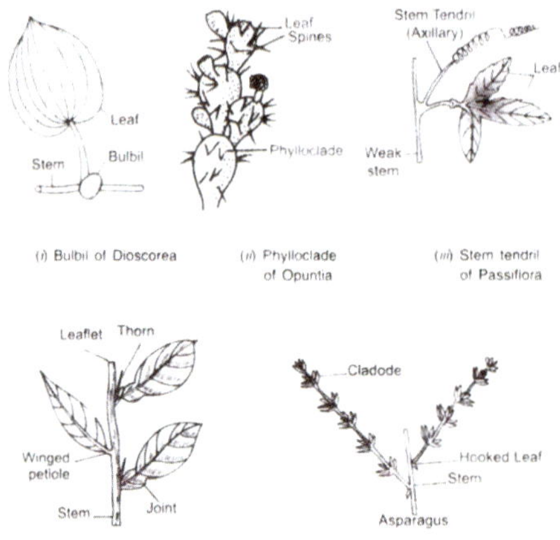

(i) Bulbil of Dioscorea (ii) Phylloclade of Opuntia (iii) Stem tendril of Passiflora
(iv) Stem thorn of Duranta (v) Cladode of Asparagus

(i) Stem tendril:

It is a leafless, spirally coiled branch formed in some climbers and helps them in climbing neighbouring objects; they may be modification of axillary bud, e.g. Passiflora.

(ii) Stem thorn:

Stem thorn is a hard, straight and pointed structure; it is a defensive organ; also helps in climbing; originates from axillary or terminal bud, e.g. Duranta.

(iii) Pylloclade:

It is a green, flattened or cylindrical stem, which takes the form and function of leaf. It contains chlorophyll and is responsible for carrying on photosynthesis. It bears succession of nodes and internodes at long or short intervals. Phylloclades are found in xerophytic plants where the leaves either grow feebly or fall off early or modified into spines e.g. Opuntia.

(iv) Cladode:

Phylloclade with one or two internodes is called cladode e.g. Asparagus. In Asparagus cladodes are needle-like, slightly flattened green structures which appear in cluster in the axil of a scaly leaf. Main stem bears leaf spines at its nodes. A scale leaf is found just above the spine. Every branch on main stem bears only scale leaves. In the axil of scale leaves cluster of cladodes appear.

(v) Bulbil:

Bulbil is the modification of vegetative or floral bud. It is swollen due to storage of food. It can function as an organ of vegetative propagation e.g. Dioscorea.

Leaf

The term leaf refers to the organ that forms the main lateral appendage on the stem of vascular plants. In general, leaves are thin, flat organs responsible for the photosynthesis of the plant. Although photosynthesis typically only occurs on the upper surface of the leaf, it can occur on both sides in some plant species. Leaves are typically comprised of a distinct upper and lower surface, stomata for gas exchange, waxy coating, hairs, and venation. Each side of the leaf differs in regarding the level to which these features are expressed. Although leaves are typically located above ground, some species have leaves which reside underground (e.g., bulb scales) or underwater (e.g., aquatic plant species). Moreover, the leaves of some plants may not be associated with photosynthesis (e.g., cataphylls). Leaves are typically oriented on a plant to avoid blocking the sunlight of the leaves situated underneath.

Parts of a Leaf

Most leaves have two main parts: (1) the blade and (2) the petiole, or leafstalk. The leaves of some kinds of plants also have a third part, called the *stipules*.

The Blade, or *lamina*, is the broad, flat part of the leaf. Photosynthesis occurs in the blade, which has many green food-making cells. Leaf blades differ from one another in several ways: (1) the types of edges, (2) the patterns of the veins, and (3) the number of blades per leaf.

Types of Edges: Almost all narrow, grasslike leaves and needles leaves have a blade with a smooth edge, as do many broadleaf plants, particularly those that are native to warm climates. The rubber plant, a common house plant, is a good example of such a plant.

The leaves of many temperate broadleaf plants have small, jagged points called teeth along the blade edge. Birch and elm trees have such leaves. Some plants have *hydathodes*, tiny valvelike structures that can release excess water from the leaf. The teeth of young leaves on many plants, including cottonwood and pin cherry trees, bear tiny glads. These glands produce liquids that protect the young leaf from plant-eating insects.

Some temperate broadleaf plants - including sassafras trees and certain mulberry and oak trees - have lobed leaves. The edge of such a leaf looks as if large bites have been taken out of it. This lobing helps heat escape from the leaf.

Patterns of the Veins. Veins carry food and water in a leaf. They also support the blade, much as the metal ribs support the fabric of an open umbrella.

In most broad leaves, the veins form a netlike pattern, with several large veins connected by smaller ones. The smallest veins supply every part of the blade with water. They also collect the food made by the green cells.

There are two main types of net-vein patterns - *pinnate* (featherlike) and *palmate* (palmlike or handlike). Pinnately veined leaves have one large central vein, called the *midrib*, which extends from the base of the blade to its tip. Other large veins branch off on each side of the midrib. The leaves of beech, birch, and elm trees have such a vein pattern. A palmately veined leaf has several main veins of about equal size, all of which extend from a common point at the base of the blade. The vein patterns of maple, sweet gum, and sycamore leaves are palmate.

Narrow leaves and needle leaves are not net-veined. Narrow leaves have a parallel-vein pattern. Several large veins run alongside one another from the base of the blade to the tip. Small crossveins connect the large veins. Needle leaves are so small that they have only one or two veins running through the center of the blade.

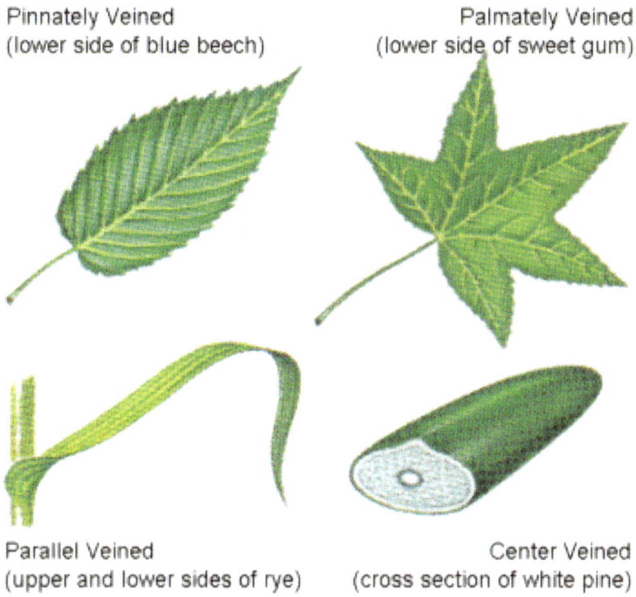

The petiole is the stemlike part of the leaf that joins the blade to the stem. Within a petiole are tiny tubes that connect with the veins in the blade. Some of the tubes carry water into the leaf. Others carry away food that the leaf has made. In many trees and shrubs, the petioles bend in such a way that the blades receive the most sunlight, thus assuring that few leaves are shaded by other leaves. The petiole also provides a flexible "handle" that enables the blade to twist in the wind and so avoid damage.

In some plants, the petioles are much larger than the stems to which they are attached. For example, the parts we eat of celery and rhubarb plants are petioles. In contrast, the leaves of some soft-stemmed plants, particularly grasses, have no petioles.

The Stipules are two small flaps that grow at the base of the petiole of some plants. In some plants, the stipules grow quickly, enclosing and protecting the young blade as it develops. Some stipules, such as those of willows and certain cherry trees, produce substances that prevent insects from attacking the developing leaf.

In many plants the stipules drop off after the blade has developed, but garden peas and a few other kinds of plants have large stipules that serve as an extra food-producing part of the leaf.

Types of Leaves

Leaf characteristics like shape, margin, hairs, etc are important external characteristics that help identify different plant species. The type of leaf is usually the characteristic of a species, and classification of leaves depend upon many different criteria.

The basic types of leaves are:

- Ferns - Leaves that have fronds.
- Conifer Leaves - These are typically needle or awl or scale shaped leaves.
- Angiosperm Leaves - They are typical, standard form of leaves that includes, stipules, a petiole and a lamina.
- Sheath Leaves - They are found in most of the grasses.

Leaves are of different types as follows based on the arrangement on the stem. They can be as follows:

- Alternate - Single leaves are attached at the nodes and they alternate in direction to a greater or lesser degree along the stem.
- Opposite - They are two structures one on each opposite side of the stem, they are attached in pairs on each node.
- Whorled - Here three or more leaves attach at each node on the stem.
- Rosulate - These leaves form a rosette.
- Rows - Leaves in this type of arrangement may be alternate or opposite in their attachment but not decussate.

Based on the divisions of the blade two basic forms of leaves can be described. They are simple leaf and compound leaf.

- Simple Leaf - It has an undivided blade. The shape of the leaf may be formed of lobes but the lobes do not reach the main vein or the midrib.
- Compound Leaf - The leaf blade is fully subdivided and each leaflet of the blase is separated along a main or secondary vein. The middle vein of a compound leaf is called a rachis.

Compound leaf can be of the following types based on the leaflets:

- Palmately Compound Leaves - They have leaflets that diverge from the end of the petiole like the fingers off the palm of the hand.
- Pinnately Compound Leaves - Here the leaflets are arranged along the main or mid-vein.
- Bipinnately Compound Leaves - Here the leaves are twice divided and the leaflets are arranged along a secondary vein.

Based on petiole leaves are of two types:

- Petiolate Leaves - They are also known as stalked leaves, these are the leaves that have a petiole. The petiole can be in different length from one plant to another.
- Sessile Leaves - These are unstalked leaves and do not possess a petiole. The lamina of the leaf expands directly from the stem.

Leaf Arrangements

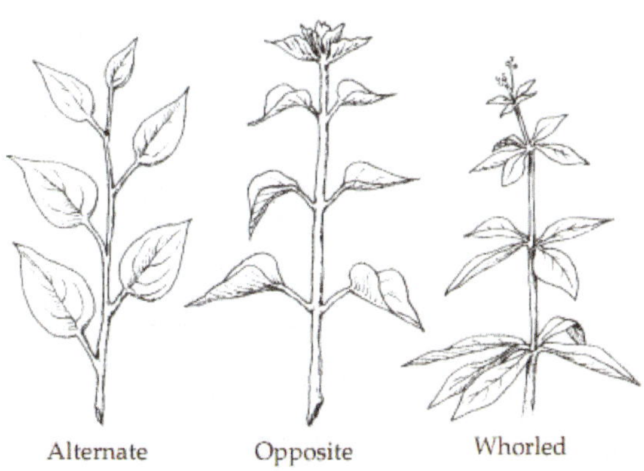

Leaf Arrangements

The study of leaf arrangements, or phyllotaxy, considers not only the descriptive classification of leaf arrangements but also theories regarding the cause of such arrangements.

The function of the arrangement of leaves (phyllotaxy) is to increase a plant's ability to carry on photosynthesis by positioning the leaves in such away as tomaximize the surface area available to intercept sunlight. Leaves may be either caulescent (on obvious stems) or acaulescent (with no obvious stems).

Flowering plants have three basic types of arrangements: alternate spiral; opposite; and whorled or verticillate. The alternate spiral arrangement is generally considered to be the most primitive condition, with the opposite and whorled conditions being derived by suppression of internode development.

There are two major hypotheses regarding the processes governing these basic arrangements. The field hypothesis of phyllotaxy posits that, as leaf primordia (newleaf cells) are created by the plant, a zone that inhibits the growth of other primordia is laid down around it, and not until the shoot tip has grown beyond that zone can a new leaf primordium be laid down. The first available space hypothesis posits that new leaves grow as soon as the plant shoot has grown out far enough to allow space for them.

The various types of leaf arrangements are usually one of the easiest vegetative characteristics to use in helping to identify vascular plants. This is especially true when leaf arrangement is combined with other characteristics, such as the presence or absence of petioles or the quality of being sessile or nonsessile. Other characteristics include the shape of the leaves and the appearance of the margins, bases, and apices types.

Alternate

grass family (Poaceae)

Alternately arranged leaves produce one leaf per node. These leaves may be on alternate sides of the stem (2-ranked or distichous), on one side of the stem (1-ranked or second), or in a spiral around the stem. If 2-ranked leaves overlap, as in some oncidium orchids and iris species, then they are referred to as equitant.

Leaves of members of the grass family (Poaceae) are distichous and alternate. Their leaves differ from most other vascular plant leaves in that they normally consist of a split tubular sheath that surrounds the stem and more or less linear blades held at right angles to the stem. They also have a small, tongue-like structure (ligule) at the junction of the sheath and blade, although in some species it may be obsolete.

Spiral

Spiral arrangements involve alternately arranged leaves in which each succeeding stem node and attached leaf is rotated slightly from the nodes below and above it. If the spiral is to the right, it is referred to as dextrorse; if to the left, it is referred to as sinistrorse.

Opposite

When two leaves occur at one node, the arrangement is called opposite. Oppositely arranged leaves may be either 2-ranked, as in Mexican heather (Cuphea hyssopifolia) in the henna family (Lythraceae), or 4-ranked or decussate, in which each succeeding pair of leaves is at right angles to the pairs above and below them. Decussate arrangement of leaves is characteristic of the mint family (Lamiaceae), themaple family (Aceraceae), and some members of the milkweed family (Asclepiadaceae), such as Asclepias viridis.

Whorled or Verticillate

When three or more leaves occur at one node, a whorled or verticillate arrangement is produced. The genera Galium and Sherardia in the madder family (Rubiaceae) are characterized by whorled leaves, as is also Isotria in the orchid family (Orchidaceae).

Rosette

carrots (Daucus carota)

Rosettes, often referred to as basal rosettes, occur in acaulescent plants, such as the common dandelion (Taraxacum officinalis) in the sunflower/aster family (Asteraceae). Acaulescent plants do have a stem, but the internodes are greatly contracted, and the leaves have a spiral alternate arrangement.

Many biennial plants, such as carrots (Daucus carota) and poison hemlock (Coniummaculatum) in the carrot family (Apiaceae), will produce a basal rosette during the first year of growth, followed by the production of a flowering stem with alternate leaves the second year.

Perfoliate

A leaf or a pair of connately fused leaves with the stem going through the center are referred to as perfoliate. Montia perfoliata and Bupleurum rotundifolium are examples of the perfoliate condition derived from a single leaf.

Silphium perfoliatum is a good example of the basal connate fusion of leaves to achieve the perfoliate condition. The upper cauline leaves of henbit (Lamium amplexicaule) in the mint family (Lamiaceae) are sessile and clasping the stem but are not actually fused.

Coniferous Leaves

The leaves of most conifers have developed with the need to minimize water loss while maximizing photosynthesis under relatively cold and dry (physiological drought) conditions where water is often not easily obtained. Needle-like leaves arranged in close, regularly spaced secund (one-sided, like a comb) divisions are referred to as pectinate or comblike.

Acicular leaves arranged in bundles or fascicles are typical of pines and spruces. In pines each fascicle is composed of two, three, four, five, six, seven, or eight divisions of needles, which form a more or less cylindrical shape if pushed together. The fascicles are spirally arranged on the tree branches.

eastern red cedar (Juniperus virginiana)

In eastern red cedar (Juniperus virginiana), the leaves are reduced to minute scales, which have an opposite decussate arrangement, giving the appearance of 4-ranks. The scales are imbricate or overlapping, much like shingles on a roof.

The leaves of the yew (Taxus) are sharp-pointed, flattened, and narrowly lance-shaped. They are spirally arranged on the branches but almost always give the appearance of being 2-ranked.

This is also true for the dawn redwood (Metasequoia) and bald cypress (Taxodium distichum), both of which are deciduous in the fall, dropping entire branchlets with the attached leaves. Yew podocarpus (Podocarpus macrophylla) in the podocarpus family (Podocarpaceae), on the other hand, has a obvious spiral arrangement of the leaves.

Function of the Leaf

As one of the most important constituents of plants, leaves have several essential functions:

Photosynthesis

The primary function of the leaf is the conversion of carbon dioxide, water, and UV light into sugar (e.g., glucose) via photosynthesis. The simple sugars formed via photosynthesis are later processed into various macromolecules (e.g., cellulose) required for the formation of the plant cell wall and other structures. Therefore, the leaf must be highly specialized to combine the carbon dioxide, water, and UV light for this process. Carbon dioxide is diffused from the atmosphere through specialized pores, termed stomata, in the outer layer of the leaf. Water is directed to the leaves via the plant's vascular conducting system, termed the xylem. Leaves are orientated to ensure maximal exposure to sunlight, and are typically thin and flat in shape to allow sunlight to penetrate the leaf to reach the chloroplasts, which are specialized organelles that perform photosynthesis. Once sugar is formed from photosynthesis, the leaves function to transport it down the plant via specialized structures called the phloem, which run in parallel to the xylem. The sugar is typically transported to the roots and shoots of the plant, to support growth.

Transpiration

Transpiration refers to the movement of water through the plant, and subsequent evaporation via the leaves. When the stomata open to accommodate the diffusion of carbon dioxide into the plant for photosynthesis, water flows out. This process also serves to cool the plant via evaporation of the water from the leaf, as well as regulate the plant's osmotic pressure.

Guttation

Guttation refers to the excretion of xylem from the edges of leaves and other vascular plants due to increased levels of water in the soil at night, when the stomata are closed. The pressure caused at the roots results in the leakage of water from the xylem out of specialized water glands at the edges of leaves.

Storage

Leaves are a primary site of water and energy storage since they provide the site of photosynthesis. Succulents are particularly adept at water storage, as evidenced by the thick leaves. Due to the high levels of nutrients and water, many animal species ingest the leaves of plants as a source of food.

Defense

Some leaves have also evolved defense mechanisms to avoid being eaten or damaged. Some examples include the spines of cacti, cones of gymnosperms, respectively. In addition, hairs found on leaves prevent water loss in dry climates and sting animals that detour herbivores (e.g., *Urticaceae*). Moreover, the waxy coatings found on leaves serve to protect against water loss, rain, and forms of contamination. Oils and other secreted substances also detract from being consumed by herbivores.

Inflorescence

Inflorescence, in a flowering plant is a cluster of flowers on a branch or a system of branches.

However, it also applies to the blooming order or pattern of those flowers, as well as how they are arranged, and more.

All flowering plants, from weeds to trees, have a specific flowering pattern called inflorescence.

There are six major types of inflorescence, divided into two categories – determinate inflorescence and indeterminate inflorescence.

Determinate inflorescence means that the youngest flowers are at the bottom of the structure, and the oldest at the top. Indeterminate inflorescence means that the youngest flowers are at the top of the structure, and the oldest flowers are at the bottom.

There are 6 major types of Inflorescence:

1. Racemose Inflorescence:

In this type of inflorescence the main axis does not end in a flower, but it grows continuously and develops flowers on its lateral sides in acropetal succession (i.e., the lower or outer flowers are older than the upper or inner ones). The various forms of racemose inflorescence may be described under three heads.

They are as follows:

(i) With the main axis elongated, i.e., (a) raceme; (b) spike; (c) spikelets; (d) catkin and (e) spadix.

(ii) With the main axis shortened, i.e., (i) corymb and (ii) umbel.

(iii) With the main axis flattened, i.e., capitulum or head.

(i) Main Axis Elongated:

(a) Raceme:

In such cases the main axis remains elongated and it bears laterally a number of stalked flowers. The lower or older flowers possess longer stalks than the upper or younger ones, e.g., radish (Raphanus sativus), mustard (Brassica campestris), etc.

When the main axis of raceme is branched and the lateral branches bear the flowers, the inflorescence is known as compound raceme or panicle, e.g., neem (Azadirachta indica), gul-mohar (Delonix regia), etc.

The main axis of the inflorescence together with the latest axes, if present, is termed as the peduncle. The stalk of the individual flower of the inflorescence is called the pedicel.

(b) Spike:

In this type of racemose inflorescence the main axis remains elongated and the lower flowers are older, i.e., opening earlier than the upper ones, as found in raceme, but here the flowers are sessile, i.e., without pedicel or stalk, e.g., amaranth (Amaranthus spp.), latjira (Achyranthes aspera), etc.

(c) Spikelets:

Each spikelet may bear one to several flowers (florets) attached to a central stalk known as rachilla. Spikeletes are arranged in a spike inflorescence which is composed of several to many spikelets which are combined in various manners on a main axis called the rachis. Some are in compound spikes (i.e., in wheat—Triticum aestivum), others are in racemes (e.g., in Festuca), while some are in panicles (e.g., in Avena).

The usual structure of spikelet is as— There is a pair of sterile glumes at the base of spikelet, the lower, outer glume called the first, and the upper, inner one called the second. Just above the glumes, there is series of florets, partly enclosed by them.

Each floret has at its base a lemma and palea. The lemma is the lower, outer bract of the floret. Usually the lemma also known as inferior palea bears a long awn as an extension of the mid-rib at the apex or back.

The floral parts borne in the axil of lemma. The palea (also known as superior palea) often with two longitudinal ridges (keels or nerves), stands between the lemma and the rachilla. Flowers and glumes are arranged on the spikelet in two opposite rows. Spikeletes are characteristic of Poaceae

(Gramineae) or Grass family, e.g., grasses, wheat, barley, oats, sorghum, sugarcane, bamboo, etc.

(d) Catkin:

This is a modified spike with a long and drooping axis bearing unisexual flowers, e.g., mulberry (Moras alba), birch (Betula spp.), oak (Quercus spp.), etc.

(e) Spadix:

This is also a modification of spike inflorescence having a fleshy axis, which remains enclosed by one or more large, often brightly coloured bracts, the spathes, e.g., in members of Araceae, Musaceae and Palmaceae. This inflorescence is found only in monocotyledonous plants.

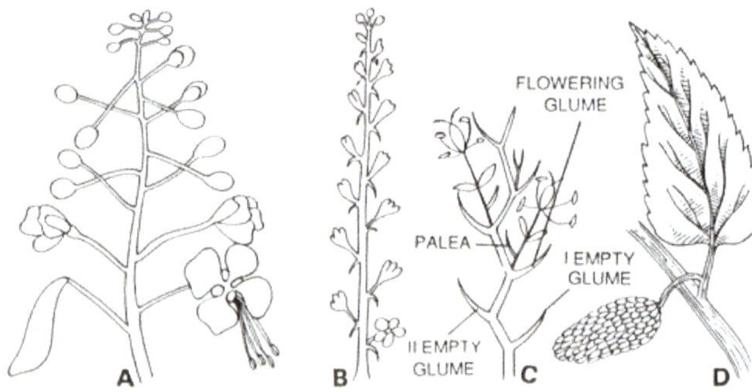

Inflorescence-racemose. A, raceme of **gul-mohar** ; B, spike ; C, spikelet of a grass; D, female catkin of mulberry.

(ii) Main Axis Shortened:

(a) Corymb:

In this inflorescence the main axis remains comparatively short and the lower flowers possess much longer stalks or pedicels than the upper ones so that all the flowers are brought more or less to the same level, e.g., in candytuft (Iberis amara).

(b) Umbel:

In this inflorescence the primary axis remains comparatively short, and it bears at its tip a group of flowers which possess pedicels or stalks of more or less equal lengths so that the flowers are seen to spread out from a common point. In this inflorescence a whorl of bracts forming an involucre is always present, and each individual flower develops from the axil of a bract.

Generally the umbel is branched and is known as umbel of umbels (compound umbel), and the branches bear flowers, e.g., in coriander (Coriandrum sativum), fennel, carrot, etc. Sometimes, the umbel is un-branched and known as simple umbel, e.g., Brahmi (Centella asiatica). This inflorescence (umbel) is characteristic of Apiaceae (Umbelliferae) family.

(iii) Main Axis Flattened:

Capitulum or Head:

In this type of inflorescence the main axis or receptacle becomes suppressed, and almost flat, and the flowers (also known as florets) are sessile (without stalk) so that they become crowded together on the flat surface of the receptacle. The florets are arranged in a centripetal manner on the receptacle, i.e., the outer flowers are older and open earlier than the inner ones.

The individual flowers (florets) are bracteate. In addition the whole inflorescence remains surrounded by a series of bracts arranged in two or three whorls.

The flowers (florets) are usually of two kinds:

(i) Ray florets (marginal strap-shaped flowers) and

(ii) Disc florets (central tubular flowers).

The capitulum (head) may also consist of only one kind of florets, e.g., only tubular florets in Ageratum or only ray or strap-shaped florets in Sonchus. A capitulum or head is characteristic of Asteraceae (Compositae) family, e.g., sunflower (Helianthus annuus), marigold (Tagetes indica), safflower (Carthamus tinctorius). Zinnia, Cosmos, Tridax, Vernonia, etc. Besides, it is also found in Acacia and sensitive plant (Mimosa pudica) of Mimosaceae family.

The capitulum inflorescence has been considered to be the most perfect. The reasons are as follows:

The individual flowers are quite small and massed together in heads, and therefore, they add to greater conspicuousness to attract the insects and flies for pollination.

At the same time there is a considerable saving of material in the construction of the corolla and other floral parts.

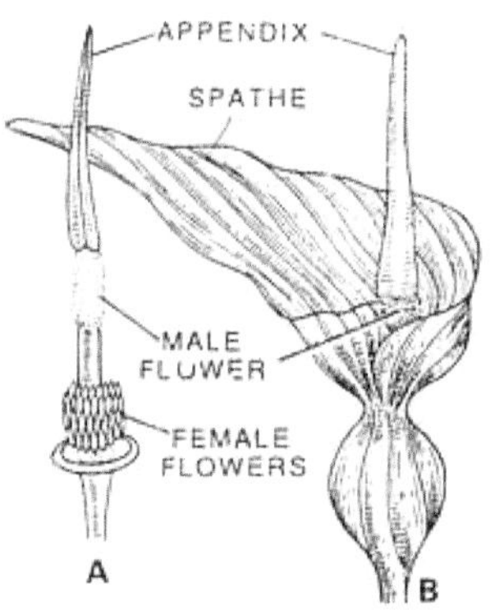

Inflorescence-spadix–A, spadix of an aroid without spathe; B, same with spathe.

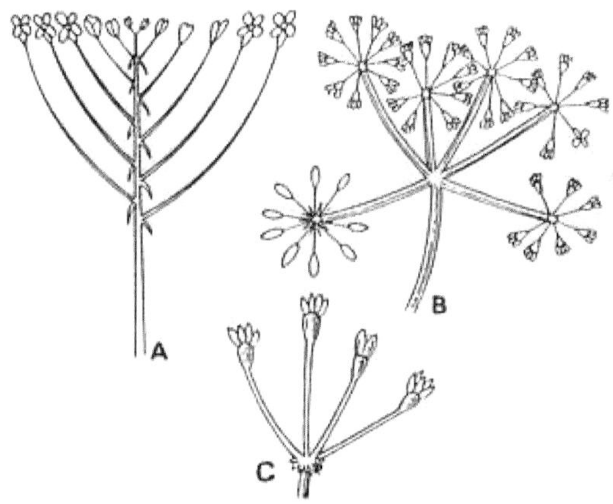

Inflorescence, A, corymb; B, compound umbel (umbel of umbels); C, simple umbel.

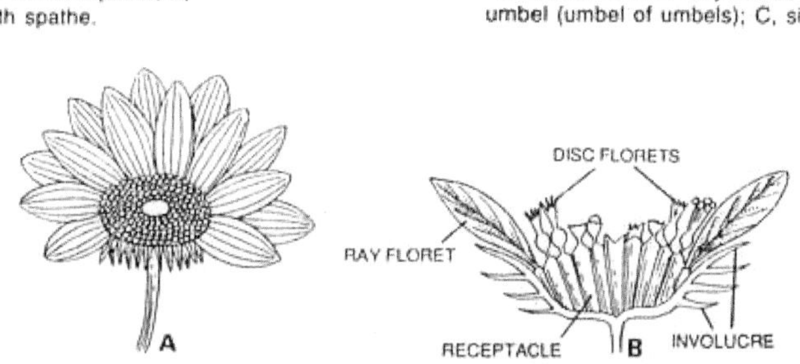

Inflorescence, Head (capitulum)—A, a head; B, a head in L.S.

A single insect may pollinate flowers in a short time without flying from one flower to another.

2. Cymose Inflorescence:

In this type of inflorescence the growth of the main axis is ceased by the development of a flower at its apex, and the lateral axis which develops the terminal flower also culminates in a flower and its growth is also ceased. The flowers may be pedicellate (stalked) or sessile (without stalk).

Here the flowers develop in basipetal succession, i.e., the terminal flower is the oldest and the lateral ones younger. This type of opening of flowers is known as centrifugal.

The cymose inflorescence may be of four main types:

(i) Uniparous or monochasial cyme;

(ii) Biparous or dichasial cyme;

(iii) Multiparous or polychasial cyme and

(iv) Cymose capitulum.

(i) Uniparous or Monochasial Cyme:

Here the main axis ends in a flower and it produces only one lateral branch at a time ending in a flower. The lateral and succeeding branches again produce only one branch at a time like the primary one.

There are three forms of uniparous cyme:

(a) Helicoid,

(b) Scorpioid, and

(c) Sympodial

(a) Helicoid Cyme:

When the lateral axes develop successively on the same side, forming a sort of helix, the cymose inflorescence is known as helicoid or one-sided cyme, e.g., in Begonia, Juncus, Hemerocallis and some members of Solanaceae.

(b) Scorpioid Cyme:

When the lateral branches develop on alternate sides, forming a zigzag, the cymose inflorescence is known as scorpioid or alternate-sided cyme, e.g., in Gossypium (cotton), Drosera (sundew), Heliotropium, Freesia, etc.

(c) Symopodial Cyme:

Sometimes, in monochasial or uniparous cyme successive axes may be at first curved or zig-zag (as in scorpioid cyme) but later on it becomes straight due to rapid growth, thus forming a central or pseudoaxis. This type of inflorescence is known as sympodial cyme as found in some members of Solanaceae (e.g., Solanum nigrum).

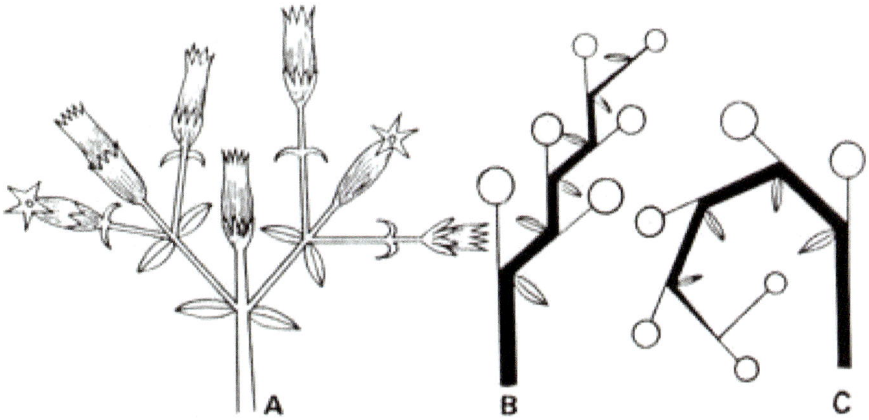

Inflorescence. Cymose—A, biparous (dichasial) cyme; B, scorpioid cyme; C, helicoid cyme.

(ii) Biparous or Dichasial Cyme:

In this type of inflorescence the peduncle bears a terminal flower and stops growing. At the same time the peduncle produces two lateral younger flowers or two lateral branches each of which terminates in a flower.

There are three flowers; the oldest one is in the centre. The lateral and succeeding branches in their turn behave in the same manner, e.g., jasmine, teak, Ixora, Saponaria, etc. This is also known as true cyme or compound dichasium.

(iii) Multiparous or Polychasial Cyme:

In this type of cymose inflorescence the main axis culminates in a flower, and at the same time it again produces a number of lateral flowers around. The oldest flower is in the centre and ends the main floral axis (peduncle). This is a simple polychasium.

The whole inflorescence looks like an umbel, but is readily distinguished from the latter by the opening of the middle flower first, e.g., Ak (Calotropis procera), Hamelia patens, etc.

(iv) Cymose Capitulum:

This type of inflorescence is found in Acacia, Mimosa and Albizzia. In such cases the peduncle is reduced or condensed to a circular disc. It bears sessile or sub-sessile flowers on it. The oldest flowers develop in the centre and youngest towards the periphery of the disc, such arrangement is known as centrifugal. The flowers make a globose head, which is also called glomerule.

3. Compound Inflorescence:

In this type of inflorescence the main axis (peduncle) branches repeatedly once or twice in racemose or cymose manner. In the former case it becomes a compound raceme and in the latter case it becomes a compound cymose inflorescence.

The main types of compound inflorescence are as follows:

1. Compound Raceme or Panicle:

In this case the raceme is branched, and the branches bear flowers in a racemose manner, e.g., Delonix regia, Azadirachta indica, Clematis buchaniana, Cassia fistula, etc.

2. Compound Umbel:

Also known as umbel of umbels. Here the peduncle (main axis) is short and bears many branches which arise in an umbellate cluster. Each such branch bears a group of flowers in an umbellate manner. Usually a whorl of leafy bracts is found at the base of branches and also at the bases of flowers arranged in umbellate way.

The former whorl of bracts is called involucre and the latter involucel. Typical examples of compound umbel are—Daucus carota (carrot), Foeniculum vulgare (fennel), Coriandrum sativum (coriander), etc.

3. Compound Corymb:

Also known as corymb of corymbs. Here the main axis (peduncle) branches in a corymbose manner and each branch bears flowers arranged in corymbs. Typical example-cauliflower.

4. Compound Spike:

Also known as spike of spikelets. The typical examples are found in Poaceae (Gramineae) family such as-wheat, barley, sorghum, oats, etc. This type has already been described under sub-head spikelets.

5. Compound Spadix:

Also known as spadix of spadices. Here the main axis (peduncle) remains branched in a racemose manner and each branch bears sessile and unisexual flowers. The whole branched structure remains covered by a single spathe. The examples are common in Palmaceae (Palmae) family.

6. Compound Head:

Also known as head of heads or capitulum of capitula. In this case many small heads form a large head. The typical example is globe thistle (Echinops). In this plant the heads are small and one-flowered and are arranged together forming a big compound head.

4. Cyathium:

This type of inflorescence is found in genus Euphorbia of family Euphorbiaceae; also found in genus Pedilanthus of the family. In this inflorescence there is a cup-shaped involucre, often provided with nectar secreting glands. The involucre encloses a single female flower, represented by a pistil, in the centre, situated on a long stalk.

This female flower remains surrounded by a number of male flowers arranged centrifugally. Each male flower is reduced to a solitary stalked stamen. It is evident that each stamen is a single male flower from the facts that it is articulated to a stalk and that it possesses a scaly bract at the base. The examples can be seen in poinsettia (Euphorbia), Pedilanthus, etc.

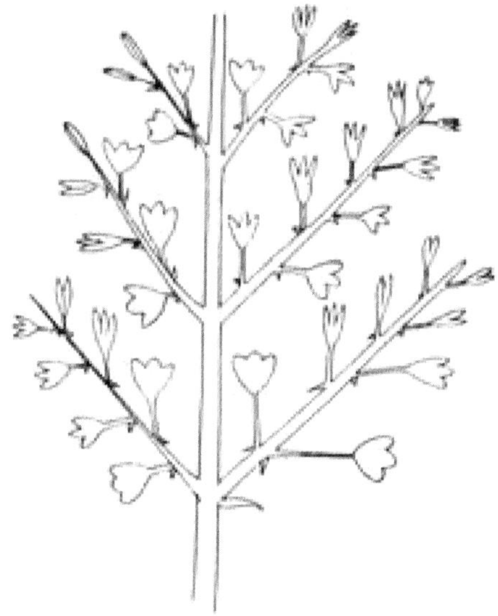

Inflorescence: Panicle.

5. Verticillaster:

This type of inflorescence is a condensed form of dichasial (biparous) cyme with a cluster of sessile or sub-sessile flowers in the axil of a leaf, forming a false whorl of flowers at the node. The first of main floral axis gives rise to two lateral branches and these branches and the succeeding branches bear only one branch each on alternate sides.

The type of inflorescence is characteristic of Lamiaceae (Labiatae) family. Typical examples, are—Ocimum, Coleus, Mentha, Leucas, etc.

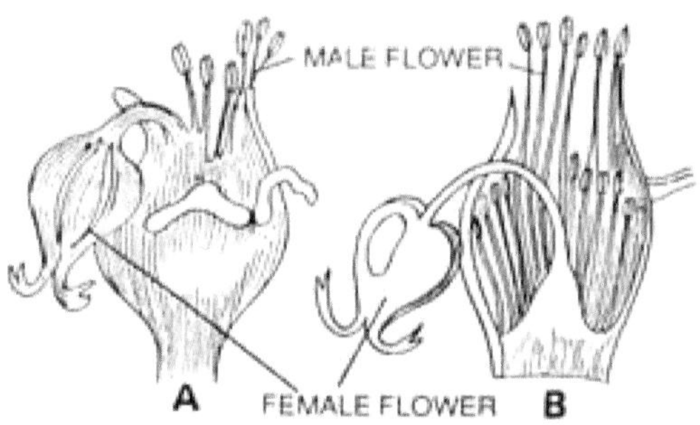

Inflorescence-special. Cyathium—A, cyathium of *Euphorbia;* B, L.S. of same.

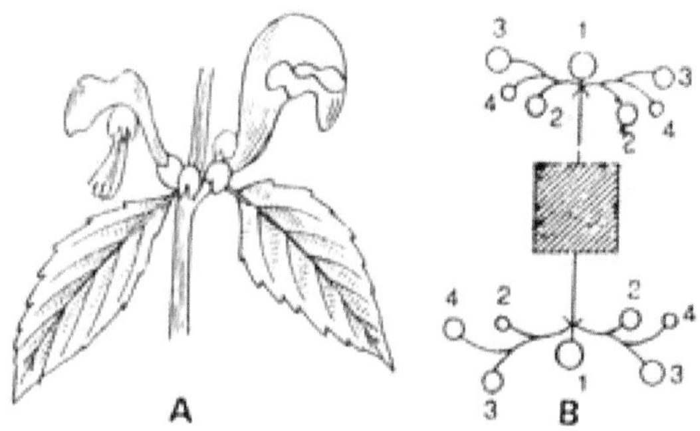

Inflorescence—special verticillaster—A verticillaster of Labiatae; B, diagram of same.

6. Hypanthodium:

In this type of inflorescence the receptacle forms a hollow cavity with an apical opening guarded by scales. Here the flowers are borne on the inner wall of the cavity. The flowers are unisexual; the female flowers develop at the base of the cavity and the male flowers towards the apical pore. The examples are found in genus Ficus of Moraceae family, e.g., Ficus carica, F. glomerata, F. benghalensis, F. religiosa, etc.

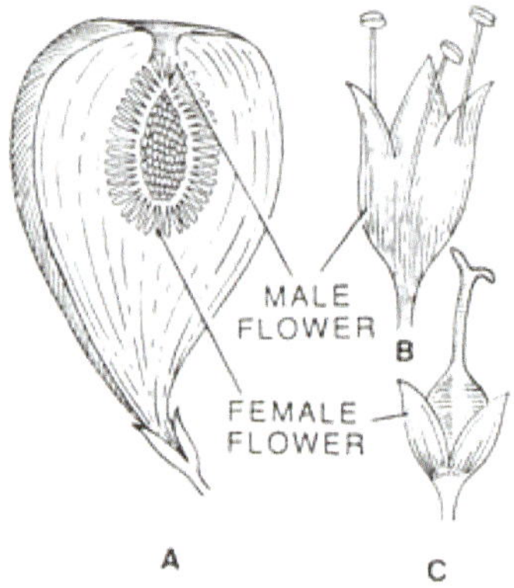

Inflorescence. A, hypanthodium of *Ficus*; B, male flower; C, female flower.

Table : Differences Between Cyathium, Verticillaster and Hypanthodium Inflorescence.

1. Cyathium	2. Verticillaster	3. Hypanthodium
The cupular structure is developed by an involucre of bracts. There is a single female flower in the centre, surrounded by numerous stalked male flowers. Each flower is represented by a single stamen, e.g., *Euphorbia*.	This is special type of cymose inflorescence. It consists of two clusters of flowers that develop from each of the two opposite axils of the leaves. Each cluster represents a dichasial cyme. Flowers are sessile and appear as a cluster around the node, *e.g.*, *Salvia, Ocimum*, etc.	The main axis develops in a cupular receptacle, with an apical opening at the apex. Both male and female flowers are present on the inner wall of cavity enclosed in a cupular structure, *e.g.*, *Ficus*.

Table : Main Types of Inflorescences

Racemose

- **Raceme.** An elongated axis bearing stalked flowers (*e.g.*, *Mustard*).
- **Spike.** Same as raceme, but flowers have no stalks, (*e.g.*, *Achyranhes*).
- **Catkin.** Spike with unisexual flowers, (*e.g.*, *Morus*).
- **Spadix.** Spike with fleshy axis enclosed by one or more large bracts, (*e.g.*, *Colocasia, Musa*, etc.)
- **Corymb.** The axis is short, and the lower flowers have longer stalks than the upper ones. Thus all flowers come to the same level, (*e.g.*, Candytuft.)
- **Umbel.** The axis is short and bears a cluster of flowers with stalks of equal length arsing from a common point, (*e.g.*, Coriander)
- **Head or Capitulum.** The main axis is a flattened, more or less convex structure on which sessile flowers (florets) are arranged in a centripetal order. The inflorescence is surrounded by prominent bracts, (*e.g.*, Sunflower, marigold, etc.)

Cymose

- **Monochasial Cyme** The main axis terminates in a flower and one lateral branch axis develops from its base which also ends in a flower, (*e.g.*, *Begonia, cotton*, etc.)
- **Dichasial Cyme.** Two lateral branches develop on either side of the terminal flower of the main axis; the lateral branches also end in a flower and may again branch similarly (*e.g.*, jasmine, *Dianthus*, etc.)
- **Polychasial Cyme.** More than two lateral branches arise from the base of the terminal flower, (*e.g.*, *Calotropis*).

Special Types

- **Hypanthodium.** The main axis forms a cup-shaped receptacle with a small opening at the top. Flowers are enclosed within the cup in cymose groups. (*e.g.* *Ficus*).
- **Cyathium.** Here the involucre forms a cup. Single female flower without perianth arises in the middle surrounded by a large number of male flowers represented by stalked stamens. *e.g.*, *Euphorbia*.
- **Verticillastor.** In this type of inflorescence, typical of plants with opposite leaves, a cyme arises in each leaf axil. The first axis ends in a flower. Two branches arise below it bearing branches in an alternating manner. Flowers are sessile and appear as a cluster around the node, *e.g.*, *Ocimum*.

Flower

A flower, also known as a bloom or blossom, is the reproductive structure found in flowering plants. Flowering plants or angiosperms are plants of the division *Magnoliophyta* and one of the two major groups of seed plants, alongside the gymnosperms.

The flower structure contains the plant's reproductive organs, and its function is to produce seeds through sexual reproduction. For the higher plants, seeds are the next generation, and serve as the primary means by which individuals of a species are dispersed across the landscape. After fertilization, portions of the flower develop into a fruit containing the seeds.

Flowers are a primary example of mutualism in nature. By attracting insects and birds for pollination, flowering plants utilize them to spread their genetic material—contained in the pollen—over great distances. At the same time, the flower provides the primary source of nutrition to the pollinating organisms. As a result, whole classes of organisms—notably bees, butterflies and hummingbirds—have evolved to feed off of flowers. In their symbiotic relationship, flowers and their pollinators have evolved together, forming an inseparable bond.

Some flowers have developed intricate methods to assure that the pollen becomes attached to the insect as it seeks for nectar. This may involve color patterns, some only seen in the ultraviolet range, or a type of tunnel that the insect has to pass through to get to the nectar. Flowers emit scents that can attract the insect of choice, even aromas like decaying meat to attract flies. Both the plant and the pollinator receive benefit—the plant is pollinated and the pollinator receives nourishment.

Some flowers, primarily those that attract birds and insects for pollination, exhibit exquisite beauty and variety in form, color, color patterns, and aroma. These flowers provide beauty and joy to humans. Flowers have been show to correlate with human well-being, including reduced stress, improved mood, enhanced self-esteem, better communication, and other benefits. Thus another symbiotic relationship has formed—between flowers and humans. People have cultivated flowers to develop unique and beautiful varieties, with colors and patterns not found in nature. To better enjoy the beauty of their blooms, they have propagated flowering plants in their gardens and greenhouses, thus increasing the success of their species.

As they fulfill their reproductive purpose while providing such tangible benefits to other living things, flowers thus provide an easy-to-understand example of the principle that every entity in creation has both an individual purpose (survival and reproduction) and a whole purpose (providing value to a higher entity or cause). Contributing to the whole is the path to the flourishing of the individual. It is this principle that underlies the harmony and unity found throughout nature, and in human society as well.

Flower Anatomy

Flowering plants are *heterosporangiate*; that is, producing two types of haploid reproductive spores. A spore is a normally haploid and unicellular reproductive mechanism produced by plants through meiosis. The *pollen* (male spores) and *ovules*(female spores) are produced in different organs, but these are together in a *bisporangiate strobilus* that is the typical flower.

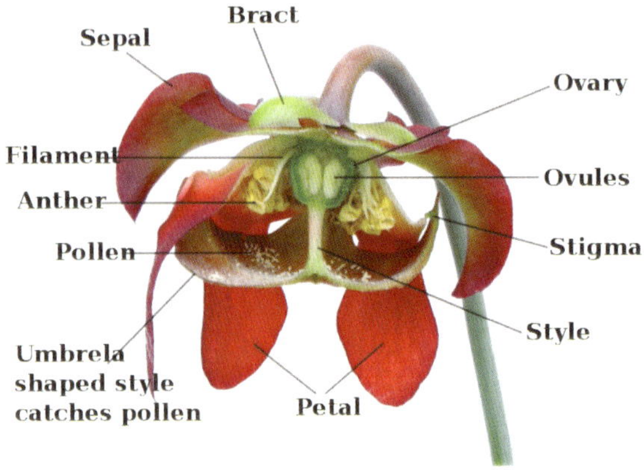

A flower is regarded as a modified stem with shortened internodes and bearing, at its nodes, structures that may be highly modified leaves. In essence, a flower structure forms on a modified shoot or *axis* with an apical meristem that does not grow continuously (growth is *determinate*). The stem is called a pedicel, the end of which is the torus or receptacle. The parts of a flower are arranged in whorls, cicles of similar parts radiating from approximately the same center, on the torus. The four main parts or whorls (starting from the base of the flower or lowest node and working upwards) are as follows:

- *Calyx* – The outer whorl of *sepals* ; typically these are green, but are petal-like in some species.
- *Corolla* – The whorl of *petals*, which are usually thin, soft and colored to attract insects that help the process of pollination.
- *Androecium* – One or two whorls of stamens, each a *filament* topped by an *anther* where pollen is produced. Pollen contains the male gametes.
- *Gynoecium* – one or more pistils. The female reproductive organ is the carpel: this contains an ovary with ovules (which contain female gametes). A pistil may consist of a number of carpels merged together, in which case there is only one pistil to each flower, or of a single individual carpel (the flower is then called apocarpous). The sticky tip of the pistil, the stigma, is the receptor of pollen. The supportive stalk, the style becomes the pathway for pollen tubes to grow from pollen grains adhering to the stigma, to the ovules, carrying the reproductive material. A single flower can have more than one carpel.

Tulip - androecium and gynoecium

Although the floral structure described above is considered the "typical" structural plan, plant species show a wide variety of modifications from this plan. These modifications have significance in the evolution of flowering plants and are used extensively by botanists to establish relationships among plant species. For example, the two subclasses of flowering plants may be distinguished by the number of floral organs in each whorl: dicotyledons typically having four or five organs (or a multiple of four or five) in each whorl and monocotyledons having three or some multiple of three. The number of carpels in a compound pistil may be only two, or otherwise not related to the above generalization for monocots and dicots.

In the majority of species, individual flowers have both pistils and stamens as described above. These flowers are described by botanists as being perfect, bisexual, or hermaphrodite—having both male and female parts.

However, in some species of plants the flowers are imperfect or unisexual—having only either male (stamens) or female (pistil) parts. In the latter case, if an individual plant is either male or female, the species is regarded as *dioecious*. However, where unisexual male and female flowers appear on the same plant, the species is considered *monoecious*.

Plant species with both functional stamens and carpels on the same flower, but that mature at different times, are called dichogamous.Depending on the stage of maturity, the single flower can be either staminate or pistillate.

Preventing Self-fertilization

Some flowers with both stamens and a pistil are capable of self-fertilization, which does increase the chance of producing seeds but limits genetic variation. The extreme case of self-fertilization occurs in flowers that always self-fertilize, such as the common dandelion.

Conversely, many species of plants have ways of preventing self-fertilization. Unisexual male and female flowers on the same plant may not appear at the same time, or pollen from the same plant may be incapable of fertilizing its ovules. The latter flower types, which have chemical barriers to their own pollen, are referred to as self-sterile or self-incompatible. Another mechanism to prevent self-pollinization is by physically separating the anthers and stigmas. When these organs are arranged so that they do not come in contact with one anther, the pollen will be more likely to transfer to a stigma of another flower.

Close-up of a Day lily flower showing six stamens and the stigma and style of a pistil

In those species that have more than one flower on an axis, the collection of flowers is termed an inflorescence. In this sense, care must be exercised in considering what is a *flower*. In botanical terminology, a single daisy or sunflower for example, is not a flower but a flower head—an inflorescence comprised of numerous small flowers (sometimes called florets). Each small flower may be anatomically as described above.

Floral Formula

A floral formula is a way to represent the structure of a flower using specific letters, numbers, and symbols. Typically, a general formula will be used to represent the flower structure of a plant family rather than a particular species. The following representations are used:

Ca = calyx (sepal whorl; e.g. Ca^5 = 5 sepals)
Co = corolla (petal whorl; e.g., $Co^{3(x)}$ = petals some multiple of three)
Z = add if *zygomorphic* (e.g., CoZ^6 = zygomorphic with 6 petals)
A = *androecium* (whorl of stamens; e.g., A^∞ = many stamens)
G = *gynoecium* (carpel or carpels; e.g., G^1 = monocarpous)

x - to represent a "variable number"
∞ - to represent "many"

A floral formula would appear something like this:

$$Ca^5Co^5A^{10-\infty}G^1$$

Flower Function: Pollination

The function of a flower is to mediate the union of male and female gametes. The process is termed pollination. Many flowers are dependent upon the wind to move pollen between flowers of the same species. Others rely on animals (especially insects) to accomplish this feat. The period of time during which this process can take place (the flower is fully expanded and functional) is called anthesis.

Grains of pollen on stigma of a lily

Many flowers in nature have evolved to attract animals to pollinate the flower, the movements of the pollinating agent contributing to the opportunity for genetic recombinations within a dispersed plantpopulation. Flowers that are insect-pollinated are called *entomophilous* (literally "insect-loving"). Flowers commonly have glands called nectaries on their various parts that attract these animals. Birds and bees are common pollinators: both having color vision, thus opting for "colorful" flowers. Some flowers have patterns, called nectar guides, that show pollinators where to look for nectar; they may be visible to us or only under ultraviolet light, which is visible to bees and some other insects. Flowers also attract pollinators by scent. Many of their scents are pleasant to our sense of smell, but not all. Some plants, such as *Rafflesia,* the titan arum, and the North American pawpaw *(Asimina triloba),* are pollinated by flies, so produce a scent imitating rotting meat.

In any case, pollinators are attracted to the plant, perhaps in search of nectar, which they eat. The arrangement of the stamens ensures that pollen grains are transferred to the bodies of the pollinator. In gathering nectar from many flowers of the same species, the pollinator transfers pollen between all of the flowers it visits.

The flowers of other species are pollinated by the wind (for example, grasses); they have no need to attract pollinators and therefore tend not to be "showy." Wind-pollinated flowers are referred to as *anemophilous*. Whereas the pollen of entomophilous flowers tends to be large-grained, sticky, and rich in protein (another "reward" for pollinators), anemophilous flower pollen is usually small-grained, very light, and of little nutritional value to insects, though it may still be gathered in times of dearth. Honeybees and bumblebees actively gather anemophilous corn (maize) pollen, though it is of seemingly little value to them.

Seed

In botany, seeds refer to the embryonic plants growing in hard or semi-hard outer coverings. The seedling stage is an essential part of propagation in seed plants. They emerge from the ripened ovule after pollen fertilization. The seed coat is developed from the ovule integuments while the embryo inside comes from the zygote. Seeds are crucial to the reproductive process of angiosperm as well as gymnosperm plants.

Angiosperm seeds are produced in fruits within the tough, fleshy interior. On the other hand, gymnosperm seeds do not require any special structure to grow and develop directly on cone bracts. With time, they become gradually enclosed by the cone scales.

Seed production varies in response to insects, climate, diseases, and the plant's internal cycle. According to botanists, both spatulate and linear gymnosperms have two different types of seeds, while dicotyledons plants have 10 varieties of seeds and monocotyledon plants boast six various types of seeds.

While most plants are propagated by seedlings, some seeds are dispersed through the wind. These include poplar, milkweed, orchids, and pine, among others. Seeds produced by Dioclea and Mucuna plants are quite buoyant. As such, they are dispersed through water as they wash up on beaches or float in ponds and rivers.

Seed Structure

A fertilized seed contains the embryo, a multicellular diploid eukaryote in its earliest stage of development, from which a new plant will grow under proper conditions. It also contains a supply of stored food and is wrapped in the seed coat or *testa*.

In angiosperms, the stored food begins as a tissue called the endosperm, which is derived from the parent plant via double fertilization. The usually triploid endosperm (containing three sets of chromosomes) is rich in oil or starch and protein.

In gymnosperms, such as conifers, the food storage tissue is part of the female gametophyte, a haploid tissue.

In some species, the embryo is imbedded in the endosperm, which the seedling will use upon germination. In others, the endosperm is absorbed by the embryo as the latter grows within the developing seed, and the cotyledons of the embryo become filled with this stored food. At maturity, seeds of these species have no endosperm. Some common plant seeds that lack an endosperm are bean, pea, oak, walnut, squash, sunflower, and radish. Plant seeds with an endosperm include all conifers and most monocotyledons (e.g., grasses and palms), and many dicotyledons (e.g. brazil nut and castor bean).

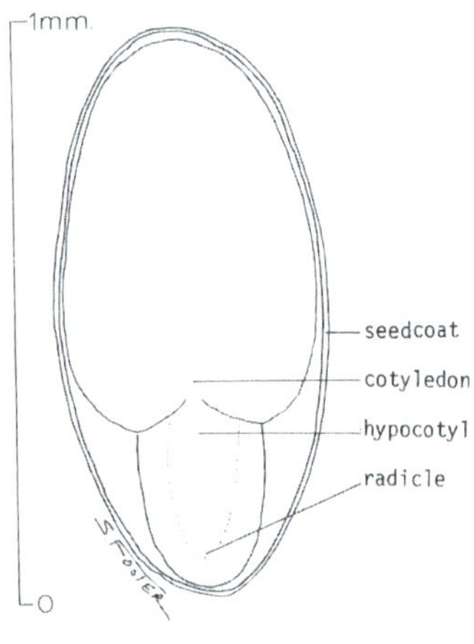

Seed of Scouler's Willow *(Salix scouleriana)*

The *seed coat* develops from tissues (called *integument*) originally surrounding the ovule. The seed coat in the mature seed can be a paper-thin layer (as for example, in the peanut) or something more substantial (as for example, thick and hard in honey locust and coconut). The seed coat helps protect the embryo from mechanical injury and from drying out.

In order for the seed coat to split, the embryo must imbibe (soak up water), which causes it to swell, splitting the seed coat. However, the nature of the seed coat determines how rapidly water can penetrate and subsequently initiate germination. For seeds with a very thick coat, scarification of the seed coat may be necessary before water can reach the embryo. Examples of scarification include gnawing by animals, freezing and thawing, battering on rocks in a stream bed, or passing through an animal's digestive tract. In the latter case, the seed coat protects the seed from digestion, while perhaps digestion weakens the seed coat such that the embryo is ready to sprout when it gets deposited (along with a bit of fertilizer) far from the parent plant. In species with thin seed coats, light may be able to penetrate into the dormant embryo. The presence of light or the absence of light may trigger the germination process, inhibiting germination in some seeds buried too deeply or in others not buried in the soil. Abscisic acid, a natural occurring compound in plants, is responsible for inducing seeds to synthesize storage proteins and plays a role in induction and maintenance of dormancy. It is usually the growth inhibitor in seeds.

The seeds of angiosperms are contained in a hard or fleshy (or with layers of both) structure called a fruit. An example of a hard fruit layer surrounding the actual seed is that of the so-called *stone* fruits (such as the peach). Gymnosperm seeds begin their development "naked" on the bracts of cones, although the seeds do become covered by the cone scales as they develop.

Types of Seeds

A Seed is primarily of two types. The two types are:

- Monocotyledonous Seed
- Dicotyledonous Seed

Structure of a Monocotyledonous Seed

A monocotyledonous seed, as the name suggests, has only one cotyledon. There is only one outer layering of the seed coat. A seed has the following parts:

- Seed Coat: In the seed of cereals such as maize, the seed coat is membranous and generally fused with the fruit wall, called Hull.
- Endosperm: The endosperm is bulky and stores food. Generally, monocotyledonous seeds are endospermic but some as in orchids are non-endospermic.
- Aleuron layer: The outer covering of endosperm separates the embryo by a proteinous layer called aleurone layer.
- Embryo: The embryo is small and situated in a groove at one end of the endosperm.
- Scutellum: This is one large and shield-shaped cotyledon.
- Embryonal axis: Plumule and radicle are the two ends.
- Coleoptile and coleorhiza: The plumule and radicle are enclosed in sheaths. They are coleoptile and coleorhiza.

Structure of a Dicotyledonous Seed

Unlike monocotyledonous seed, a dicotyledonous seed, as the name suggests, has two cotyledons. It has the following parts:

- Seed coat: This is the outermost covering of a seed. The seed coat has two layers, the outer testa and the inner tegmen.
- Hilum: The hilum is a scar on the seed coat through which the developing seed was attached to the fruit.

- Micropyle: It is a small pore present above the hilum.
- Embryo: It consists of an embryonal axis and two cotyledons.
- Cotyledons: These are often fleshy and full of reserve food materials.
- Radicle and plumule: They are present at the two ends of the embryonal axis.
- Endosperm: In some seeds such as castor, the endosperm formed as a result of double fertilisation, is a food storing tissue. In plants such as bean, gram and pea, the endosperm is not present in the matured seed. They are known as non-endospermous.

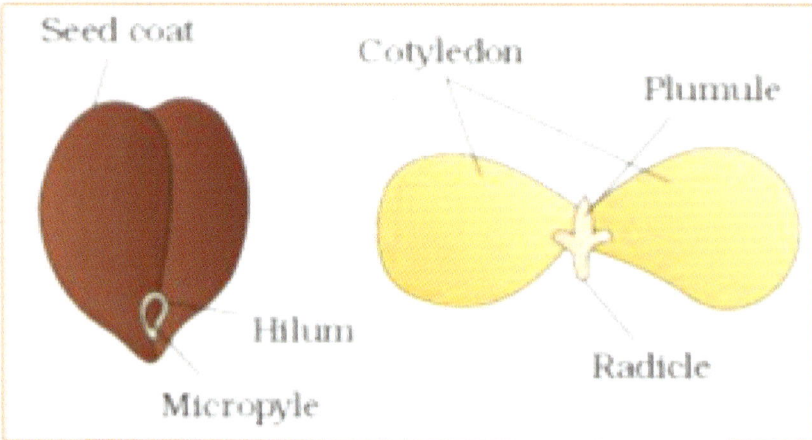

Structure of dicotyledonous seed

Functions

Seeds serve several functions for the plants that produce them. Key among these functions are nourishment of the embryo, dispersal to a new location, and dormancy during unfavorable conditions. Seeds fundamentally are means of reproduction, and most seeds are the product of sexual reproduction which produces a remixing of genetic material and phenotype variability on which natural selection acts.

Embryo Nourishment

Seeds protect and nourish the embryo or young plant. They usually give a seedling a faster start than a sporeling from a spore, because of the larger food reserves in the seed and the multicellularity of the enclosed embryo.

Dispersal

Unlike animals, plants are limited in their ability to seek out favorable conditions for life and growth. As a result, plants have evolved many ways to disperse their offspring by dispersing their seeds. A seed must somehow "arrive" at a location and be there at a time favorable for germination and growth. When the fruits open and release their seeds in a regular way, it is called dehiscent, which is often distinctive for related groups of plants; these fruits include capsules, follicles, legumes, silicles and siliques. When fruits do not open and release their seeds in a regular fashion, they are called indehiscent, which include the fruits achenes, caryopsis, nuts, samaras, and utricles.

By Wind (Anemochory):

Dandelion seeds are contained within achenes, which can be carried long distances by the wind

The seed pod of milkweed (*Asclepias syriaca*)

- Some seeds (e.g., pine) have a wing that aids in wind dispersal.
- The dustlike seeds of orchids are carried efficiently by the wind.
- Some seeds (e.g. milkweed, poplar) have hairs that aid in wind dispersal.

Other seeds are enclosed in fruit structures that aid wind dispersal in similar ways:

- Dandelion achenes have hairs.
- Maple samaras have two wings.

By water (hydrochory)

- Some plants, such as *Mucuna* and *Dioclea*, produce buoyant seeds termed sea-beans or drift seeds because they float in rivers to the oceans and wash up on beaches.

By animals (zoochory)

- Seeds (burrs) with barbs or hooks (e.g. acaena, burdock, dock) which attach to animal fur or feathers, and then drop off later.

- Seeds with a fleshy covering (e.g. apple, cherry, juniper) are eaten by animals (birds, mammals, reptiles, fish) which then disperse these seeds in their droppings.
- Seeds (nuts) are attractive long-term storable food resources for animals (e.g. acorns, hazelnut, walnut); the seeds are stored some distance from the parent plant, and some escape being eaten if the animal forgets them.

Myrmecochory is the dispersal of seeds by ants. Foraging ants disperse seeds which have appendages called elaiosomes (e.g. bloodroot, trilliums, acacias, and many species of Proteaceae). Elaiosomes are soft, fleshy structures that contain nutrients for animals that eat them. The ants carry such seeds back to their nest, where the elaiosomes are eaten. The remainder of the seed, which is hard and inedible to the ants, then germinates either within the nest or at a removal site where the seed has been discarded by the ants. This dispersal relationship is an example of mutualism, since the plants depend upon the ants to disperse seeds, while the ants depend upon the plants seeds for food. As a result, a drop in numbers of one partner can reduce success of the other. In South Africa, the Argentine ant (*Linepithema humile*) has invaded and displaced native species of ants. Unlike the native ant species, Argentine ants do not collect the seeds of *Mimetes cucullatus* or eat the elaiosomes. In areas where these ants have invaded, the numbers of *Mimetes* seedlings have dropped.

Dormancy

Seed dormancy has two main functions: the first is synchronizing germination with the optimal conditions for survival of the resulting seedling; the second is spreading germination of a batch of seeds over time so a catastrophe (e.g. late frosts, drought, herbivory) does not result in the death of all offspring of a plant (bet-hedging). Seed dormancy is defined as a seed failing to germinate under environmental conditions optimal for germination, normally when the environment is at a suitable temperature with proper soil moisture. This true dormancy or innate dormancy is therefore caused by conditions within the seed that prevent germination. Thus dormancy is a state of the seed, not of the environment. Induced dormancy, enforced dormancy or seed quiescence occurs when a seed fails to germinate because the external environmental conditions are inappropriate for germination, mostly in response to conditions being too dark or light, too cold or hot, or too dry.

Seed dormancy is not the same as seed persistence in the soil or on the plant, though even in scientific publications dormancy and persistence are often confused or used as synonyms.

Often, seed dormancy is divided into four major categories: exogenous; endogenous; combinational; and secondary. A more recent system distinguishes five classes: morphological, physiological, morphophysiological, physical, and combinational dormancy.

Exogenous dormancy is caused by conditions outside the embryo, including:

- Physical dormancy or hard seed coats occurs when seeds are impermeable to water. At dormancy break, a specialized structure, the 'water gap', is disrupted in response to environmental cues, especially temperature, so water can enter the seed and germination can occur. Plant families where physical dormancy occurs include Anacardiaceae, Cannaceae, Convulvulaceae, Fabaceae and Malvaceae.

- Chemical dormancy considers species that lack physiological dormancy, but where a chemical prevents germination. This chemical can be leached out of the seed by rainwater or snow melt or be deactivated somehow. Leaching of chemical inhibitors from the seed by rain water is often cited as an important cause of dormancy release in seeds of desert plants, but little evidence exists to support this claim.

Endogenous dormancy is caused by conditions within the embryo itself, including:

- In morphological dormancy, germination is prevented due to morphological characteristics of the embryo. In some species, the embryo is just a mass of cells when seeds are dispersed; it is not differentiated. Before germination can take place, both differentiation and growth of the embryo have to occur. In other species, the embryo is differentiated but not fully grown (underdeveloped) at dispersal, and embryo growth up to a species specific length is required before germination can occur. Examples of plant families where morphological dormancy occurs are Apiaceae, Cycadaceae, Liliaceae, Magnoliaceae and Ranunculaceae.
- Morphophysiological dormancy includes seeds with underdeveloped embryos, and also have physiological components to dormancy. These seeds, therefore, require a dormancy-breaking treatments, as well as a period of time to develop fully grown embryos. Plant families where morphophysiological dormancy occurs include Apiaceae, Aquifoliaceae, Liliaceae, Magnoliaceae, Papaveraceae and Ranunculaceae. Some plants with morphophysiological dormancy, such as *Asarum* or *Trillium* species, have multiple types of dormancy, one affects radicle (root) growth, while the other affects plumule (shoot) growth. The terms "double dormancy" and "two-year seeds" are used for species whose seeds need two years to complete germination or at least two winters and one summer. Dormancy of the radicle (seedling root) is broken during the first winter after dispersal while dormancy of the shoot bud is broken during the second winter.
- Physiological dormancy means the embryo, due to physiological causes, cannot generate enough power to break through the seed coat, endosperm or other covering structures. Dormancy is typically broken at cool wet, warm wet or warm dry conditions. Abscisic acid is usually the growth inhibitor in seeds, and its production can be affected by light.
    - o Drying, in some plants, including a number of grasses and those from seasonally arid regions, is needed before they will germinate. The seeds are released, but need to have a lower moisture content before germination can begin. If the seeds remain moist after dispersal, germination can be delayed for many months or even years. Many herbaceous plants from temperate climate zones have physiological dormancy that disappears with drying of the seeds. Other species will germinate after dispersal only under very narrow temperature ranges, but as the seeds dry, they are able to germinate over a wider temperature range.
- In seeds with combinational dormancy, the seed or fruit coat is impermeable to water and the embryo has physiological dormancy. Depending on the species, physical dormancy can be broken before or after physiological dormancy is broken.
- Secondary dormancy is caused by conditions after the seed has been dispersed and occurs in some seeds when nondormant seed is exposed to conditions that are not favorable to

germination, very often high temperatures. The mechanisms of secondary dormancy are not yet fully understood, but might involve the loss of sensitivity in receptors in the plasma membrane.

The following types of seed dormancy do not involve seed dormancy, strictly speaking, as lack of germination is prevented by the environment, not by characteristics of the seed itself:

- Photodormancy or light sensitivity affects germination of some seeds. These photoblastic seeds need a period of darkness or light to germinate. In species with thin seed coats, light may be able to penetrate into the dormant embryo. The presence of light or the absence of light may trigger the germination process, inhibiting germination in some seeds buried too deeply or in others not buried in the soil.
- Thermodormancy is seed sensitivity to heat or cold. Some seeds, including cocklebur and amaranth, germinate only at high temperatures (30°C or 86°F); many plants that have seeds that germinate in early to midsummer have thermodormancy, so germinate only when the soil temperature is warm. Other seeds need cool soils to germinate, while others, such as celery, are inhibited when soil temperatures are too warm. Often, thermodormancy requirements disappear as the seed ages or dries.

Not all seeds undergo a period of dormancy. Seeds of some mangroves are viviparous; they begin to germinate while still attached to the parent. The large, heavy root allows the seed to penetrate into the ground when it falls. Many garden plant seeds will germinate readily as soon as they have water and are warm enough; though their wild ancestors may have had dormancy, these cultivated plants lack it. After many generations of selective pressure by plant breeders and gardeners, dormancy has been selected out.

For annuals, seeds are a way for the species to survive dry or cold seasons. Ephemeral plants are usually annuals that can go from seed to seed in as few as six weeks.

Germination

Germinating sunflower seedlings

Seed germination is a process by which a seed embryo develops into a seedling. It involves the reactivation of the metabolic pathways that lead to growth and the emergence of the radicle or seed root and plumule or shoot. The emergence of the seedling above the soil surface is the next phase of the plant's growth and is called seedling establishment.

Three fundamental conditions must exist before germination can occur. (1) The embryo must be alive, called seed viability. (2) Any dormancy requirements that prevent germination must be overcome. (3) The proper environmental conditions must exist for germination.

Seed viability is the ability of the embryo to germinate and is affected by a number of different conditions. Some plants do not produce seeds that have functional complete embryos, or the seed may have no embryo at all, often called empty seeds. Predators and pathogens can damage or kill the seed while it is still in the fruit or after it is dispersed. Environmental conditions like flooding or heat can kill the seed before or during germination. The age of the seed affects its health and germination ability: since the seed has a living embryo, over time cells die and cannot be replaced. Some seeds can live for a long time before germination, while others can only survive for a short period after dispersal before they die.

Seed vigor is a measure of the quality of seed, and involves the viability of the seed, the germination percentage, germination rate and the strength of the seedlings produced.

The germination percentage is simply the proportion of seeds that germinate from all seeds subject to the right conditions for growth. The germination rate is the length of time it takes for the seeds to germinate. Germination percentages and rates are affected by seed viability, dormancy and environmental effects that impact on the seed and seedling. In agriculture and horticulture quality seeds have high viability, measured by germination percentage plus the rate of germination. This is given as a percent of germination over a certain amount of time, 90% germination in 20 days, for example. 'Dormancy' is covered above; many plants produce seeds with varying degrees of dormancy, and different seeds from the same fruit can have different degrees of dormancy. It's possible to have seeds with no dormancy if they are dispersed right away and do not dry (if the seeds dry they go into physiological dormancy). There is great variation amongst plants and a dormant seed is still a viable seed even though the germination rate might be very low.

Environmental conditions affecting seed germination include; water, oxygen, temperature and light.

Three distinct phases of seed germination occur: water imbibition; lag phase; and radicle emergence.

In order for the seed coat to split, the embryo must imbibe (soak up water), which causes it to swell, splitting the seed coat. However, the nature of the seed coat determines how rapidly water can penetrate and subsequently initiate germination. The rate of imbibition is dependent on the permeability of the seed coat, amount of water in the environment and the area of contact the seed has to the source of water. For some seeds, imbibing too much water too quickly can kill the seed. For some seeds, once water is imbibed the germination process cannot be stopped, and drying then becomes fatal. Other seeds can imbibe and lose water a few times without causing ill effects, but drying can cause secondary dormancy.

Repair of DNA Damage

During seed dormancy, often associated with unpredictable and stressful environments, DNA damage accumulates as the seeds age. In rye seeds, the reduction of DNA integrity due to damage is associated with loss of seed viability during storage. Upon germination, seeds of *Vicia faba* undergo DNA repair. A plant DNA ligase that is involved in repair of single- and double-strand breaks during seed germination is an important determinant of seed longevity. Also, in Arabidopsis seeds, the activities of the DNA repair enzymes Poly ADP ribose polymerases (PARP) are likely needed for successful germination. Thus DNA damages that accumulate during dormancy appear to be a problem for seed survival, and the enzymatic repair of DNA damages during germination appears to be important for seed viability.

Inducing Germination

A number of different strategies are used by gardeners and horticulturists to break seed dormancy.

Scarification allows water and gases to penetrate into the seed; it includes methods to physically break the hard seed coats or soften them by chemicals, such as soaking in hot water or poking holes in the seed with a pin or rubbing them on sandpaper or cracking with a press or hammer. Sometimes fruits are harvested while the seeds are still immature and the seed coat is not fully developed and sown right away before the seed coat become impermeable. Under natural conditions, seed coats are worn down by rodents chewing on the seed, the seeds rubbing against rocks (seeds are moved by the wind or water currents), by undergoing freezing and thawing of surface water, or passing through an animal's digestive tract. In the latter case, the seed coat protects the seed from digestion, while often weakening the seed coat such that the embryo is ready to sprout when it is deposited, along with a bit of fecal matter that acts as fertilizer, far from the parent plant. Microorganisms are often effective in breaking down hard seed coats and are sometimes used by people as a treatment; the seeds are stored in a moist warm sandy medium for several months under nonsterile conditions.

Stratification, also called moist-chilling, breaks down physiological dormancy, and involves the addition of moisture to the seeds so they absorb water, and they are then subjected to a period of moist chilling to after-ripen the embryo. Sowing in late summer and fall and allowing to overwinter under cool conditions is an effective way to stratify seeds; some seeds respond more favorably to periods of oscillating temperatures which are a part of the natural environment.

Leaching or the soaking in water removes chemical inhibitors in some seeds that prevent germination. Rain and melting snow naturally accomplish this task. For seeds planted in gardens, running water is best—if soaked in a container, 12 to 24 hours of soaking is sufficient. Soaking longer, especially in stagnant water, can result in oxygen starvation and seed death. Seeds with hard seed coats can be soaked in hot water to break open the impermeable cell layers that prevent water intake.

Other methods used to assist in the germination of seeds that have dormancy include prechilling, predrying, daily alternation of temperature, light exposure, potassium nitrate, the use of plant growth regulators, such as gibberellins, cytokinins, ethylene, thiourea, sodium hypochlorite, and others. Some seeds germinate best after a fire. For some seeds, fire cracks hard seed coats, while in others, chemical dormancy is broken in reaction to the presence of smoke. Liquid smoke is often used by gardeners to assist in the germination of these species.

Sterile Seeds

Seeds may be sterile for few reasons: they may have been irradiated, unpollinated, cells lived past expectancy, or bred for the purpose.

Evolution and Origin of Seeds

The origin of seed plants is a problem that still remains unsolved. However, more and more data tends to place this origin in the middle Devonian. The description in 2004 of the proto-seed *Runcaria heinzelinii* in the Givetian of Belgium is an indication of that ancient origin of seed-plants. As with modern ferns, most land plants before this time reproduced by sending spores into the air, that would land and become whole new plants.

The first "true" seeds are described from the upper Devonian, which is probably the theater of their true first evolutionary radiation. With this radiation came an evolution of seed size, shape, dispersal and eventually the radiation of gymnosperms and angiosperms and monocotyledons and dicotyledons. The seed plants progressively became one of the major elements of nearly all ecosystems.

Fruit

Fruit is the fleshy or dry ripened ovary of a flowering plant, enclosing the seed or seeds. Thus, apricots, bananas, and grapes, as well as bean pods, corn grains, tomatoes, cucumbers, and (in their shells) acorns and almonds, are all technically fruits. Popularly, however, the term is restricted to the ripened ovaries that are sweet and either succulent or pulpy.

What is an apple? Learn what creates the signature taste and smell of one of the world's favorite fruits.

Botanically, a fruit is a mature ovary and its associated parts. It usually contains seeds, which have developed from the enclosed ovule after fertilization, although development without fertilization, called parthenocarpy, is known, for example, in bananas. Fertilization induces various changes in a flower: the anthers and stigma wither, the petals drop off, and the sepals may be shed or undergo modifications; the ovary enlarges, and the ovules develop into seeds, each containing an embryo plant. The principal purpose of the fruit is the protection and dissemination of the seed.

Fruits are important sources of dietary fibre, vitamins (especially vitamin C), and antioxidants. Although fresh fruits are subject to spoilage, their shelf life can be extended by refrigeration or by the removal of oxygen from their storage or packaging containers. Fruits can be processed into juices, jams, and jellies and preserved by dehydration, canning, fermentation, and pickling. Waxes, such as those from bayberries (wax myrtles), and vegetable ivory from the hard fruits of a South American palm species (*Phytelephas macrocarpa*) are important fruit-derived products. Various drugs come from fruits, such as morphine from the fruit of the opium poppy.

Types of Fruits

Fruits may be true or false.

(i) True Fruits:

A true fruit or eucarp is a mature or ripened ovary, developed after fertilization, e.g., Mango, Maize, Grape etc.

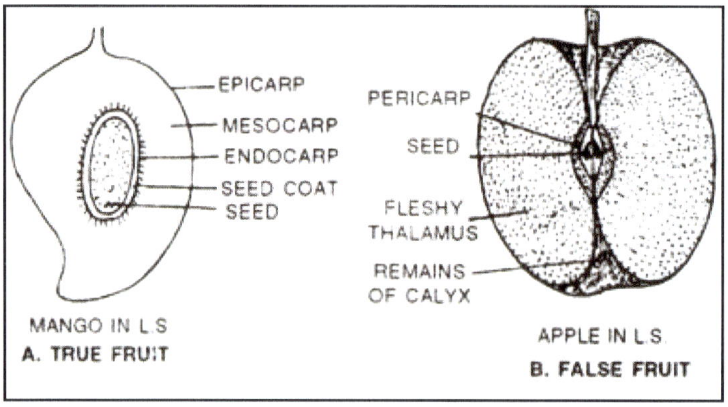

Figure: Morphology of true and false fruits

(ii) False Fruits:

A false fruit or pseudo-carp is derived from the floral parts other than ovary, e.g., peduncle in cashew-nut, thalamus in apple, pear, gourd and cucumber; fused perianth in mulberry and calyx in Dillenia (Or. Ou). Jack fruit and pine apple are also false fruits as they develop from the entire inflorescence. False fruits are also called spurious or accessory fruits.

(iii) Parthenocarpic fruits:

These are seedless fruits that are formed without fertilization, e.g., Banana. Now a day many seedless grapes, oranges and water melones are being developed by horticulturists. Pomology is a branch of horticulture that deals with the study of fruits and their cultivation.

Morphology of a Typical Fruit:

A fruit consists of pericarp and seeds. Seeds are fertilized and ripened ovules. The pericarp develops from the ovary wall and may be dry or fleshy. When fleshy, pericarp is differentiated into outer epicarp, middle mesocarp and inner endocarp.

On the basis of the above mentioned features, fruits are usually classified into three main groups:

(1) Simple

(2) Aggregate and

(3) Composite or Multiple fruits.

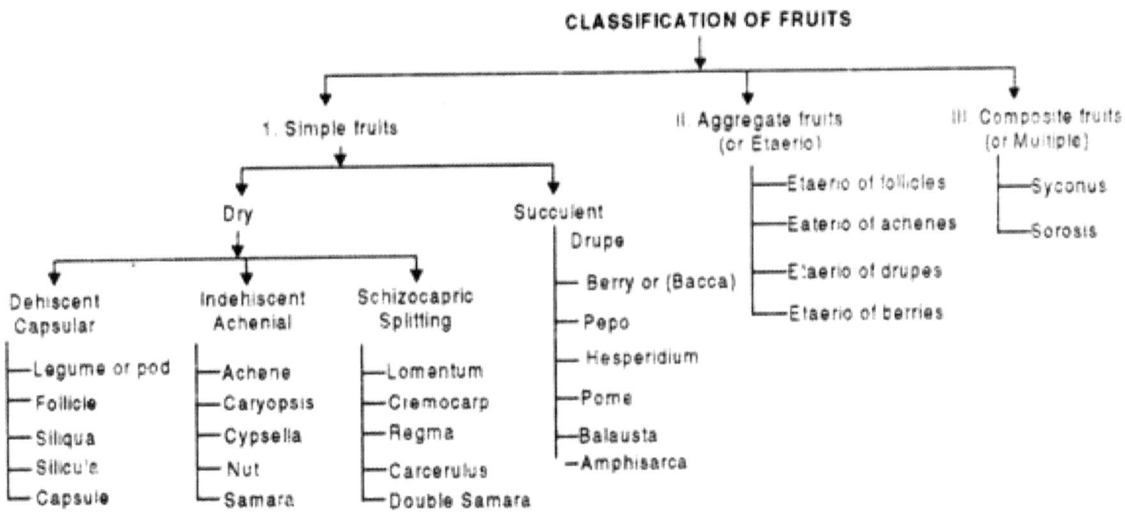

1.

Simple Fruits:

When a single fruit develops from a single ovary of a single flower, it is called a simple fruit. The ovary may belong to a monocarpellary simple gynoecium or to a polycarpellary syncarpous gynoecium. There are two categories of simple fruits—dry and fleshy.

Simple fruits are of two types:

1. Dry Fruits:

These fruits are not fleshy, and their pericarp (fruit wall) is not distinguished into three layers.

2. Succulent Fruits (Fleshy fruits):

In these fruits pericarp is distinguished into epicarp, mesocarp and endocarp. Mesocarp is fleshy or fibrous. These fruits are indehiscent, and seeds are liberated after the decay of the flesh.

1. Dry Fruits:

Three types of dry fruits are distinguishable:

(A) Dehiscent Fruits (Capsular Fruits):

Characteristic of these fruits is that their pericarp rupture after ripening and the seeds are disseminated.

(B) Indehiscent Fruits (Achenial Fruits):

As their name indicates, pericarp of such fruits does not rupture on ripening and the seeds remain inside.

(C) Schizocarpic Fruits (Splitting Fruits):

These fruits fall in between the above-mentioned two categories. Here, the fruit on ripening divides into one-seeded segments or mericarp; but the mericarps remain un-ruptured.

(A) Dehiscent Fruits (Capsular Fruits):

Depending on the mode of dehiscence, these fruits can be divided into the following five classes:

1. Legume or Pod:

Legume develops from a superior, monocarpellary, unilocular ovary. At maturity, the fruit dehisces along both the sutures i.e. ventral as well as dorsal. It is characteristic of family Leguminosae (Pea, Gram etc).

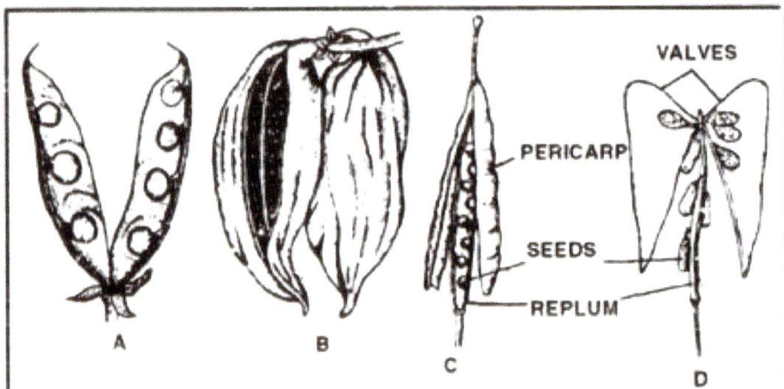

Figure : Dehiscent fruits. A. Legume of Pea. B. Follicles of Calatropis. C. Siliqua of Brassica. D. Silicula of Capsella.

2. Follicle:

It is similar to legume but it dehisces only along the ventral suture, e.g. Larkspur, Calatropis, Michelia, Vinca.

3. Siliqua:

Siliqua develops from a bicarpellary, syncarpous, superior ovary which is unilocular but becomes bilocular due to a false septum called replum. It is an elongated fruit in which dehiscence occurs along both the sutures from base to apex and the seeds attached to the replum get exposed. Example-Brassica (Mustard).

4. Silicula:

A short and flattened siliqua is called silicula. It is almost as broad as long. Examples: Iberis amara (Candytuft), Capsella bursa-pastoris (Shepherd's purse).

5. Capsule:

It is a simple dry many seeded dehiscent fruit developing from a multi-carpellary syncarpous ovary.

On the basis of dehiscence capsules are of the following types:

(i) Porocidal:

The dehiscence occurs through pores as in Poppy (Papaver).

(ii) Pyxis:

This is a special name given to a capsule when the dehiscence is transverse so that the top comes off as a lid as if exposing a box of seeds, e.g., Celosia (Cock's comb), Amaranth us, Chalfweed.

(iii) Loculicidal:

The dehiscence occurs by longitudinal slits which open into the loculi, e.g., 1 dy's finger (Abelmoschus).

(iv)Septicidal:

The dehiscence line appears along the septa, e.g.. Linseed, Cotton to the central axis ,eg. Datura.

(v) Sentifragai:

Tin- broken parts separate exposing the seeds attached to the central axis, e.g Datum.

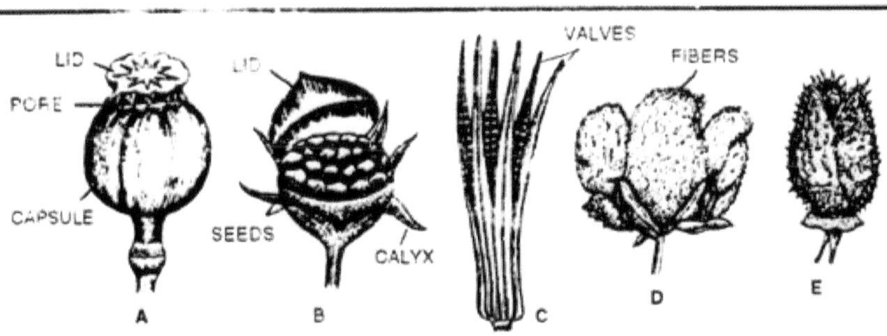

Figure 7.3. Capsule fruits **A.** Porocidal capsule of Poppy. **B.** Pyxis of Celosia **C.** Loculicidal capsule of Abelmoschus. **D.** Septicidal capsule of Gossypium. **E.** Septifragal capsule of Datura.

(B) Indehiscent or Achenial Fruits:

Achenial fruits are simple, indehiscent, single seeded having a thin, dry, woody or leathery pericarp.

There are five common types of achenial fruits:

1. Achene

The pericarp of the fruit is free from the testa of the seed. The seed is attached to the pericarp only at one point. It develops from superior monocarpellary pistill having unilocular and uniovuled ovary, e.g., Mirabilis jalapa, but more commonly achenes occur in the form of aggregate fruits as in Ranunculus and Clematis etc.

2. Caryopsis

It is similar to achene except that in this case pericarp and testa are inseparably fused as in cereals. It is a characteristic future of family Gramineae. Example—Wheat, Maize etc.

3. Cypsela

It is a characteristic feature of family Compositae. The fruit wall is free from testa and a typical feature of the fruit is the presence of a pappus having a crown of hair like processes which helps in wind-dispersal. The fruit develops from bicarpellary, syncarpous, interior ovary having a single basal ovule, e.g., Sonchus, Dandelion etc.

4. Samara

It develops from a monocarpellary pistil with a superior, unilocular and uniovuled ovary. The pericarp is expanded in the form of wings which help in dispersal. Example—Holoptelea and Elm.

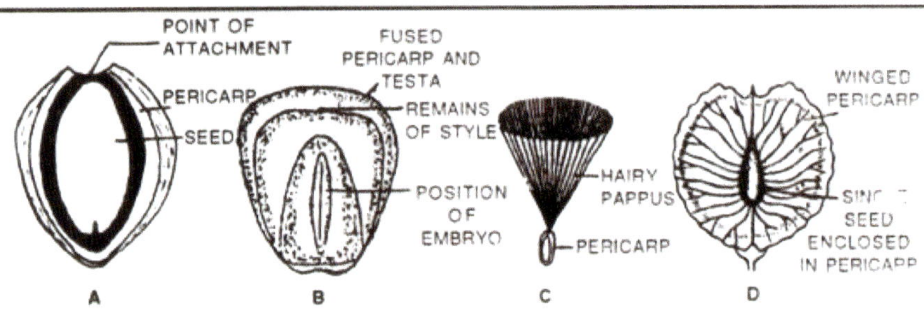

Figure : Achenial fruits. **A.** Achene of Mirabilis (L.S.). **B.** Caryopsis of Maize (external view). **C.** Cypsela of Sonchus. **D.** Samara of Elm.

5. Nut:

The pericarp is harder and leathery or woody. It may develop from a simple or compound pistil with superior or inferior, uniovuled ovary. Examples—Quercus (Oak), Litchi and Cashew-nut Trapa etc. In case of Litchi pericarp is hard and leathery. The edible part is aril which is an outgrowth of testa from the micropylar end and becomes juicy.

(C) Schizocarpic or Splitting Fruits:

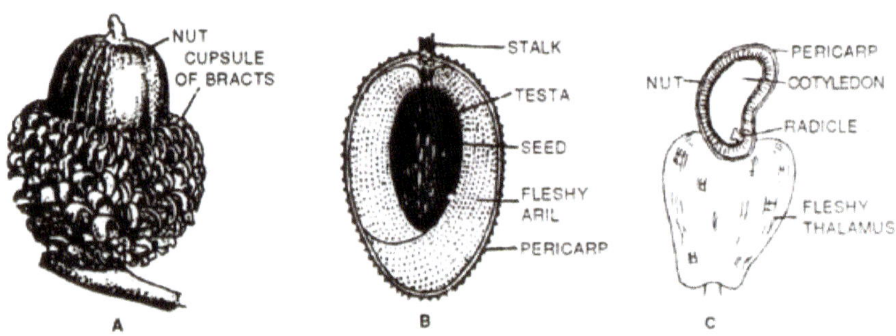

Figure : Nuts **A.** Nut of OAK. **B.** Nut of Litchi. **C.** Nut of Cashewnut

These fruits maybe considered intermediate between achenial (being indehiscent) and capsular (being many seeded) fruits. The fruit breaks up into a number of indehiscent single-seeded segments called mericarps from which seeds are liberated only when pericarp gets rotten. In some cases one- seeded parts of the fruit are dehiscent and are called Cocci.

Schizocarpic fruits are of following 5 types:

1. Lomentum:

The fruit is constricted between the seeds and usually breaks up into segments containing one or more seeds, e.g. Mimosa, Acacia arabica. In case of radish, the fruit is lomentaceous siliqua.

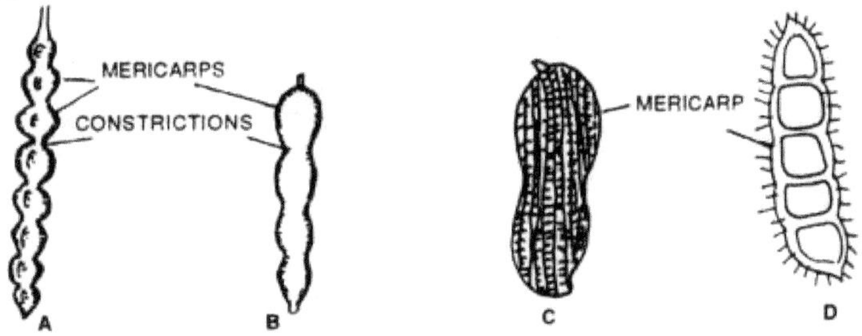

Figure: Lomentums **A.** Acacia. **B.** Radish. **C.** Groundnut. **D.** Mimosa

2. Compound Samara:

This is a type of two or more-chambered fruit derived from a syncarpous (i.e., compound) ovary. The pericarp is extended in the form of wings and at maturity the fruit breaks up % into single seeded mericarps. e.g., EIm (Holoptelea), maple.

3. Cremocarp:

This is a two-seeded fruit derived from bicarpellary, syncarpous, inferior, bilocular and uniovuled ovary. It is a typical fruit of family umbelliferae. The two mericarps split along the central axis or carpophore to which they remain attached. Persistent style and stylopodium are present e.g. Coriander.

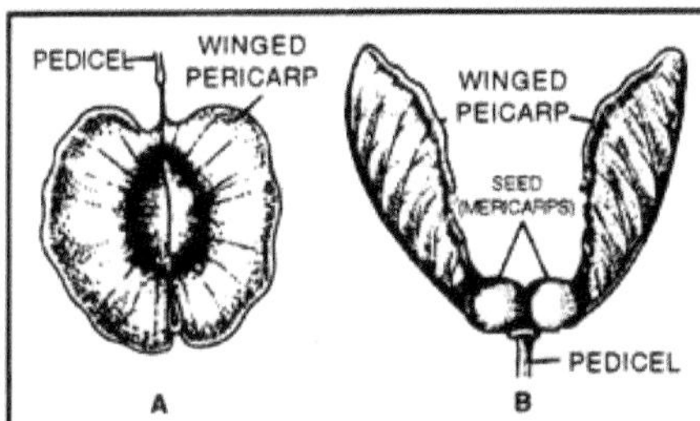

Figure : Compound Samara
A. Holoptelea, **B.** Acer (Maple)

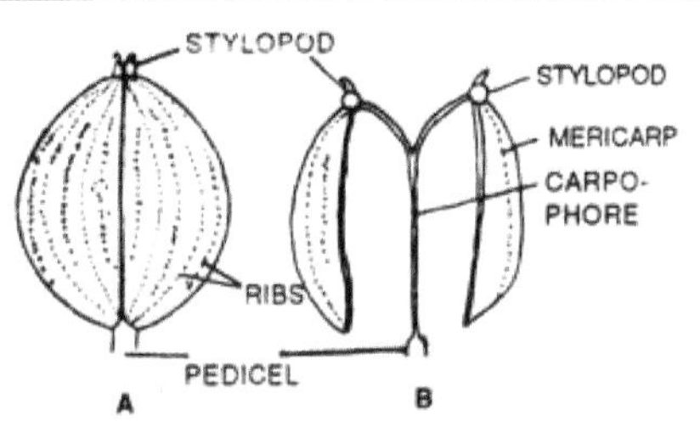

Figure: Cremocarp of Coriander
A. Complete **B.** Mericarps Separated

4. Carcerulas:

This fruit is derived from superior, syncarpous pistil, multilocular with axile placentation. The fruit splits into many mericarps. e.g. Hollyhock (Althaea rosea), Salvia, Ocimum.

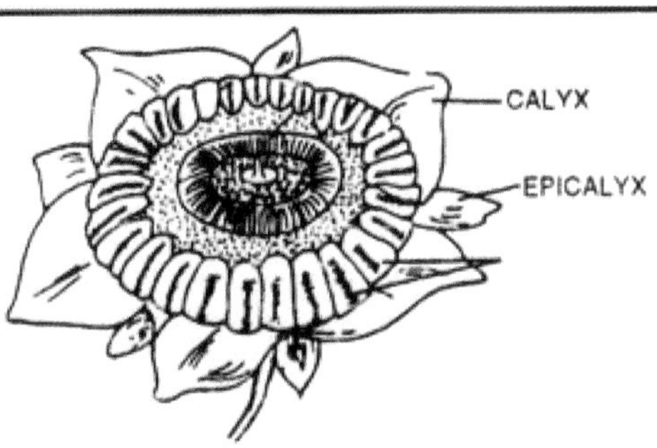

Figure : Carcerulus of Althaea (Hollyhock).

5. Regma:

It is derived from polycarpellary pistil which splits into as many Cocci (dehiscent segments) as there are carpels. Regma of castor breaks up into three cocci as it is derived from tricarpellary syncarpous pistil. Similarly, regma of Geranium breaks into five cocci as it is derived from five carpels.

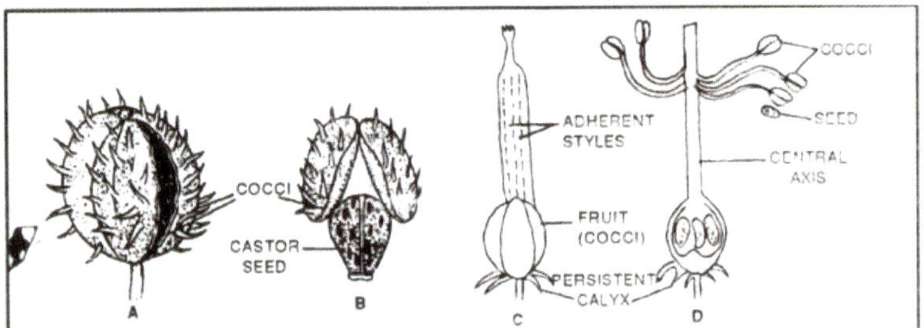

Figure : Regma A. Regma of castor breaking into three cocci. B. Coccus releasing a seed. C. Regma of Geranium. D. Dehisced Regma of Germanium

2. Succulent or Fleshy Fruits:

These are simple fruits with fleshy pericarp. The simple succulent fruits are of 3 types – drupe, pome and berrie.

1. Drupe:

The pericarp or fruit wall is differentiated into thin epicarp (skin) fleshy mesocarp and stony endocarp.Hence.it is also called as stone fruit, e.g., Mango, Coconut, Peach, Almond, Trapa etc. In mango, mesocarp is juicy and edible. In coconut mesocarp is fibrous and edible part is endocarp. In almond, epicarp and mesocarp get peeled off and only hard endocarp can be seen in marketed fruits. The edible part is cotyledons.

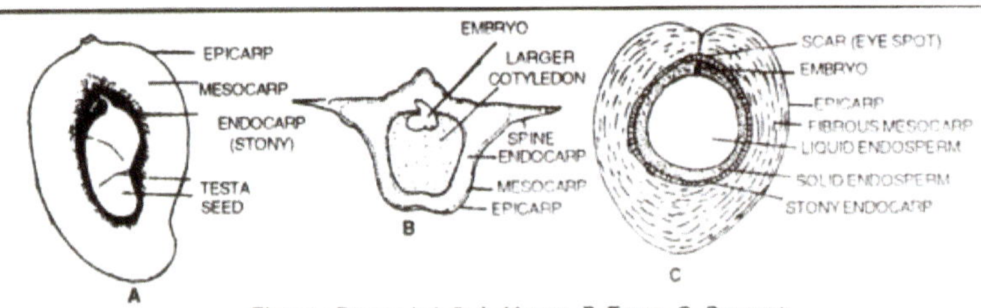

Figure : Drupes in L.S. A. Mango, B. Trapa, C. Coconut

2. Pome:

It is a simple, fleshy but false fruit as it is surrounded by a fleshy thalamus which is edible while actual fruit lies within, e.g., apple, pear, loquat etc.

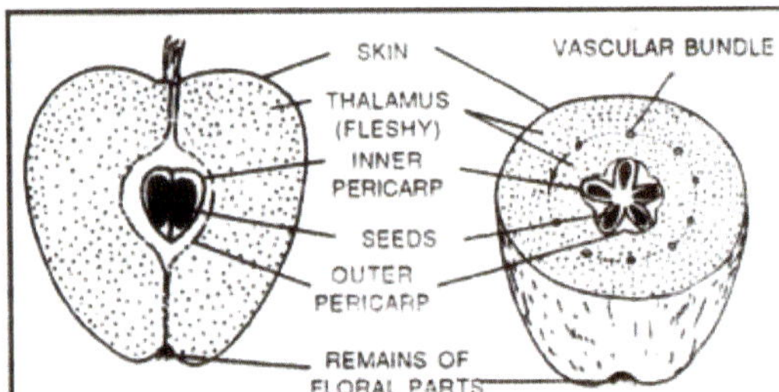

Figure : Pome of Apple A. L.S., B. T.S.

3. Berry and Bacca:

Berry is a fleshy fruit in which there is no hard part except the seeds. Pericarp may be differentiated into epicarp, mesocarp and endocarp. One or other of these layers may form pulp in which seeds are embedded which generally gets detached from the placenta.

The fruits derived from superior ovary are called superior or true berries as in brinjal, grape, tomato. False berries are derived from inferior ovary and thalamus and pericarp are fused as in banana and guava etc. In case of bannana epicarp and thalamus are peeled off, mesocarp and endocarp with embedded unripe seeds forms the edible part. In case of Date, epicarp and mesocarp are edible while papery and thin endocarp is thrown away along with the seed.

There are some fruits which show variations from the normal berry:

(i) Pepo:

This develops from inferior ovary which is unilocular or falsely trilocular having parietal placentation. The seeds remain attached to placenta. The outer ring is very hard as in Cucurbits.

(ii) Hesperidium:

It develops from polycarpellary, syncarpous, superior, multilocuiar ovary with axile placentation. Epicarp forms the leathery peeling, mesocarp is in the form of fibres while the endocarp projects inwards forming distinct chambers from which juicy ingrowths in the form of hair arise which form the edible part, eg. Citrus (Orange, Lemon).

(iii) Amphisarca:

It is derived from polycarpellary, syncarpous, multilocuiar and superior ovary. In this case, epicarp is woody. The placenta and inner layers of pericarp become pulpy and edible in which the seeds are scattered. The testa is muclilagenous, e.g., Aegle marmelose.

(iv) Balusta:

It is a berry with an outer hard rind formed of epicarp and a part of mesocarp. The inward foldings of mesocarp form chambers. Each chamber is lined by papery endocarp which encloses a group of seeds. The seeds are covered by edible juicy testa. e.g., Pomranate.

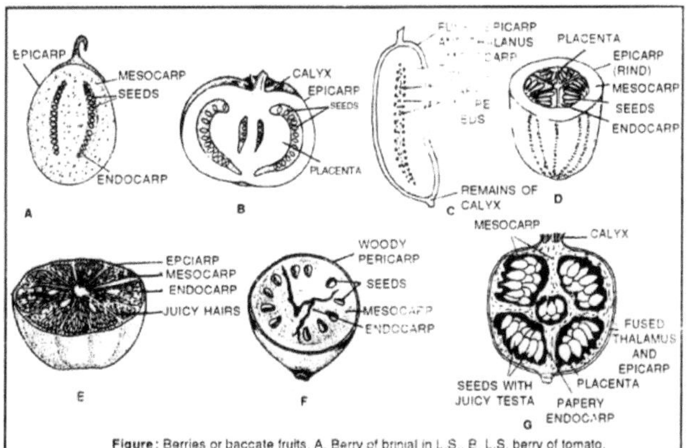

Figure: Berries or baccate fruits. A. Berry of brinjal in L.S., B. L.S. berry of tomato, C. L.S. berry of banana, D.T.S. pepo of cucumber, E.T.S. Hesperidium of Lemon, F.T.S. amphisaraca of wood apple (Aegle), G.L.S. balusta of Pomegranate

II. Aggregate Fruits:

Flowers with polycarpellary and apocarpous gynoecium give rise to a number of fruitlets as there are a number of free ovaries, each giving rise to one fruitlet. Sometimes, these fruitlets coalesce together appearing to be a single fruit but in many other cases, the fruitlets remain free from one another forming etaerio of fruitlets. An aggregate fruit is named according to the nature of fruitlets.

1. Etaerio of achenes:

Aggregate of achenes are found in Fragaria (strawberry), Rose, Ranunculus, Nelumbium (lotus) etc. Here each fruitlet is an achene; and achenes are hairy. In rose (Rosa), many achenes are present on a saucer (cup) – shaped thalamus. In lotus (Nelumbium), thalamus becomes spongy and some achenes are embedded in it. In strawberry [Fragaria), the thalamus is fleshy and becomes red on maturation and is the edible part.

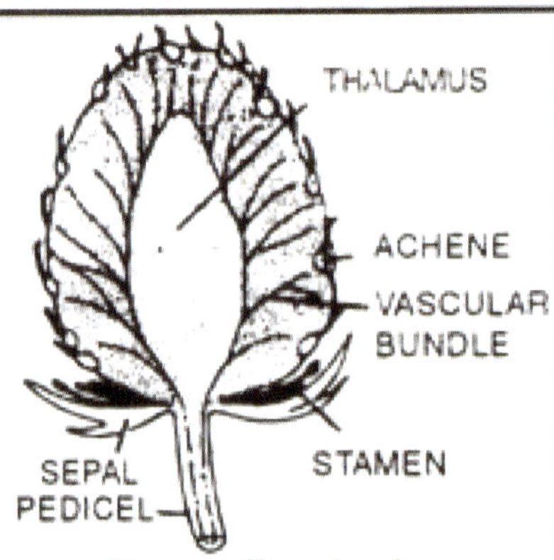

Figure : Etaerio of achenes of strawberry.

2. Etaerio of follicles:

Etaerio of follicles can be seen in Aconitum, Catotropis, Crypiostegia etc. In Aconitum three fruitlets from each flower while two fruitlets (follicles) develop from one flower in Calotroiis, Cryptostegia and Michelia.

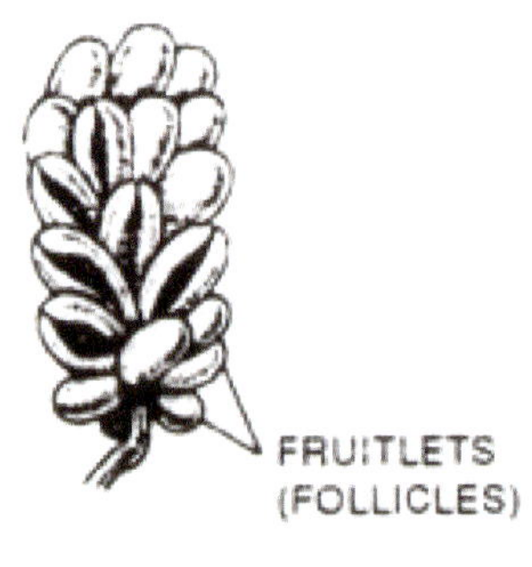

Figure : Etaerio of follicles of *Michelia*.

3. Etaerio of samaras

It can be studied in Ailanthus where many winged samaras develop from one flower.

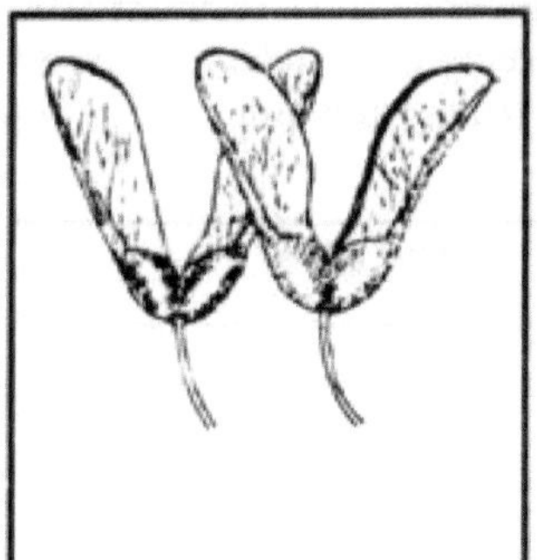

Figure : Etaerio of samaras of *Alianthus*.

4. Etaerio of berries:

In Artabotrys berries occur in a bunch. In Anona squomosa (Custard apple) the berries become very fleshy and being crowded together on a thick thalamus form a complex single fruit. The apices of berries fuse together forming something like a common rind.

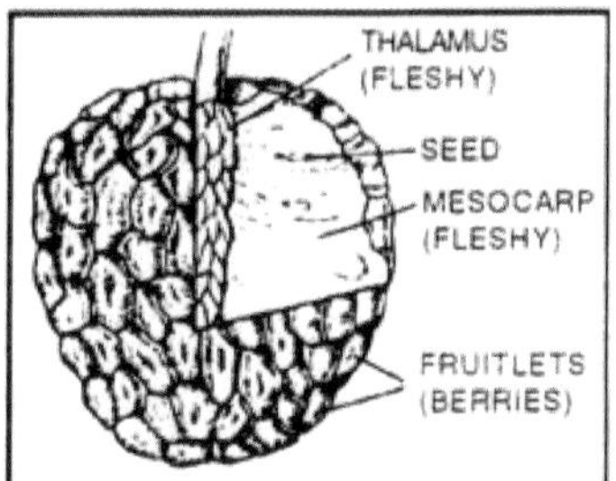

F·ure : Etaerio of berries of custard apple.

5. Etaerio of drupes

It is an aggregate of small drupes or drupelets developing from different carpels of a flower, and arranged collectively on fleshy thalamus, e.g. Rubus idaeus.

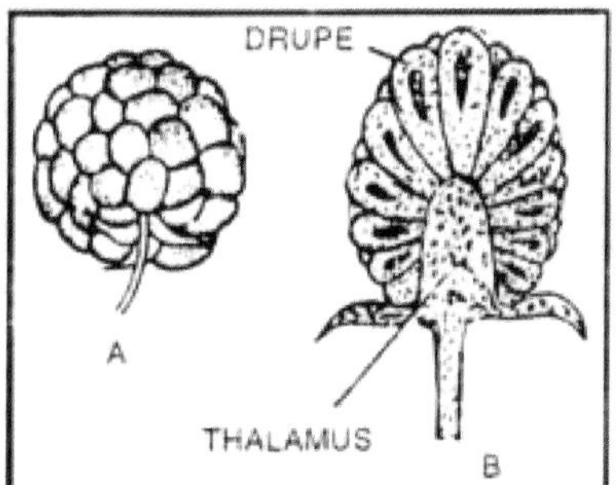

Figure : A. Etaerio of drupes of *Rubus* (Raspberry). B. Section of the same.

III. Composite Fruits:

A fruit developing from a complete inflorescence is called a multiple or a composite fruit.

There are two main types of composite fruits:

1. Sorosis:

This type of fruit is found in Mulberry, Pineapple and Jack fruit (kathal). These fruits are derived from catkin, spike and spadix type of inflorescence.

Mulberry (Morus indica) fruit develops from catkin in which fleshy perianth encloses dry achenes.

In jack fruit, thick club-shaped peduncle has the flowers arranged on it. The fertile fruits have juicy, edible perianth lobes and the bracts form more or less juicy chaffs around them. The spines on the tough rind represent the stigmas of the carpels. Each seed is covered by a membranous testa. In Pineapple (Ananas sativus), the ovaries are not so conspicuous, edible portion being formed by peduncle, perianth and bracts. Each polygonal area on the surface represents a flower. This fruit develops from an intercalary spike.

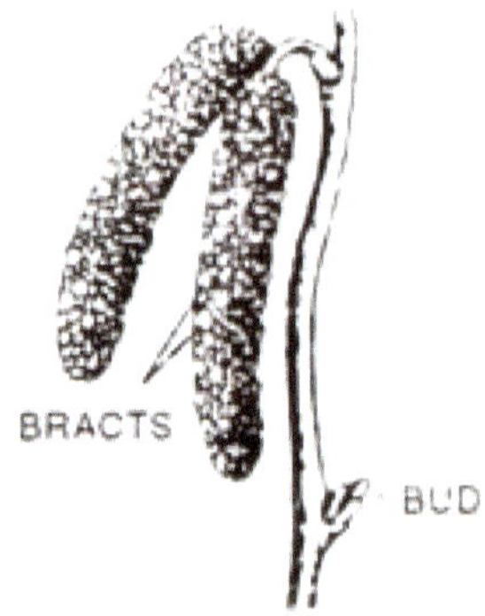

Figure : Sorosis of Mulberry

2. Syconus:

This fruit develops from the hypanthodium type of inflorescence and is characteristic of Ficus. In fig, Banyan etc. female flowers within the closed receptacle (which becomes fleshy) of the inflorescence develop into achenes giving rise to a multiple fruit of achenes.

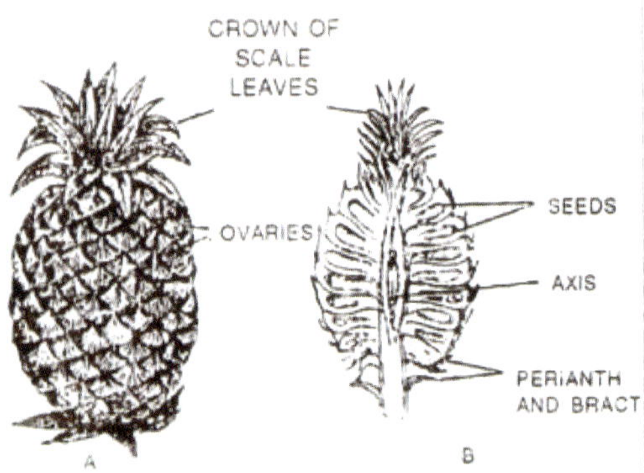

Figure: A Sorosis of pine-apple. B. V.S. of the same.

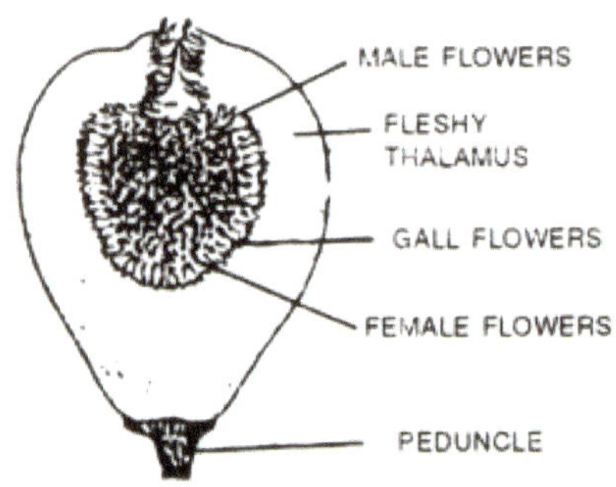

Figure : Syconus

Edible Parts of Some Common Fruits

Common/ English name	Botanical name	Type	Edible parts
	I. Simple Fruits		
Pea	*Pisum sativum*	Legume	Seeds
Lady's finger/ Okra	*Abelmoschus esculentus syn. Hibiscus esculentus*	Capsule	Entire fruit
Wheat	*Triticum aestivum*	Caryopsis	Entire Fruit
Corn/Maize	*Zea mays*	Caryopsis	Entire fruit
Cashew nut	*Anacardium occidentale*	Nut	Cotyledons and fleshy thalamus
Litchi	*Litchi chinensis*	Nut	Aril
Water chestnut	*Trapa bispinosa*	Nut	Seeds
Ground nut	*Arachis hypogaea*	Lomentum	Seeds
Coriander	*Coriandrum sativum*	Cremocarp	Entire fruit
Mango	*Mangifera indica*	Drupe	Fleshy Mesocarp
Coconut	*Cocos nucifera*	Drupe	Endosperm or entire seed
Indian Jujube	*Zizyphus jujuba*	Drupe	Drupe Epicarp and mesocarp
Almond	*Prunus amygdalus*	Drupe	Seeds
Walnut	*Juglans regia*	Nut	Cotyledons
Apple	*Pyrus malus*	Pome	Fleshy thalamus
Pear	*Pyrus communis*	Pome	Fleshy thalamus
Tomato	*Lycopersicum esculentum*	Berry	Pericarp and placentae
Brinjal	*Solanum melongena*	Berry	Pericarp and placentae
Grape	*Vitis vinifera*	Berry	Pericarp and placentae
Date plam	*Phoenix dactylifera*	Berry	Pericarp
Banana	*Musa paradisiaca var sapientum*	Berry	Mesocarp and endocarp
Guava	*Psidium guajava*	Berry	Pericarp, placentae and thalamus
Betel nut	*Areca catechu*	Berry	Seeds
Bottle gourd	*Lagenaria siceraria*	Pepo	Mesocarp, endocarp and young seeds
Cucumber	*Cucumis sativus*	Pepo	Mesocarp, endocarp and young seeds
Watermelon	*Citrullus lanatus*	Pepo	Mesocarp, endocarp and shelled seeds
Muskmelon	*Cucumis melo*	Pepo	Mesocarp, endocarp and shelled seeds
Lime	*Citrus aurantifolia*	Hesperidium	Glandular hairs arising from endocarp
Loose skined orange	*Citrus reticulata*	Hesperidium	Glandular hairs arising from endocarp
Pomegranate	*Punica granatum*	Balausta	Succulent testa
Bengal quince	*Aegle marmelos*	Amphisarca	Inner fleshy layer of pericarp and placentae
	II. Aggregate Fruits		
Lotus	*Nelumbo nucifera*	Etaerio of Achenes	Fleshy thalamus and seeds
Strawberry	*Fragaria vesca*	or Etaerio of Achenes	Fleshy thalamus and seeds
Custard apple	*Annona squamosa*	Etaerio of Berries	Inner layer of pericarp and thalamus
	III Multiple Composite Fruits		
Mulberry	*Morus alba and M.nigra*	Sorosis	Succulent perianth and fleshy axis
Pine apple	*Ananas comosus*	Sorosis	Fleshy axis, bracts, fused perianth and pericarp
Jack fruit	*Artocarpus heterophyllus*	Sorosis	Fleshy axis, bracts, perianth and seeds
Fig	*Ficus carica*	Syconus	Fleshy receptacle or thalamus

Plant Pigments

Pigments are responsible for many of the beautiful colors we see in the plant world. Dyes have often been made from both animal sources and plant extracts. Some of the pigments found in animals have also recently been found in plants.

Bilirubin is responsible for the yellow color seen in jaundice sufferers and bruises, and is created when hemoglobin (the pigment that makes blood red) is broken down. Recently this pigment has also been found in plants, specifically in the orange fuzz on seeds of the white Bird of Paradise tree. The bilirubin in plants doesn't come from breaking down hemoglobin. (In animals hemoglobin is broken down to heme, and then converted to bilirubin.) Chlorophyll molecules have similar ring structures to those of heme, and it appears that breaking down chlorophyll can also yield bilirubin - or almost; the breakdown products require just one more step to produce bilirubin. It is fascinating to realize that the process of degradation starts the same way in plants as it does in our own bodies.

Major Plant Pigments and their Occurrence

Pigment	**Common types**	**Where they are found**	**Examples of typical colors**
Chlorophylls	Chlorophyll	Green plants	Green
Carotenoids	Carotenes and xanthophylls (e.g. astaxanthin)	Bacteria. Green plants (masked by chlorophyll), vegetables like carrots, mangoes and so on. Some birds, fish and crustaceans absorb them through their diets	Oranges, reds, yellows, pinks
Flavonoids	Anthocyanins, aurones, chalcones, flavonols and proanthocyanidins	Produce many colors in flowers. Common in plants such as berries, eggplant, and citrus fruits. Present in certain teas, wine, and chocolate	Yellow, red, blue, purple
Betalains	Betacyanins and betaxanthins	Flowers and fungi	Red to violet, also yellow to orange

Chlorophylls

Chlorophyll is green, and is responsible for the green color of foliage and leaves. More importantly, by enabling plants to produce oxygen during photosynthesis, it is critical to sustaining our life on earth. Chlorophyll has structural features similar to heme. Bilirubin, which produces a yellow color, has recently been found in plants.

Red, orange, and yellow plants, as well as other organisms, generally rely on carotenoids for their vivid colors.

Carotenoids

Carotene is a pigment that absorbs blue and indigo light, and that provides rich yellows and oranges. The distinctive colors of mango, carrots, fallleaves, and yams are due to various forms of carotene, as is the yellow of butter and other animal fats. This pigment is important to our diet, as the human body breaks down each carotene molecule to produce two vitamin A molecules.

Lycopene, canthaxanthin, and astaxanthin share a similar structure to carotene. The red tones of tomatoes, guava, red grapefruit, papaya, rosehips, and watermelon indicate the presence of lycopene.

Canthaxanthin produces the pink colors of flamingos, some crustaceans, salmon, and trout. In its synthetic form, it is used to ensure captive flamingos retain their coloring, as a red food colorant, and even as a tanning aid. Astaxanthin provides the red colors of cooked salmon, red bream, trout, lobster, and shellfish. In a live animal, astaxanthin is combined with a protein and is blackish in color. When boiled, the protein breaks down to unmask the true "lobster red" of astaxanthin.

Flavonoids

Flavonoids are the yellow plant pigments seen most notably in lemons, oranges, and grapefruit. The name stems from the Latin word "flavus," which means yellow. Flavonoids in flowers and fruit provide visual cues for animal pollinators and seed dispersers to locate their targets. Flavonoids are located in the cytoplasm and plastids. Many of the foods that we eat, including dark chocolate, strawberries, blueberries, cinnamon, pecans, walnuts, grapes, and cabbage, contain flavonoids. These chemicals lower cholesterol levels, and many have antioxidant properties. Anthocyanins and proanthocyanidins, and the reddish-brown pigment theaflavin found in tea, act to create color, while most other flavonoids are visible only under UV light.

Flavonoids include red, purple, or blue anthocyanins, as well as white or pale yellow compounds such as rutin, quercetin, and kaempferol.

Anthocyanins play a role in the colors of ripening fruit. They are found in most other plant parts and in most genera. Anthocyanin pigments take their color from the range of red, purple, or blue, depending on their pH. Blueberries, cranberries, and bilberries are rich in anthocyanins, as are the berries of the Rubus genus (including black raspberry, red raspberry and blackberry), blackcurrants, cherries, eggplant peel, black rice, Concord and muscadine grapes, red cabbage, and violet petals. Anthocyanins are partly responsible for the red and purple colors of some olives.

Proanthocyanidins are linked to the beige color of the broad bean seed coat, and also to shades of black, red, brown, and tan. Apples, pine bark, cinnamon, grape seed, cocoa, grape skin, and the grapes used to make most red wines all contain proanthcyanidin.

The yellow colors of flavonoid pigments can be found as chalcones (found in flowers and the organs of plants), aurones (found in flowers and some bark, wood, or leaves) and flavonols.

Betalains

Like carotenoids and flavonoids, betalains also seem to play an important role in attracting animals to flowers and fruit, and produce a similar range of colors. The betalains consist of two subgroups, red-violet (betacyanin) and yellow to orange (betaxanthin) pigments. They only occur in a few plant families, and always independently of anthocyanins.

Betacyanins are established food colorants. Betalains give rise to the distinctive deep red of beetroot. The composition of different betalain pigments can vary, giving rise to breeds of beetroot that are yellow or other colors, in addition to the familiar deep red. The betalains in beets include betanin, isobetanin, probetanin, and neobetanin (the red to violet ones are known collectively as *betacyanin*). Other pigments contained in beet are indicaxanthin and vulgaxanthins (yellow to orange pigments known as *betaxanthins*). Betalains cause the crimson of Amaranthus flowers (class of Caryophylalles).

Interestingly, betalains are only found in one sub-group of flowering plants (Caryophylalles or Centrospermae). Bougainvillea, certain cacti, and amaranth are all examples of this family. These plants lost, or never acquired, genes for the synthesis of other plant pigments. Genes for the synthesis of betalains appear in unrelated fungi (such as Amanita muscaria) as violet and yellow pigments.

Betalains are close in structure and in their synthesis to the animal pigment group melanins, and to eulamelanins in particular.

References

- Galili G; Kigel J (1995). "Chapter One". Seed development and germination. New York: M. Dekker. ISBN 0-8247-9229-7
- What-is-a-taproot-4125537: thespruce.com, Retrieved 19 June 2018
- Igor Kosinki (2007). "Long-term variability in seed size and seedling establishment ofMaianthemum bifolium". Plant Ecology. 194 (2): 149–156. doi:10.1007/s11258-007-9281-1
- Difference-between-taproot-and-fibrous-root, science: differencebetween.net, Retrieved 11 May 2018
- The Seed Biology Place, Gerhard Leubner Lab, Royal Holloway, University of London, archived from the original on 24 September 2015, Retrieved 13 October 2015
- Primary-and-secondary-functions-of-roots-plants-70311: biologydiscussion.com, Retrieved 10 July 2018

- Vleeshouwers L.M., Bouwmeester H.J., Karssen C.M. (1995). "Redefining seed dormancy: an attempt to integrate physiology and ecology". Journal of Ecology. 83: 1031–1037. doi:10.2307/2261184
- Stems-in-plants-function-types-and-anatomy-2601: biologydiscussion.com, Retrieved 31 March 2018
- Rost, Thomas L.; Weier, T. Elliot; Weier, Thomas Elliot (1979). Botany: a brief introduction to plant biology. New York: Wiley. p. 319. ISBN 0-471-02114-8
- 6-major-types-of-inflorescence-with-diagrams-botany-20048: biologydiscussion.com, Retrieved 10 April 2018
- Shannon DA; Isaac L; Brockman FE (February 1996). "Assessment of hedgerow species for seed size, stand establishment and seedling height". Agroforestry Systems. 35 (1): 95–110. doi:10.1007/BF02345331

Chapter 7

Plant Reproduction

Plant reproduction refers to the production of offspring in plants, achieved either through sexual or asexual reproduction. Sexual reproduction involves the fusion of gametes and results in new progeny that are genetically different from the parent or parents. Asexual reproduction does not require a fusion of gametes and the offspring is genetically identical to the parent. This chapter closely examines the key concepts of sexual and asexual reproduction to provide an extensive understanding of plant reproduction.

Plant reproduction is the production of new individuals from one or more parent plants. This can be accomplished by sexual or asexual means.

Asexual reproduction in plants involves a variety of widely disparate methods for producing new plants identical in every respect to the parent. Sexual reproduction, on the other hand, depends on a complex series of basic cellular events, involving chromosomes and their genes that take place within an elaborate sexual apparatus evolved precisely for the development of new plants in some respects different from the two parents that played a role in their production.

Physiology of Plant Reproduction

The maturation of sporophytes and gametophytes, as manifested by their ability to produce spores and gametes, respectively, involves both internal and environmental factors. With respect to the former, the organism must have completed a certain minimum period of vegetative development before environmental factors are able to stimulate formation of spores and gametes.

The liverwort Marchantia polymorpha

Among environmental factors affecting reproduction, the duration, intensity, and quality of light, as well as temperature, have primary roles; for example, the liverwort *Marchantia polymorpha* continues in the vegetative state indefinitely under daily fluorescent illumination of 16 hours. Control plants exposed to daily incandescent lighting of 16 hours become sexually mature after 30 days. Addition of sucrose hastens the development of the sexual phase, an indication that chemical factors also play a role. In hornworts (e.g., *Anthoceros* and *Phaeoceros*), antheridia develop under daily light periods of 4–12 hours; none develop when the plants are illuminated for longer periods.

Temperature also affects sexual maturation. In mosses, for example, initiation of sex organs in bacteria-free laboratory cultures of *Funaria* occurs at 10°C (50°F) when cultures are illuminated 6, 12, or 20 hours daily. On the other hand, the moss *Polytrichum* is seemingly not affected by duration of light but forms sex organs best at 21°C (70°F).

Among vascular cryptogams little is known about the various environmental factors that affect development of sex organs in nature, and, except for certain ferns and horsetails, experimental studies of controlled laboratory cultures are lacking. It has been shown that laboratory cultures of gametophytes of certain horsetails derived from single spores produce sex organs 40–60 days after the spores germinate. In horsetails generally, conditions favoring vegetative growth evoke a preponderance of initially female gametophytes; less favorable conditions induce the production of male gametophytes.

Considerable information is available about the physiology of reproduction in ferns, especially with respect to the gametophyte generation. Spore germination occurs if adequate moisture is present and in temperatures between 15 and 35°C (59 and 95°F). The spores germinate and the young gametophytes do best under neutral or slightly acid conditions. Most fern spores require light for germination. Some fern spores remain viable for as long as 20 years.

Spores germinating in darkness or in the absence of blue wavelengths of light remain in the protonemal, or filamentous, condition. Growth to the mature gametophyte requires light of blue wavelengths in most species.

An interesting aspect of fern reproductive physiology is the discovery that antheridial production is under hormonal control. Several types of antheridium-evoking substances (hormones) have been recognized, and there is some evidence that they are specific. The sperm of ferns are attracted chemically to the vicinity of the maturing archegonia, where they become trapped in extruded mucilage. Although only one sporophyte usually develops on a gametophyte, as many as five may be induced to form if the gametophyte is repeatedly exposed to sperm.

The preceding paragraphs, dealing with vascular cryptogams, have emphasized the influence of environmental factors on the expression of sexual reproduction; the remaining account deals with the angiosperms. Because the gametophytic phase of angiosperms is abbreviated and because it is parasitic on the sporophyte, the maturation of the sporophyte, as manifested by flowering, has received more intensive study in angiosperms than has that of the gametophyte.

Angiosperms that complete their life cycles within one growing season (in the temperate zone) are known as annuals. Perhaps the shortest angiospermous life cycle known is that of *Plantago insularis*, a southern California species, that can grow from seed to the production of its own seed in four to six weeks under domestication; the cycle of most annuals, however, is longer. A number of angiosperms are biennial in the temperate zone: they grow vegetatively for one season, but their flowering and seed production are delayed until a second growing season, after which the plants die (e.g., beets, carrots). Still other angiosperms are perennial: they continue growing, flowering, and producing seeds for a number of growing seasons (e.g., irises, roses, oaks, eucalypti).

It has been demonstrated that duration and wavelength of light are of paramount importance in controlling the flowering of angiosperms. On the basis of extensive experimental studies, flowering plants have been classified as "long-day," "short-day," or "day-neutral" with respect to their requirements of light for flowering. What they actually perceive is the length of the night, using their phytochrome system, which employs blue pigments in photoperiodic processes. Botanists use flashes of red or "far-red" light to interrupt the dark period, which allows them to classify various plants with respect to their photoperiodism.

Temperature also plays an important role in flowering. Thus, members of the cabbage family can be grown without flowering when high enough temperatures are maintained. A number of perennial and biennial plants (bulbous plants, beets) require a prior period of low temperature before they flower. In addition to internal (i.e., genetic factors), therefore, temperature and light are of paramount importance for many plants in evoking maturation.

Several other physiological aspects of reproduction in angiosperms are noteworthy. Once transported to the stigma, the germination of the pollen grains is markedly affected by the chemical composition of the stigmatic exudate; e.g., pollen of an unrelated species will not germinate on the stigma. There is evidence also that the directional growth of the pollen tubes through the style toward the ovular micropyle is stimulated by a chemical substance produced by the style, ovules, and probably other tissues of the pistil.

It is also clear that both pollination and fertilization have physiological effects on fruit production. In a number of cases, pollination alone (not followed by fertilization) is sufficient stimulus to evoke enlargement of the pistil to form a seedless (parthenocarpic) fruit. This phenomenon occurs in banana and in certain varieties of citrus fruits, grapes, and cucumbers. Parthenocarpy can also be induced by exposure of the stigma to indoleacetic acid, naphthoxyacetic acid, indole butyric acid, naphthalene-acetic acid, or other synthetic hormones. The hormone gibberellin is effective in producing seedless grapes and is the active component in preparations used to prevent premature dropping of fruit.

Also significant in the reproduction of flowering plants are the phenomena of self-sterility and self-incompatibility, which prevail in certain plants with perfect flowers (having both stamens and pistils). Pollen from one flower or from another flower of the same plant or from a plant with identical genetic constitution, when applied to the stigma, fails to germinate or grows so slowly that fertilization does not occur. For example, sweet cherries are self-incompatible; a number of groups of genetically intrasterile types are known, but cross-pollination of trees of different genetic groups results in fruit production. Self-incompatibility occurs also in certain strains of garden plants and in plum, apple, pear, and tobacco.

Self-sterility, usually associated with differences in chromosome number, often occurs in hybrids from widely divergent parents. This is true of cultivated varieties of blackberry and raspberry. Such chromosomal imbalance creates irregularities in meiosis during formation of the pollen, which results in infertility.

Sexual Reproduction

Sexual reproduction in plants requires separate male and female parts whose separate sex cells come together and fuse or unite. This unification produces a seed that if cultivated under the proper conditions, will grow into a unique offspring.

In flowering plants (angiosperms), flowers are the reproductive structure. A stamen produces gametes called pollen grains, which attach to a pistil, in which the female gametes (ovules) are located. Here, the female gamete is fertilized and develops into a seed. The ovary, which produced the gamete, then grows into a fruit, which surrounds the seed(s). Plants may either self-pollinate or cross-pollinate.

That plants employ many different strategies to engage in sexual reproduction was used, from just a structural perspective, by Carolus Linnaeus to propose a system of classification of flowering plants. Later this subject received attention from Christian Konrad Sprengel who described plant sexuality as the "revealed secret of nature" and, for the first time, understood the biotic and abiotic interactions of the pollination process. Charles Darwin's theories of natural selection are based on his work.

Flowers, the reproductive structures of angiosperms, are more varied than the equivalent structures of any other group of organisms, and flowering plants also have an unrivaled diversity of sexual systems. But sexuality and the significance of sexual reproductive strategies is no less important in all of the other plant groups. The breeding system is the single most important determinant of the mating structure of non-clonal plant populations. The mating structure in turn controls the amount and distribution of genetic variation.

Androecium and Gynoecium are known to be the conceptive structures of a flower.

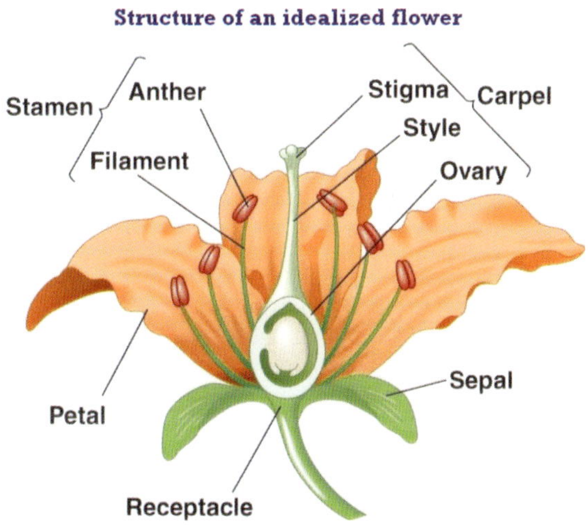

The male conceptive structure is Androecium. It comprises of whorl of stamens.

Stamen: It has slim stalk known as filament. At its tip is contained a bilobed structure called anther. Usually these are bilobed anthers containing two theca (dithecous) isolated by longitudinal groove. The transverse section of any anther shows four sided micro sporangia two in every flap. Microsporangia further form the pollen sac. Contained in the pollen sac are the pollen grains.

Structure of Microsporangia

Microsporangium is encompassed by 4 layers:

- Epidermis: External single layer of defensive cells.
- Endothecium: Endothecium is a single celled layer possessing the character of dehiscence.

It discharges pollen dust. The character of dehiscence is because of the hygroscopic nature of endothecium.

- Middle layers: It is characterized by three to four layers of cells.
- Tapetum: It is the deepest layer of micro sporangia that provides nourishment to maturing microspores.

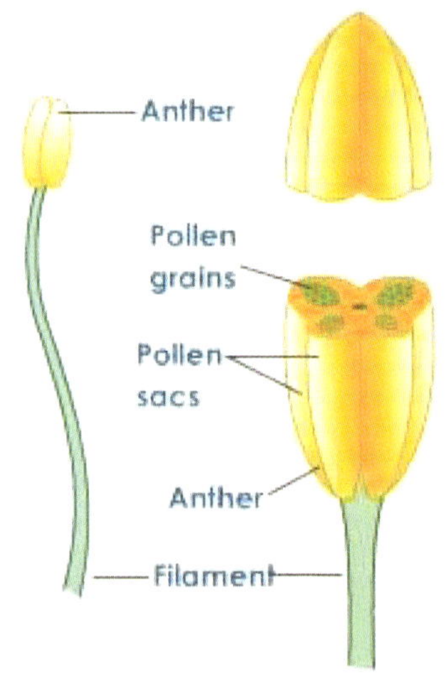

Structure of Anther

Microsporangia genesis: It is the procedure of development of microspores from dust mother cells through the process of meiotic cell division. The microspores framed are a group of four cells called Microspore quadruplicate. As there is further development in the anther the moisture from the anther also dries out. The microspores separate and form into dust grains.

Pollen grains: Pollen grains or dust grains are round in shape. It consists of two layers: external called Exine and internal called Intine. The exine is comprised of a natural compound sporopollenin. It is hard and resistance natural compound. No one protein is known to debase this protein. The exine consists of a typical pore called germ pore.

As the dust grain develops, it contains two cells: a greater Vegetative Cell and minute Generative Cell. The generative cell isolates mitotically into two male gametes.

Gynoecium

The female regenerative part of the flower is the gynoecium. It comprises of pistil. Gynoecium is termed in many ways:

- Monocarpellary: Gynoecium comprises of single pistil. For Example: Pea, Bean.
- Multicarpellary: Gynoecium comprises of more than one pistil.
  - o Syncarpous: Gynoecium with combined pistil. For Example: Tomato, cucumber.
  - o Apocarpous: Gynoecium with free pistil. For Example: Lotus vinca.

Pistil: Each pistil has three sections - the Stigma, style and ovary.

- Stigma: At the tip of pistil is the stigma that provides a landing stage for dust grains.
- Style: The style is the lengthened slender part underneath the stigma.
- Ovary: The basal lump part of the pistil is ovary.

The hole inside the ovary is called ovarian hole (locule). The megasporangia usually called ovule emerge from the placenta. The quantity of ovules inside an ovary might be one (mango, paddy, wheat) or numerous (orchids, watermelon, papaya).

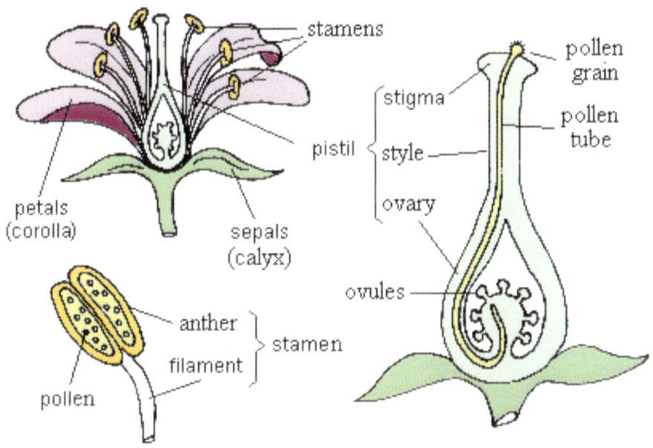

Gynoecium diagram in a flower

Gynoecium diagram

Megasporangium (ovule): The ovule is a minute structure connected to the placenta with a stalk known as Funicle. The ovule's body wires with funicle in the area called hilum.

Each ovule has maybe a couple of defensive envelopes called integuments. Integuments encompass the ovule aside from at the tip. The minute opening at the tip is called Micropyle. Chalaza is a structure present opposite to the micropylar end. It denotes the basal part of the ovule.

The integument encloses nucellus, which is a mass of cells. These cells have abundant nourishment materials. Inside the nucleus is the embryo sac or female gametophyte. An ovule by and large has a single embryo sac shaped from a megaspore through reduction division.

Megasporogenesis

The procedure of arrangement of megaspores from the megaspore mother cell is known as Megasporogenesis. Ovules for the most part separate a single megaspore mother cell (MMC) in the micropylar area of the nucellus. The MMC experiences meiotic division.

The dominant part of flowering plants, one of the megaspores is utilitarian while the other three deteriorate. This functional megaspore forms into the female gametophyte.

The nucellus is by nature diploid. MMC is diploid; the utilitarian haploid, female gametophyte is haploid.

The core of the utilitarian megaspore divides mitotically to shape two nuclei. These two move to the inverse shafts, forming the 2-nucleate embryo sac. Two more successive mitotic atomic divisions result in the arrangement of 8-nucleate phases of the embryo sac.

One nucleus from every shaft comes to center of embryo sac. They intertwine to frame secondary nucleus. The three nuclei at the chalazal end shape the antipodal.

The three nuclei at the micropylar end sort out in to egg apparatus. They are recognized in to two Synergids with an egg between them. The angiosperm embryo sacs, at development, however 8 nucleate but is 7-celled.

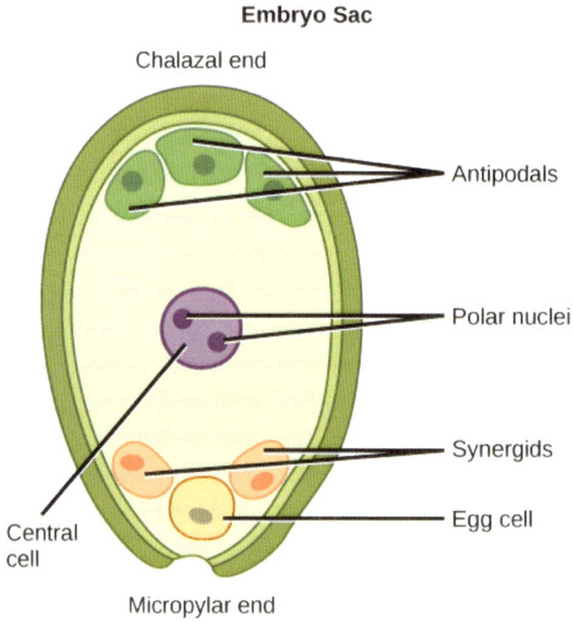

Pollination

Transfer of dust grains to the stigma of a pistil is called pollination.

Type of Pollination

- Autogamy: Transfer of dust grains from the anther to the stigma of a similar flower is called autogamy. In such flower anthers and the stigma lie near each other with the goal that self-pollination can happen.
- Chasmogamous flowers and Cleistogamous flowers: Some plants produce two sorts of flowers. For Example: Viola (Regular Pansy), Oxalis, and Commelina. Flowers with uncovered anthers and stigma are known to be chasmogamous flower. Flowers which don't open are known as Cleistogamous Flowers.
- Geitonogamy: Transfer of dust grains from the anther to the stigma of the diverse flower but of the similar plant is called Geotonogamy. Hereditarily it is like autogamy on the grounds that the dust grains originate from the same plant.
- Xenogamy: Transfer of dust grains from anther to the stigma of an alternate plant but of the same species.

Agents of Pollination: Plants utilize two abiotic (Wind and Water) and one biotic (Animals) operators to accomplish pollination. Majority of plants utilize biotic operators for pollination. Just a little extent of plants utilizes abiotic operators.

- Anemophily: The pollination happened by the impact of wind as an operator is called Anemophily.
- Hydrophily: The pollination happened by the impact of water as an operator in hydrophytes is called Hydrophily.
- Zoophily: The pollination happened by the impact of creatures as an operator is called Zoophily.

On the basis of various sorts of creatures involved in pollination, they are grouped into:

- Entamophily: The pollination happened by the impact of insects as an operator is called Entamophily.
- Ornithophily: The pollination happened by the impact of birds as a specialist is called Ornithophily.
- Chirapterophily: The pollination happened by the impact of bats as a specialist is called Chirapterophily.
- Anemophily: It is the name given to the pollination by wind. Little flowers are pollinated by anemophily. These plants produce huge measure of dust grains. For Example: Sugar Stick, Grasses, Bamboo, Rice, and so on. The anemophilous flowers are small and never shaded in bright color. They are unscented and never create nectar. Pollen grains are little, dry and light. Stigmas are bushy and spread to trap dust grains floating in air.
- Hydrophily: It is the pollination by water. Not all aquatic plants experiences hydrophily. When the pollination happens underneath the water level it is called Hypohydrogamic. It found in plants like Hydrilla, Ceratophyllum. When the pollination happens at the water level is called Epihydrogamic. It is found in plants like Vallisneria.

In some other aquatic plants like water hyacinth (Eichhornia), water lily flower develops out of water and pollination happens by wind or insects. In vallisneria, the sub-consolidated aquatic plant, the male flowers opens at the surface of water. The discharged dust grains coasts on water flow. The pedicel of female flower stretches till it achieves the water surface to contact with dust grains.

- Zoophily: It is the pollination by creatures like bats, birds, insects, wasps, honey bees and so on. Insects pollinate majority of plants. The entamophilous flowers have some trademark highlights:
    - o The flowers are fragrant and overlook scent. For Example: Jasmine, Rose.
    - o The flowers are colorful to attracted insects.
    - o The nectar organ produces nectar to encourage visiting insects.
    - o The dust grains are sticky or spiny. It attaches to the group of visiting insects.
    - o To catch dust grains, the stigma has to be sticky.

Yucca plant is pollinated by moth. The moth and yucca can't finish their existence without each other. The moth drills an opening in ovary to lay its eggs. At that point it gathers dust grains from a few flowers and pushes in the honor end of stigma. After fertilization and seed improvement, the hatchlings of moth eat developing seeds. The unconsumed seeds are scattered to spread. Accordingly, moth can't sustain life without moth and yucca can't imitate sexually without moth.

Inbreeding: The fertilization that happens within the single plant is called inbreeding. Larger part of plants is androgynous. The autogamy (self fertilization) and geitonogamy brings about inbreeding. The significant disadvantage of self fertilization is, the offspring gets feeble and causes diminished wellness in population. It is called inbreeding wretchedness. To stay away from these, plants create numerous structures that demoralize self fertilization.

Outbreeding: The fertilization happens between two unique plants of same species is called out breeding. The Xenogamy (cross-pollination) brings about outbreeding. The plants accomplished out-breeding by numerous systems. The Unisexual Flowers are delivered in same plant or distinctive plant. In monoecious plants with unisexual flowers, similar to castor, cucurbita, maize it maintains a strategic distance from autogamy. In diecious plants like papaya, mulberry, it brings about xenogamy.

Self-Sterility or Incompatibility: In this dust of the flower has no fertilizing impact on stigma of a similar flower. For Example: Passiflora, Potato.

Dichogamy: In this male and female sex organ develops at various times. Ex: China rose, lady's finger, jasmine, Custard apple.

Herkogamy: The anther and stigma are put at various positions so that pollens are not able to achieve stigma of same flower.

Artificial Hybridization

Pollinating the dust grains of one flower to the next flower of various plants of same species is known as artificial hybridization. It is one of the significant approaches of harvest improvement program. In this strategy, the required dust grains are utilized for pollination. The stigma is shielded from contamination with other dust grains. It is accomplished by castration and bagging.

Emasculation: It is the expulsion of anthers/stamens from flower bud before it develops.

Emasculation

Bagging: It is the covering of undermined flower with margarine paper to avert contamination with different pollens.

Different methods of Bagging

Pollen Pistil Interaction: It is the capacity of pistil to perceive the pollen of same species. It experience post pollination occasion that prompts to fertilization. If wrong pollen grain (different species) is arrived on stigma, pistil rejects by preventing dust germination. The acknowledgment and dismissal is interceded by the compound part of dust grain interacting with pistil.

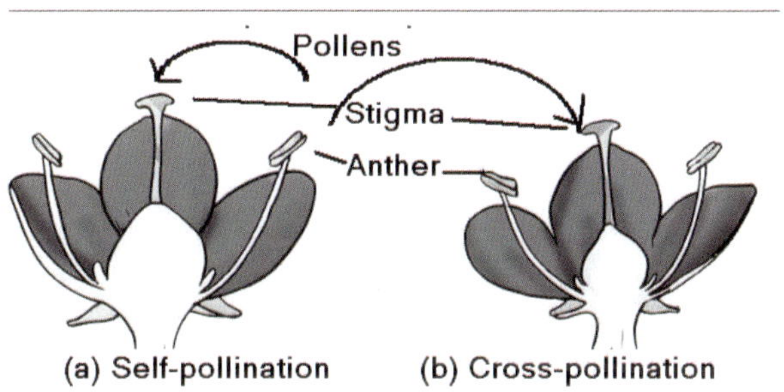

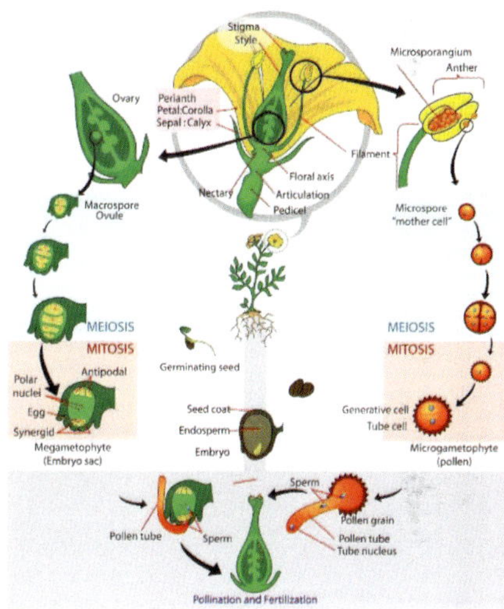

Pollination and Fertilization

Fertilization

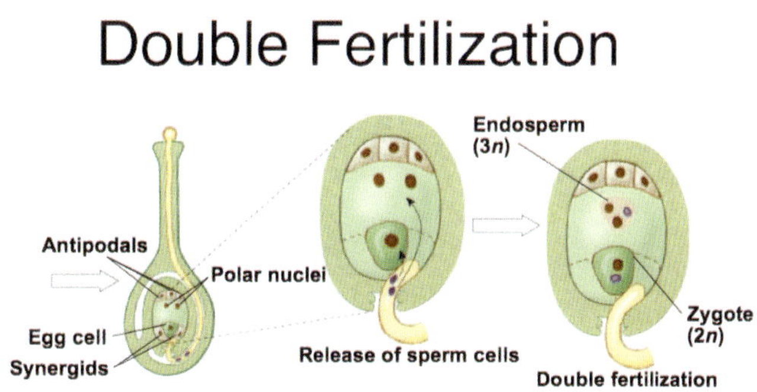

The fertilization in angiosperms is named as double fertilization and triple fusion. As the pollen-grain develops as a rule it contains one vegetative cell and two generative cells that form into male gamete. In a few plants dust grain sheds as two cell condition. In this, generative cell partitions into two male gametes as pollen tube shapes.

As the dust grain arrives on stigma, it begins to germinate. The pollen tube create leaves germ pore. It develops through the tissue of stigma, style and infiltrates the ovule.

The section of pollen tube in to ovule is from a distinct place and is distinguished into:

1. Porogamy: It is the passage of dust tube through micropyle.
2. Chalazogamy: It is the passage of dust tube through chalaza.
3. Misogamy: It is the section of dust tube through integument or funicle.

As pollen tube enters the ovule, the tip breaks inside the embryo sac leading to the release of two male gametes.

One of the male gamete wires with egg and structures diploid zygote. Another male gamete wires with diploid secondary nucleus to frame triploid endosperm.

The fusion of two polar nuclei (to frame secondary diploid nucleus) and male gamete is named as Triple Fusion. The two fusions as zygote development and triple fusion are named as Double Fertilization.

Post-fertilization Stages

It is the series of occasions that happens after double fertilization. It includes:

1. Endosperm development.
2. Embryo improvement.
3. Development of ovule into seed.
4. Development of ovary into fruit.

1. Endosperm: The endosperm cell, shaped by triple fusion under goes rehashed mitotic cell division to frame triploid endosperm tissue. It gives sustenance to the developing embryo. Three unique sorts of endosperms are accounted as follows:

   Nuclear endosperm: In this endosperm cell nucleus experiences free nuclear division. Every one of these nuclei is pushed to the fringe forming central vacuole. The cytokines is brings about the development of endosperm tissue. Example: coconut. The vacuole is loaded with liquid.

   Cellular endosperm: In this endosperm cellular division experiences cytokinesis. Example: Datura.

   Helobial endosperm: In first endosperm cell division two unequal cells shapes. One cell experiences cellular division and other under go free nuclear division. Example: monocots.

In pea, groundnut, beans endosperm is totally devoured by developing embryo. In a few seeds like groundnut, castor endosperm perseveres developed seeds.

2. Embryo improvement

   In Dicot: The zygote separates transversely in to upper suspensor cell exhibited at micropyle area and lower embryonal cell exhibited at chalazal district. The suspensor cell partitions and structures 7 – 10 cells called suspensor. The upper cell of suspensor extends and appends to the embryo sac. The embryonal cell separates into eight cells. The lower four cells creates in to form one plumule and two cotyledons. The above four cells close to the suspensor creates in to form hypocotyl and radicle. The completely created embryo of dicotyledons has an embryonal hub that separated into plumule, two cotyledons and a radicle. The lower end of radicle (root tip) is secured with root cap. In the starting embryo is globular in shape. At the point when two cotyledons separate it frames heart shape.

   In monocots: The zygote isolates transversely in to upper suspensor cell display at micropyle locale and lower embryonal cell exhibited at chalazal district. The embryonal cell partitions into globular embryo. It frames a huge cotyledon and plumule. Plumule is covered by coleoptile. The single cotyledon is called scutellum. At lower end of embryonal pivot radicle creates. It is secured by a sheath called coleorhiza.

3. Seed: The final result of sexual reproduction is the formation of seed. It is the ovule that is fertilized. Seed comprises of seed coat, cotyledon and embryonal hub. The cotyledons are swollen because of their capacity of hold sustenance. The seeds are either albuminous or non-albuminous. In albuminous seed endosperm is not totally utilized by embryo. Example: wheat, maize, castor, sunflower. In a few seeds like dark pepper the nucellus is additionally present. The leftover persistent nucellus is called Perisperm. In non albuminous seed endosperm is totally utilized by embryo. Example: Pea, Groundnut. The seed coat is developed from the integument of ovule. The micropyle remains as a pore in the seed coat. It encourages the section of oxygen and water in to seed during germination. Seeds become scarce by loosing water. The embryo enters dormancy, the phase of inactivity. As the conditions become favorable again it begins germination.

4. Fruit: The wall of ovary forms into organic product wall called pericarp. It encompasses the seed. In apple, strawberry the fruit is developed from the thalamus. Subsequently these are termed false fruit. The fruits created from unfertilized ovary are called Parthenocarpic Fruits. Parthenocarpic fruits are seedless.

Apomixes: Apomixes is the process of formation of seed without the fertilization of ovule and the fruits created from unfertilized ovule are termed apomictic fruits. The descendants of hybrid isolate their characters and not continue the hybrid character till a great many generations. Consequently the ranchers ought to purchase hybrid seeds each year. It makes broad. The apomixes is performed in hybrid plants to maintain a strategic distance from the isolation of characters. The farmers can utilize hybrid seeds to raise new yield each year.

Polyembryony: The nearness of more than one embryo in seed is known as Polyembryony. Example: Seeds of citrus fruits.

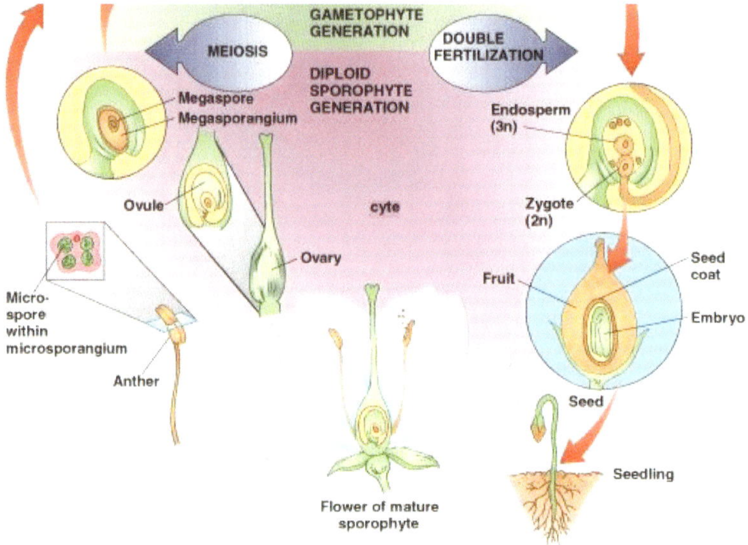

Process of Fertilization in plants

Asexual Reproduction

Roots: Different types of stems allow for asexual reproduction

(a) The corm of a garlic plant looks similar to (b) a tulip bulb, but the corm is solid tissue, while the bulb consists of layers of modified leaves that surround an underground stem. Both corms and bulbs can self-propagate, giving rise to new plants. (c) Ginger forms masses of stems called rhizomes that can give rise to multiple plants. (d) Potato plants form fleshy stem tubers. Each eye in the stem tuber can give rise to a new plant. (e) Strawberry plants form stolons: stems that grow at the soil surface or just below ground and can give rise to new plants.

Many plants are able to propagate themselves using asexual reproduction. This method does not require the investment required to produce a flower, attract pollinators, or find a means of seed dispersal. Asexual reproduction produces plants that are genetically identical to the parent plant because no mixing of male and female gametes takes place. Traditionally, these plants survive well under stable environmental conditions when compared with plants produced from sexual reproduction because they carry genes identical to those of their parents.

Plants have two main types of asexual reproduction: vegetative reproduction and apomixis. Vegetative reproduction results in new plant individuals without the production of seeds or spores. Many different types of roots exhibit vegetative reproduction. The corm is used by gladiolus and garlic. Bulbs, such as a scaly bulb in lilies and a tunicate bulb in daffodils, are other common examples of this type of reproduction. A potato is a stem tuber, while parsnip propagates from a taproot. Ginger and iris produce rhizomes, while ivy uses an adventitious root (a root arising from a plant part other than the main or primary root), and the strawberry plant has a stolon, which is also called a runner.

Some plants can produce seeds without fertilization. Either the ovule or part of the ovary, which is diploid in nature, gives rise to a new seed. This method of reproduction is known as apomixis.

An advantage of asexual reproduction is that the resulting plant will reach maturity faster. Since the new plant is arising from an adult plant or plant parts, it will also be sturdier than a seedling. Asexual reproduction can take place by natural or artificial (assisted by humans) means.

Natural and Artificial Methods of Asexual Reproduction in Plants

Plants can undergo natural methods of asexual reproduction, performed by the plant itself, or artificial methods, aided by humans.

Natural Methods of Asexual Reproduction

Natural methods of asexual reproduction include strategies that plants have developed to self-propagate. Many plants, such as ginger, onion, gladioli, and dahlia, continue to grow from buds that are present on the surface of the stem. In some plants, such as the sweet potato, adventitious roots or runners (stolons) can give rise to new plants. In *Bryophyllum* and kalanchoe, the leaves have small buds on their margins. When these are detached from the plant, they grow into independent plants; they may also start growing into independent plants if the leaf touches the soil. Some plants can be propagated through cuttings alone.

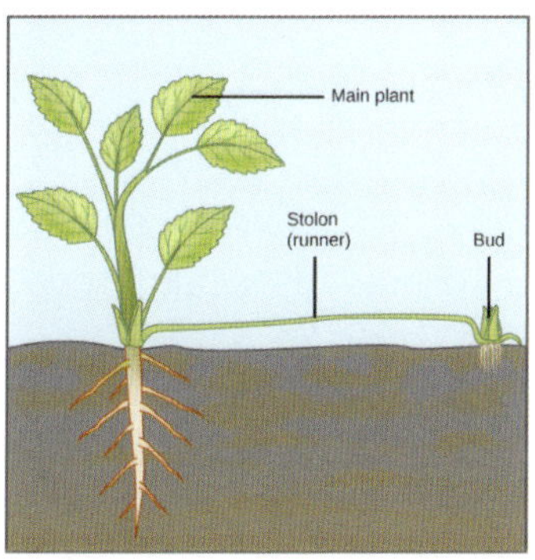

Runners: asexual reproduction: A stolon, or runner, is a stem that runs along the ground. At the nodes, it forms adventitious roots and buds that grow into a new plant.

Artificial Methods of Asexual Reproduction

Artificial methods of asexual reproduction are frequently employed to give rise to new, and sometimes novel, plants. They include grafting, cutting, layering, and micro propagation.

Vegetative Propagation

Vegetative propagation, or vegetative plant propagation, refers to the process of plant reproduction whereby a fragment of a parent plant (a cutting) is taken in order to produce more plants (clones). These fragments, known as cuttings, are able to take root and grow into full plants without much interference from the gardener.

In most cases, the vegetation used in vegetative plant propagation is derived from the roots, stems, and leaves of a gardener's healthiest, highest-yielding plant, commonly referred to as the mother plant.

Many indoor growers refer to vegetative plant propagation simply as cloning, or taking cuttings. It is considered to be an asexual method of plant propagation.

Many plants can reproduce naturally via vegetative plant propagation, but it is also possible for botanists to artificially induce vegetative propagation. In fact, this is a common practice for indoor growers looking for swifter growth cycles and more predictable plant growth in their growrooms.

Because vegetative propagation essentially produces natural clones of the parent plant, the new plants contain the same genetic material. This enables botanists to continuously reproduce the same traits that are desirable in the original plant, provided that the growing conditions are favorable.

As a result, vegetative propagation is commonly used for commercial purposes, enabling planters to maintain both a consistent taste and quality.

Another advantage of vegetative propagation is that plants can skip the seedling phase, which enables farmers to save money and time. For example, instead of taking around three years to reach maturity, a young tea plant cutting can easily reach maturity in a matter of months with vegetative propagation.

On the other hand, vegetative propagation has been shown to negatively impact the plant's biodiversity while putting both the original and new plant at risk of certain kinds of diseases.

Plants may be propagated artificially by vegetative means by working on natural principles.

(I) Natural Methods:

Among seed plants, vegetative propagation by natural methods is very common. Frequently, vegetative organs such as roots, stems and leaves bear adventitious buds and bring about the formation of new plants. These structures are also called as vegetative propagates.

The role played by various vegetative parts are:

A. Vegetative propagation by roots:

The roots of some plants develop adventitious buds on them e.g., Dalbergia sissoo (Shisham), Guava, poplars, Albizia lebbeck, Murraya, etc. Some tuberous adventitious roots besides possessing adventitious buds also contain sufficient quantities of, food, e.g., Dahlia and Sweet potato.

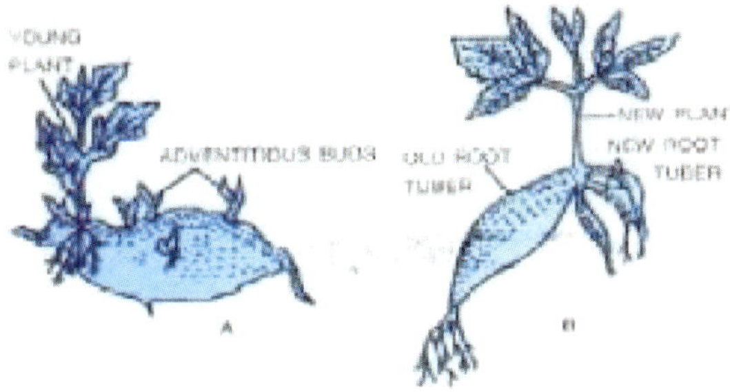

A-B Vegitable reproduction by roots
A. Adventitouse buds growing into new shoots in tuberous root of sweet potato:
B. Old root tuber of Dahlia growing into new plant.

These buds sprout under suitable conditions. These sprouts may be separated and planted. In shisham (Tahli), young fast growing, shoots will arise from the roots around the cut stumps of trees.

B. Vegetative propagation by stems:

Straggling and branched rhizomes reproduce by vegetative reproduction e.g. Banana, Ginger Turmeric. Decay of older parts isolates the newly formed branches. The latter henceforth leads an independent life. Corms like Colocasia (Kachalu), Freesia, Crocus, etc., have sufficient stored food and also bear many adventitious buds. A bulb also bears a number of buds, e.g., Garlic and Narcissus, Onion.

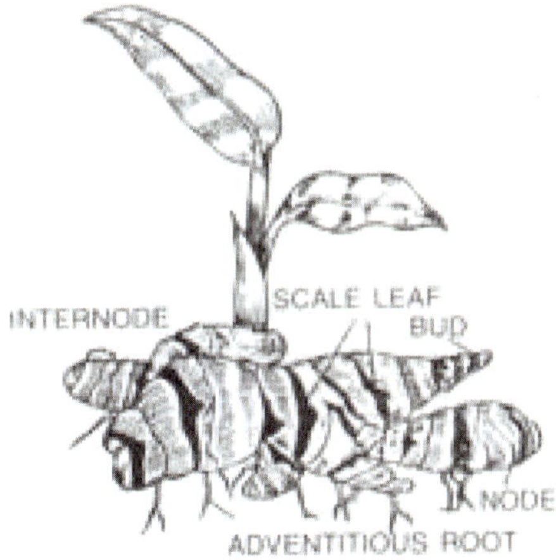

Rhizome of ginger

(i) Stem tuber:

Stem tubers are found in Potato and Artichoke. Tuber of potato is a swollen apical part of an underground stem branch and bears a number of nodes or eyes. Each eye bears one or more buds. New plants are produced from the buds on the eyes. The potato crop is raised by tubers and not by seeds.

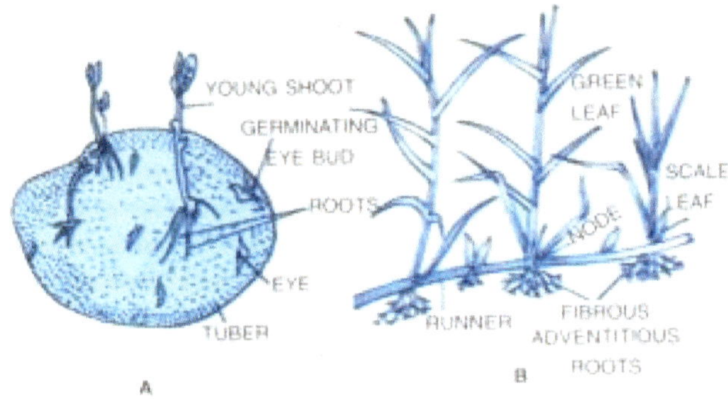

A. Aerial shoots arising from the buds in the regions of nodes or eyes in potato stem tuber.
B. A runner of grass.

(ii) Runner:

Aerial weak stem e.g., in lawn grass Cyanodon (Doob grass); Mint and Wood sorrel (Oxalis) when touches the ground, gives off adventitious roots at nodes. When older parts of plant die, the branches separate from the parent plant and form independent plants.

C. Vegetative propagation by leaf:

Some plants produce adventitious buds on their leaves e.g., Bryophyllum, Begonia, Streptocarpus, Kalanchoe and Saintpaulia. In Bryophyllum notched margins of succulent leaves bear adventitious buds.

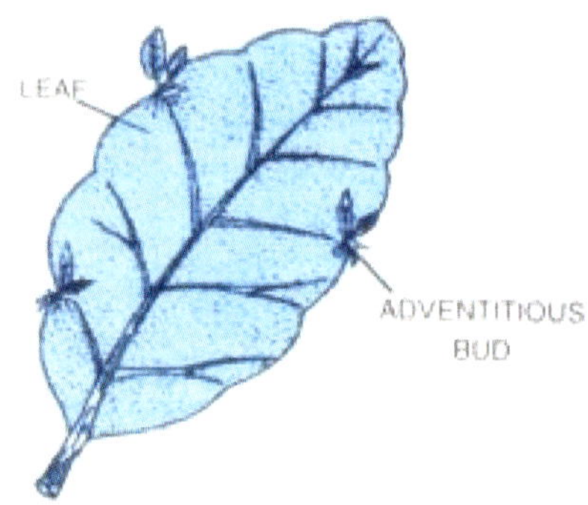

A leaf of *Bryophyllum* showing formation of new plants from marginal adventitious buds

These buds usually remain dormant, when the leaf is attached with plant. However, the leaves when in contact with moist soil develop new plantlets along the margins. However in some species of Bryophyllum (B. diagremontianum), plantlets develop along the margins of intact leaves.

II. Artificial Methods:

Gardeners and horticulturists have employed the various methods of vegetative propagation adopted by plants in nature. These practices constitute artificial means of vegetative propagation.

These methods are meant for:

(a) Quick production of new plants.

(b) Combining good qualities of two different varieties.

Plants raised through vegetative propagation are the exact genetic copies of the parents, showing the same characters like of their parents. Such a genetically uniform population raised from a single plant by vegetative methods is known as clone.

Artificial methods of vegetative propagation provide excellent methods of propagating the desirable varieties of plants economically, with least attention and comparatively in shorter time.

Some common and important artificial or horticultural methods of vegetative propagation are:

A. Cutting:

A cutting is separated portion of root, stem or leaf. It is very common method employed for artificial vegetative propagation. Sugar cane, Roses, Citrus, Duranta, Grapes, Cocoa, Bougainvillea, carnations and many other plants are largely and rapidly propagated by stem cuttings. A cutting is put into the soil.

Care is taken that nodes which were lower in parent plant (morphologically) are put in the soil, while the morphologically higher nodes are kept up. Adventitious roots are given off at the lower nodes.

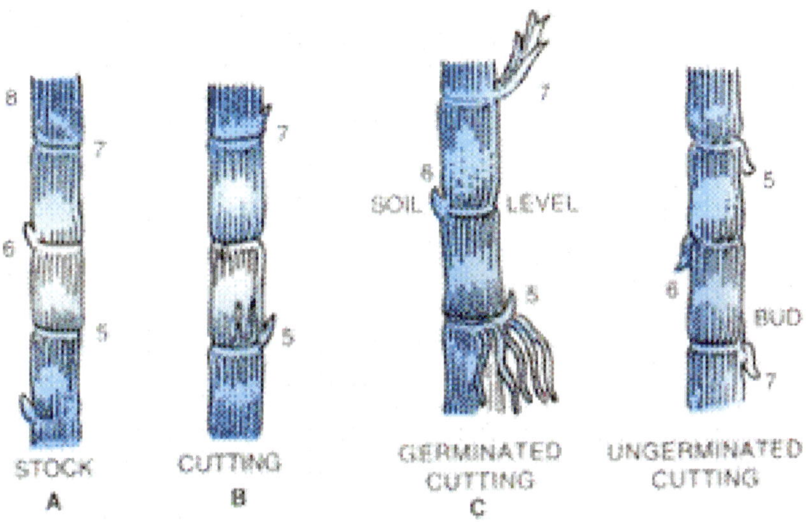

Cutting technique

Nowadays special growth hormones like IAA (Indole acetic acid), IBA (Indole butyric acid), NAA (Naphthalene acetic acid) are applied in dilute quantities to the lower end of stem cuttings. Such application of hormones leads to the quick formation of adventitious roots. Favorable time for planting of cuttings is the spring.

B. Grafting:

Grafting is performed between two plants of closely related varieties. The plant of which the root system is to be taken is called stock and the other of which shoot system is to be used is known as scion.

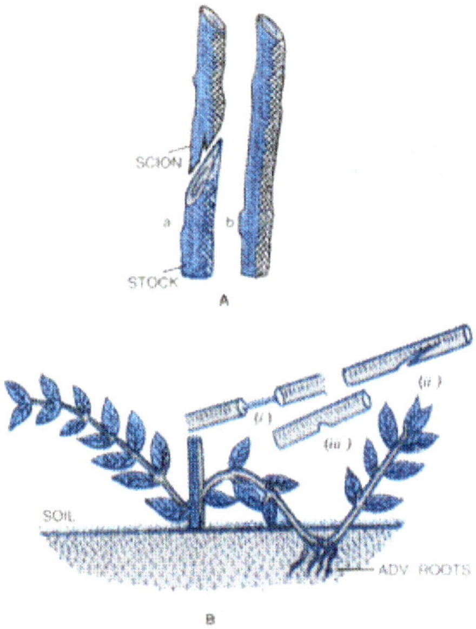

A. Vegetative propagation by Grafting;
B. Layering; (i) Ringing; (ii) Notching; (iii) Tonguing.

Stock usually bears strong root system (e.g., Desi mango) and scion has a desired variety of flowers or fruits (e.g., Dussehri mango), but with a weak root system. The success of the grafting mostly depends upon the union of the cambium of scion and stock which results in organic connection between them.

The cut surfaces should be held together tightly by wrapping, nailing or some such method. The grafting operations should be done at proper time of the year. All shoots sprouting from the stock should be removed otherwise they will not permit the scion to grow.

C. Budding:

In budding, a bud with small portion of surrounding bark is taken from the stem of a desired plant. It is then inserted into the small slit made in the bark of stock. Both are tied together and not allowed to dry. Buds of stock are not permitted to sprout. This method is widely practiced in rose, plums, etc.

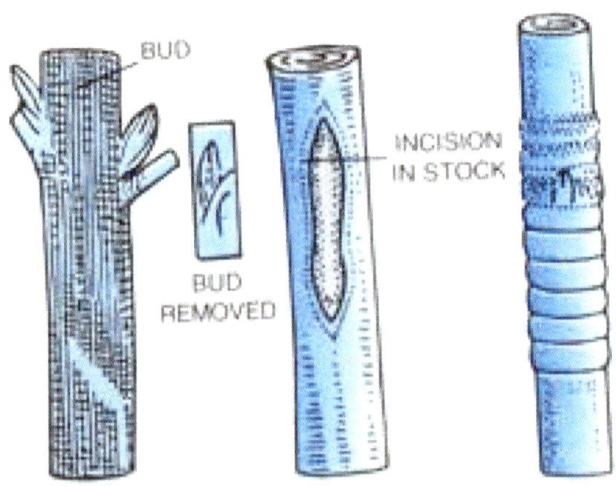

Budding

D. Layering:

In this method, stem branch chosen for propagation (layer) remains attached to parent till it has produced adventitious roots. In ground layering, stem is capable of bending (herbs) and the selected branch is bent down to the soil. Portion touching the soil is cleared of leaves.

An injury is made on it by cutting a shallow ring around (ringing) or by giving an oblique upward cut on bent portions of the layer at the node (tonguing) or by giving a V-shaped cut at the Lower side of bent portion of layer (notching). Injured parts can be treated with auxins (hormones) to stimulate the formation of adventitious roots e.g., jasmine. Strawberry, Clematis.

In air layering, bark of the chosen branch is cut by girdling or ringing near the base. This injury initiates root formation. The cut area is treated with auxins (hormones). Then a ball of Sphagnum moss is placed around the cut portion. It is kept moist and enclosed in a polythene bag.

When the roots appear, the stem is cut below the level of roots and planted. This method is also popularly called gootee. A longitudinal cut can also be made in branch and wet moss can be placed under the flap. This method of artificial vegetative propagation is widely used in plants like Litchi, Citrus, Pomegranate, Guava, Croton etc.

E. Bulbil:

Sometimes buds become fleshy e.g., Agave. It takes part in propagation of plant.

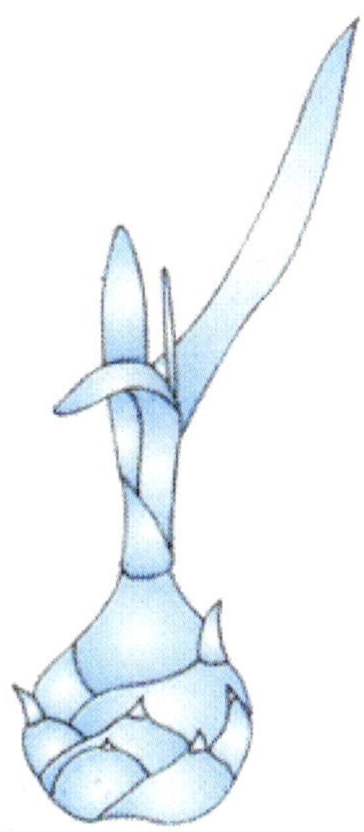

Bulbil of Agave

Sporogenesis

Sporogenesis is the production of spores in biology. The term is also used to refer to the process of reproduction via spores. Reproductive spores were found to be formed in eukaryotic organisms, such as plants, algae and fungi, during their normal reproductive life cycle. Dormant spores are formed, for example by certain fungi and algae, primarily in response to unfavorable growing conditions. Most eukaryotic spores are haploid and form through cell division, though some types are diploid or dikaryons and form through cell fusion.

Reproduction via Spores

Reproductive spores are generally the result of cell division, most commonly meiosis (e.g. in plant sporophytes). Sporic meiosis is needed to complete the sexual life cycle of the organisms using it.

In some cases, sporogenesis occurs via mitosis (e.g. in some fungi and algae). Mitotic sporogenesis is a form of asexual reproduction. Examples are the conidial fungi *Aspergillus* and *Penicillium*, for which mitospore formation appears to be the primary mode of reproduction. Other fungi, such as ascomycetes, utilize both mitotic and meiotic spores. The red alga *Polysiphonia* alternates between mitotic and meiotic sporogenesis and both processes are required to complete its complex reproductive life cycle.

In the case of dormant spores in eukaryotes, sporogenesis often occurs as a result of fertilization or karyogamy forming a diploid spore equivalent to a zygote. Therefore, zygospores are the result of sexual reproduction.

Reproduction via spores involves the spreading of the spores by water or air. Algae and some fungi (chytrids) often use motile zoospores that can swim to new locations before developing into sessile organisms. Airborne spores are obvious in fungi, for example when they are released from puffballs. Other fungi have more active spore dispersal mechanisms. For example, the fungus *Pilobolus* can shoot its sporangia towards light. Plant spores designed for dispersal are also referred to as diaspores. Plant spores are most obvious in the reproduction of ferns and mosses. However, they also exist in flowering plants where they develop hidden inside the flower. For example, the pollen grains of flowering plants develop out of microspores produced in the anthers.

Reproductive spores grow into multicellular haploid individuals or sporelings. In heterosporous organisms, two types of spores exist: microspores give rise to males and megaspores to females. In homosporous organisms, all spores look alike and grow into individuals carrying reproductive parts of both genders.

Formation of Reproductive Spores

Sporogenesis occurs in reproductive structures termed sporangia. The process involves sporogenous cells (sporocytes, also called spore mother cells) undergoing cell division to give rise to spores.

In meiotic sporogenesis, a diploid spore mother cell within the sporangium undergoes meiosis, producing a tetrad of haploid spores. In organisms that are heterosporous, two types of spores occur: Microsporangia produce male microspores, and mega sporangia produce female megaspores. In megasporogenesis, often three of the four spores degenerate after meiosis, whereas in microsporogenesis all four microspores survive.

In gymnosperms, such as conifers, microspores are produced through meiosis from microsporocytes in microstrobili or male cones. In flowering plants, microspores are produced in the anthers of flowers. Each anther contains four pollen sacs, which contain the microsporocytes. After meiosis, each microspore undergoes mitotic cell division, giving rise to multicellular pollen grains (six nuclei in gymnosperms, three nuclei in flowering plants).

Megasporogenesis occurs in megastrobili in conifers (for example a pine cone) and inside the ovule in the flowers of flowering plants. A megasporocyte inside a mega sporangium or ovule undergoes meiosis, producing four megaspores. Only one is a functional megaspore whereas the others stay dysfunctional or degenerate. The megaspore undergoes several mitotic divisions to develop into a female gametophyte (for example the seven-cell/eight-nuclei embryo sac in flowering plants).

Mitospore Formation

Some fungi and algae produce mitospores through mitotic cell division within a sporangium. In fungi, such mitospores are referred to as conidia.

Formation of Dormant Spores

Some algae, and fungi form resting spores made to survive unfavorable conditions. Typically, changes in the environment from favorable to unfavorable growing conditions will trigger a switch from asexual reproduction to sexual reproduction in these organisms. The resulting spores are protected through the formation of a thick cell wall and can withstand harsh conditions such as drought or extreme temperatures. Examples are chlamydospores, teliospores, zygospores, and myxospores. Similar survival structures produced in some bacteria are known as endospores.

Chlamydospore and Teliospore Formation

Chlamydospores are generally multicellular, asexual structures. Teliospores are a form of chlamydospore produced through the fusion of cells or hyphae where the nuclei of the fused cells stay separate. These nuclei undergo karyogamy and meiosis upon germination of the spore.

Zygospore, Oospore and Auxospore Formation

Zygospores are formed in certain fungi (zygomycota, for example *Rhizopus*) and some algae (for example *Chlamydomonas*). The zygospore forms through the isogamic fusion of two cells (motile single cells in *Chlamydomonas*) or sexual conjugation between two hyphae (in zygomycota). Plasmogamy is followed by karyogamy, therefore zygospores are diploid (zygotes). They will undergo zygotic meiosis upon germinating.

In oomycetes, the zygote forms through the fertilization of an egg cell with a sperm nucleus and enters a resting stage as a diploid, thick-walled oospore. The germinating oospore undergoes mitosis and gives rise to diploid hyphae which reproduce asexually via mitotic zoospores as long as conditions are favorable.

In diatoms, fertilization gives rise to a zygote termed auxospore. Besides sexual reproduction and as a resting stage, the function of an auxospore is the restoration of the original cell size, as diatoms get progressively smaller during mitotic cell division. Auxospores divide by mitosis.

Endospore Formation

The term sporogenesis can also refer to endospore formation in bacteria, which allows the cells to survive unfavorable conditions. Endospores are not reproductive structures and their formation does not require cell fusion or division. Instead, they form through the production of an encapsulating spore coat within the spore-forming cell.

Parts of Spore

There are many parts of the spore 'plant'. The case that holds all the spores is called a sporangium.

Fragmentation

Among plants, fragmentation is the regularly found type of asexual reproduction. Fragmentation is a very common type of vegetative reproduction in plants. Fragmentation occurs when a shoot that is rooted becomes detached from the main group. In plants, there are different other mechanisms. There are several other known mechanisms of natural fragmentation in plants.

Specialized Structures (Reproductive): There are few plants which could form adventitious plantlets on their leaves, which would later detach to form individually grown plants. Certain other ones produce organs like turions and bulbils.

Even in nonvascular plants, the process of fragmentation is a common phenomenon like mosses and liverworts. In case of moss leaves or stems, they are carried by the wind, animals or even water. Once the moss fragment reaches an environment that is suitable, it would root itself to form a new plant.

Fragmentation is used by people artificially to propagate by grafting, cutting, layering, and division. People use fragmentation to artificially propagate many plants via division, layering, cuttings, grafting, via storage organs like corms, rhizomes and tubers, and micro propagation.

Examples

Mosses can reproduce themselves by fragmentation. When the tops of liverworts such as Marchantia are broken off, each fragmented top grows into a new plant. Potato tubers, when broken apart and planted, form a whole new plant from each fragment. Many weeds can regenerate completely just from fragments of stems left in the ground, which is why weeds need to be pulled up instead of cut down.

Uses

Gardeners and farmers use fragmentation to their benefit when they start new plants from the cuttings of old plants. They also use it when they graft parts of one plant onto another. Scientists can even grow whole plants out of individual cells taken from other plants. This allows them to rapidly produce thousands of clones of the original plant in a short amount of time.

Advantages

One of the chief advantages of reproduction by fragmentation is that it can rapidly produce many new plants without the waiting and uncertainty of pollination and sexual reproduction. Marchantia liverworts fragment and regrow so easily that the process can be initiated simply by having raindrops strike the tops of the plants. It can also help plants reproduce in times of drought. They need water to allow fertilization to occur. During drought, sexual reproduction becomes more difficult. Fragmentation also allows gardeners to make many identical copies of plants that have particularly desirable looks or other characteristics.

Disadvantages

Reproduction by fragmentation carries its disadvantages as well, such as a limiting of genetic variability. If a mutation occurs at any point during the process, then all subsequent generations will carry the mutation. This leaves them much more vulnerable to rapid widespread die-offs. If the parent plant is particularly vulnerable to a disease or predator, then all of the offspring of that plant will carry the same vulnerability, and the entire population may be quickly killed by the same problem.

Parthenogenesis

Parthenogenesis is a type of asexual reproduction in which a female gamete or egg cell develops into an individual without fertilization. Many plants are capable of reproducing by parthenogenesis.

Process of Parthenogenesis

There are two main ways in which parthenogenesis occurs. One method is by apomixis, where egg cells are produced by mitosis. In apomictic parthenogenesis, the female sex cell (oocyte) replicates by mitosis producing two diploid cells. These cells have the full compliment of chromosomes needed to develop into an embryo. The resulting offspring are clones of the parent cell. Organisms that reproduce in this manner include flowering plants.

The other main method of parthenogenesis is through automixis. In automictic parthenogenesis, egg cells are produced by meiosis. Normally in oogenesis (egg cell development), the resulting daughter cells are divided unequally during meiosis. This asymmetrical cytokinesis results in one large egg cell (oocyte) and smaller cells called polar bodies. The polar bodies degrade and are not fertilized. The oocyte is haploid and only becomes diploid after it is fertilized by male sperm. Since automictic parthenogenesis does not involve males, the egg cell becomes diploid by fusing with one of the polar bodies or by duplicating its chromosomes and doubling its genetic material. Since the resulting offspring are produced by meiosis, genetic recombination occurs and these individuals are not true clones of the parent cell.

Advantages and Disadvantages of Parthenogenesis

Parthenogenesis is an adaptive strategy to ensure the reproduction of organisms when conditions are not favorable for sexual reproduction. Asexual reproduction can be advantageous to organisms that must remain in a particular environment and in places where mates are scarce. Numerous offspring can be produced without "costing" the parent a great amount of energy or time. A disadvantage of this type of reproduction is the lack of genetic variation. There is no movement of genes from one population to another. Due to the fact that environments are unstable, populations that are genetically variable will be able to adapt to changing conditions better than those that lack genetic variation.

References

- Plant-reproductive-system, science: britannica.com, Retrieved 19 June 2018
- Sexual-reproduction-in-flowering-plants: askiitians.com, Retrieved 21 May 2018
- Asexual-reproduction, boundless-biology: lumenlearning.com, Retrieved 11 June 2018
- Vegetative-propagation-1970: maximumyield.com, Retrieved 31 March 2018
- Vegetative-reproduction-methods-in-plants-natural-and-artificial-methods-plants-26735: yourarticlelibrary.com, Retrieved 11 May 2018
- Parthenogenesis-373474: thoughtco.com, Retrieved 24 April 2018

Permissions

All chapters in this book are published with permission under the Creative Commons Attribution Share Alike License or equivalent. Every chapter published in this book has been scrutinized by our experts. Their significance has been extensively debated. The topics covered herein carry significant information for a comprehensive understanding. They may even be implemented as practical applications or may be referred to as a beginning point for further studies.

We would like to thank the editorial team for lending their expertise to make the book truly unique. They have played a crucial role in the development of this book. Without their invaluable contributions this book wouldn't have been possible. They have made vital efforts to compile up to date information on the varied aspects of this subject to make this book a valuable addition to the collection of many professionals and students.

This book was conceptualized with the vision of imparting up-to-date and integrated information in this field. To ensure the same, a matchless editorial board was set up. Every individual on the board went through rigorous rounds of assessment to prove their worth. After which they invested a large part of their time researching and compiling the most relevant data for our readers.

The editorial board has been involved in producing this book since its inception. They have spent rigorous hours researching and exploring the diverse topics which have resulted in the successful publishing of this book. They have passed on their knowledge of decades through this book. To expedite this challenging task, the publisher supported the team at every step. A small team of assistant editors was also appointed to further simplify the editing procedure and attain best results for the readers.

Apart from the editorial board, the designing team has also invested a significant amount of their time in understanding the subject and creating the most relevant covers. They scrutinized every image to scout for the most suitable representation of the subject and create an appropriate cover for the book.

The publishing team has been an ardent support to the editorial, designing and production team. Their endless efforts to recruit the best for this project, has resulted in the accomplishment of this book. They are a veteran in the field of academics and their pool of knowledge is as vast as their experience in printing. Their expertise and guidance has proved useful at every step. Their uncompromising quality standards have made this book an exceptional effort. Their encouragement from time to time has been an inspiration for everyone.

The publisher and the editorial board hope that this book will prove to be a valuable piece of knowledge for students, practitioners and scholars across the globe.

Index